Towards Mechanism-based Treatments for Fragile X Syndrome

Towards Mechanism-based Treatments for Fragile X Syndrome

Special Issue Editors

Daman Kumari
Inbal Gazy

MDPI • Basel • Beijing • Wuhan • Barcelona • Belgrade

Special Issue Editors
Daman Kumari
National Institute of Diabetes
USA

Inbal Gazy
National Institute of Diabetes
USA

Editorial Office
MDPI
St. Alban-Anlage 66
4052 Basel, Switzerland

This is a reprint of articles from the Special Issue published online in the open access journal *Brain Sciences* (ISSN 2076-3425) from 2018 to 2019 (available at: https://www.mdpi.com/journal/brainsci/special_issues/Fragile_X_Syndrome)

For citation purposes, cite each article independently as indicated on the article page online and as indicated below:

LastName, A.A.; LastName, B.B.; LastName, C.C. Article Title. *Journal Name* **Year**, *Article Number*, Page Range.

ISBN 978-3-03921-505-8 (Pbk)
ISBN 978-3-03921-506-5 (PDF)

Cover image courtesy of Elisa A. Waxman, Children's Hospital of Philadelphia, Philadelphia, PA, USA.

Contents

About the Special Issue Editors

Daman Kumari is a Staff Scientist in the Laboratory of Cell and Molecular Biology at the National Institutes of Health. Dr. Kumari has made several important contributions to the understanding of repeat-mediated *FMR1* gene silencing in FXS. Her current research is focused on developing treatment strategies for FXS that are based on *FMR1* gene reactivation. Dr. Kumari serves on the Editorial Board of *Genetic Testing and Molecular Biomarkers, Brain Sciences*, and *Journal of Autism and Developmental Research* and is an ad hoc peer reviewer for many scientific journals.

Inbal Gazy received her Ph.D. in Human Genetics from the Hebrew University of Jerusalem, Israel. She is currently a Postdoctoral Fellow in the Laboratory of Cell and Molecular Biology at the National Institutes of Health. Dr. Gazy has contributed to the understanding of molecular mechanisms that are responsible for several genetic diseases. Her recent work focuses on the molecular mechanisms and consequences of the unusual repeat expansion mutation responsible for fragile X-related disorders.

Editorial

Towards Mechanism-Based Treatments for Fragile X Syndrome

Daman Kumari * and Inbal Gazy

Section on Gene Structure and Disease, Laboratory of Cell and Molecular Biology, National Institute of Diabetes, Digestive and Kidney Diseases, National Institutes of Health, Bethesda, MD 20892, USA
* Correspondence: damank@niddk.nih.gov

Received: 13 August 2019; Accepted: 14 August 2019; Published: 16 August 2019

Fragile X syndrome (FXS) is the most common heritable form of intellectual disability, as well as the most common known monogenic cause of autism spectrum disorder (ASD), affecting 1 in 4000–8000 people worldwide. Almost 30 years ago, in 1991, the causative mutation for FXS was identified to be a CGG-repeat expansion in a gene named *Fragile X Mental Retardation-1* (*FMR1*) located on the X chromosome [1,2]. At that time, it also became clear that methylation of a CpG island proximal to the repeats results in loss of gene expression and disease pathology [3–7]. While many decades have elapsed since these early discoveries, the underlying mechanisms involved in the expansion mutation, as well as the resulting gene silencing, still remain elusive.

Soon after the identification of *FMR1*, a knockout mouse model was developed to better understand the role of *FMR1* and its protein product, FMRP, in FXS [8]. Work from this mouse model, as well as other models, implicated FMRP activity in brain development, synaptic plasticity and neuronal transmission circuits [9,10]. This body of work led to clinical trials attempting to correct the deficient pathways in the brain of FXS patients [11]. However, all these clinical trials, although based on successful preclinical studies, did not show beneficial effects in FXS patients. Thus, currently there is no cure or effective treatment for FXS, and all the available interventions are focused on managing patient symptoms.

Confronted with the gaps in knowledge and lack of treatments for FXS patients, the scientific community decided to revisit the approaches and practices used, implementing changes both in the preclinical and the clinical arenas. This special issue addresses some of the changes that are being made in the field towards finding effective treatments for FXS.

As reviewed by Kumari et al. [12], there are two major avenues currently being pursued for development of effective treatments: (1) Targeting pathways altered in the absence of FMRP in the brain, and (2) Restoring FMRP expression. The fact that clinical trials focused on the restoration of altered pathways have not shown much promise to date, together with the development of new techniques and approaches in the biomedical field, has led many scientific groups to revisit the restoration of FMRP as a potential treatment approach for FXS. Several strategies are currently being investigated as potential ways to restore FMRP expression in patients. One such strategy is gene therapy, wherein, FMRP could be produced from an exogenous DNA introduced into patients' cells. Hampson et al. [13] discuss this approach and the current work using viral vectors to introduce the FMRP coding sequence into patients. While studies using adeno-associated viral vector FMRP therapy in the *Fmr1* KO mouse model are showing promising results, additional work needs to be done in this area before such treatments can be considered for clinical trials.

Given that the loss of *FMR1* expression leads to the absence of FMRP in the majority of FXS patients, and that epigenetic silencing is reversible, another approach for restoring FMRP would be to reactivate the silenced *FMR1* gene. Understanding the molecular mechanism underlying *FMR1* silencing in FXS is key for the better identification of targets for its reactivation. In this regard mouse models cannot be used as they fail to recapitulate the repeat mediated *FMR1* gene silencing seen in humans. Thus, human cell lines are currently the most widely used model for such research. In this special issue, Abu Diab et al. [14] describe the use of pluripotent stem cells as a model for investigating both the timing and mechanism of gene silencing. They also describe how such models have been used to understand the mechanism of repeat expansion. Kumari et al. [12] expand further on the use of cultured cells as a model system to investigate gene silencing. The authors describe what has been learnt from these models about the pathways involved in silencing and how this knowledge can enable us to develop mechanism-targeted drugs for gene reactivation. They further describe how unbiased screening in cell culture models can be used as an alternative to identify small molecules that target silencing pathways and could potentially restore *FMR1* expression in FXS patients.

The third strategy discussed in this special issue for the restoration of FMRP expression in patients utilizes the CRISPR-Cas9 gene-editing technology. Yrigollen et al. [15] discuss the use of CRISPR-Cas9 as an alternative method for the reactivation of the *FMR1* gene via two different approaches: the use of Cas9 to (1) facilitate the recruitment of either DNA demethylases or transcriptional activators to the *FMR1* gene, or (2) delete the CGG repeats, which will hopefully lead to the loss of methylation. While such approaches are very appealing, they have many limitations such as their efficacy, off-target effects and efficiency of delivery. Thus, as with the previously described approaches, further study is required before they can be considered for therapeutic uses.

Another avenue for therapeutic targeting is to reduce, if not eliminate, the pathological expansions of the CGG repeats. To identify such targets, understanding the mechanism underlying the CGG-repeat expansion mutation is crucial. Moreover, better understanding of the expansion mechanism can also enable clinicians to better assess disease risk in patients who display high variability in the extent of expansions as well as the penetrance and manifestation of disease. Zhao et al. [16] address the current knowledge from an *Fmr1* KI mouse model, and the potential implications for humans. Saré et al. [17] present data on a different mouse model, investigating the function of FXR2P, an FMRP paralog, as a possible modulator of FXS phenotypes. Applying the lessons learnt from different mouse models to develop research in humans has the potential to increase our understanding of disease risk and repeat instability, develop better diagnostic tools for use in the clinic, identify outcome measures for clinical trials and discover new targets for drugs.

Regardless of the therapeutic approach, when moving from preclinical studies to clinical trials, objective, measurable and reliable molecular biomarkers that differentiate healthy controls from patients are necessary for evaluating the efficacy of the drugs used in the clinical trials. The failure of clinical trials in finding drugs that can be used specifically for FXS has drawn attention to the need for finding better outcome measures in FXS clinical trials. Such need is highlighted by the fact that placebo response in FXS clinical trials is strong, which can result in difficulties in assessing positive responses. Zafarullah et al. [18] list in their review the currently known candidate molecular biomarkers, and Pal et al. [19] focus on one such measure, protein synthesis, as a potential biomarker. While there are a few potential candidates, to date none of these biomarkers have been shown to be robust, reliable or accurate, or can be measured in an accessible tissue (such as blood). Therefore, there is still an urgent need to find appropriate biomarkers for the assessment of drug response in FXS clinical trials in parallel to advancing novel drug discovery.

Fragile X syndrome is a multi-symptomatic disorder and in addition to intellectual disability and ASD, there are many associated behavioral symptoms such as anxiety, depression and others. As discussed above, scientists are trying to find drugs that will be able to cure FXS, yet, until then, treatments for FXS are focused on ameliorating specific symptom/s of patients. However, what symptom/s should be the focus when developing treatments and designing clinical trials for FXS? There is an increased appreciation in the field of the importance of involving patients and their families in creating priorities for treatments. Weber and colleagues [20] created a survey for family members/caretakers as well as patients to identify such priorities. They found that learning difficulties, anxiety and behavioral problems are major concern areas that should be taken into consideration when developing treatments for FXS. They also point out that when designing clinical trials for treating these, or any other, symptoms there is a need to take into consideration both the age and the gender of patients to better target the treatment and assess outcomes.

Bartholomay et al. [21] expand on the differences between genders and draw attention to the fact that the scientific community mainly focused their research and clinical interventions on affected males, as females tend to exhibit milder symptoms of FXS. In this review, they present preliminary data from their prospective longitudinal study to identify factors that contribute to the overall functional outcomes in girls with FXS and may represent potential targets of intervention in this patient group.

The study by Bartholomay et al. [21] emphasizes the fact that we still have much to do to understand all the deficits in this multi-symptomatic syndrome. Language and communication skills are known to be significantly delayed in patients with FXS. However, very little work in the field has focused on language development and abilities during infancy and toddlerhood, due to the difficulties in assessing these early on in development. Reisinger et al. [22] describe their pilot study for the evaluation of early language development using the LENA (Language ENvironment Analysis) automated vocal analysis system. Consistent with previous literature, they found that caregivers of infants with FXS vocalize less around their children when compared to those of typically developing matched controls. Because language acquisition and cognitive development have been found to correlate with the amount of language in a child's environment, this might indicate that a simple and effective intervention such as an increase in the FXS children's caregiver's verbal responses may positively affect the child's language development.

Deficits in executive functions, cognitive abilities that support adaptive goal-directed behavior, are also a characteristic of FXS. In this special issue, Schmitt et al. [23] present a literature summary of executive function deficits in FXS patients. Given that deficits in executive functions have negative effects on FXS patients as well as their families, the authors emphasize the importance of better understanding the underlying affected processes and the identification of good outcome measures to develop treatments for improving these functions in FXS patients.

Another important factor when developing treatments is the time of intervention. It is believed that in a neurodevelopmental disease such as FXS, early intervention is critical for achieving the maximal therapeutic effect. The work by Reisinger et al. [22] described above, addresses language and communication skills as one example for the potential benefits of early intervention. A major confounding factor for early intervention is the age of disease diagnosis. Currently, the average age of diagnosis of FXS is three years, and usually later in affected females, which delays the onset of treatments. Okoniewski et al. [24] address this major concern and discuss screening approaches such as voluntary newborn screening (NBS) that can enable early diagnosis. They describe the creation of one such expanded NBS, called Early Check, for FXS in North Carolina. The potential benefits of such screening programs extend beyond the ability for early intervention, such as the ability of long-term follow-up and the collection of natural history data that can be used to better understand the development of the disease and its risk factors. This is especially important for documenting the relative risk of developmental differences and the identification of biological or environmental predictors of worse outcomes in infants with a premutation allele.

A major challenge for finding better drugs for FXS is the ability to translate the preclinical studies into successful clinical trials. To address this issue and provide guidelines to enhance the success of clinical trials, the Clinical Trials Committee (CTC) was formed in 2015. The CTC is made up of FXS experts as well as family stakeholders as a one-stop point of contact to consult and give input on any interventional trials planned for FXS patients. In this special issue, members of the CTC describe the changes that need to be implemented and factors to be considered when designing future clinical trials [25].

For optimal outcomes, changes should also be implemented in drug and clinical trial designs carried out within the pharmaceutical industry, a major driver of drug development. Lee et al. [26] discuss drug development from the industry point of view, and describe the modifications taking place within the industry to improve clinical trials. One of the major changes is the involvement of

FXS families and patients and the consideration of their insights when designing trials. This can have a positive effect not only on the outcomes but also in the recruitment and engagement of patients in clinical trials. This, together with reaching out to the scientific community (such as the CTC), the use of better outcome measures, reducing potential placebo responses and considering the heterogeneity of the condition can all lead to the development of drugs that have a meaningful impact on FXS patients' lives.

We hope that addressing the issues raised in this special issue will result in new studies that will help fill the knowledge gaps and identify objective outcome measures for successful clinical trials. This will ultimately advance the field in its search for effective treatments, or even a cure for FXS.

Conflicts of Interest: The authors declare no conflict of interest.

References

1. Kremer, E.J.; Pritchard, M.; Lynch, M.; Yu, S.; Holman, K.; Baker, E.; Warren, S.T.; Schlessinger, D.; Sutherland, G.R.; Richards, R.I. Mapping of DNA instability at the fragile X to a trinucleotide repeat sequence p(CCG)n. *Science* **1991**, *252*, 1711–1714. [CrossRef] [PubMed]
2. Verkerk, A.J.; Pieretti, M.; Sutcliffe, J.S.; Fu, Y.H.; Kuhl, D.P.; Pizzuti, A.; Reiner, O.; Richards, S.; Victoria, M.F.; Zhang, F.P.; et al. Identification of a gene (FMR-1) containing a CGG repeat coincident with a breakpoint cluster region exhibiting length variation in fragile X syndrome. *Cell* **1991**, *65*, 905–914. [CrossRef]
3. Bell, M.V.; Hirst, M.C.; Nakahori, Y.; MacKinnon, R.N.; Roche, A.; Flint, T.J.; Jacobs, P.A.; Tommerup, N.; Tranebjaerg, L.; Froster-Iskenius, U.; et al. Physical mapping across the fragile X: Hypermethylation and clinical expression of the fragile X syndrome. *Cell* **1991**, *64*, 861–866. [CrossRef]
4. Heitz, D.; Rousseau, F.; Devys, D.; Saccone, S.; Abderrahim, H.; Le Paslier, D.; Cohen, D.; Vincent, A.; Toniolo, D.; Della Valle, G.; et al. Isolation of sequences that span the fragile X and identification of a fragile X-related CpG island. *Science* **1991**, *251*, 1236–1239. [CrossRef] [PubMed]
5. Oberle, I.; Rousseau, F.; Heitz, D.; Kretz, C.; Devys, D.; Hanauer, A.; Boue, J.; Bertheas, M.F.; Mandel, J.L. Instability of a 550-base pair DNA segment and abnormal methylation in fragile X syndrome. *Science* **1991**, *252*, 1097–1102. [CrossRef] [PubMed]
6. Pieretti, M.; Zhang, F.P.; Fu, Y.H.; Warren, S.T.; Oostra, B.A.; Caskey, C.T.; Nelson, D.L. Absence of expression of the FMR-1 gene in fragile X syndrome. *Cell* **1991**, *66*, 817–822. [CrossRef]
7. Vincent, A.; Heitz, D.; Petit, C.; Kretz, C.; Oberle, I.; Mandel, J.L. Abnormal pattern detected in fragile-X patients by pulsed-field gel electrophoresis. *Nature* **1991**, *349*, 624–626. [CrossRef]
8. The Dutch-Belgian Fragile X Consortium. Fmr1 knockout mice: A model to study fragile X mental retardation. *Cell* **1994**, *78*, 23–33.
9. Bassell, G.J.; Warren, S.T. Fragile X syndrome: Loss of local mRNA regulation alters synaptic development and function. *Neuron* **2008**, *60*, 201–214. [CrossRef]
10. Sidorov, M.S.; Auerbach, B.D.; Bear, M.F. Fragile X mental retardation protein and synaptic plasticity. *Mol. Brain* **2013**, *6*, 15. [CrossRef]
11. Gross, C.; Hoffmann, A.; Bassell, G.J.; Berry-Kravis, E.M. Therapeutic Strategies in Fragile X Syndrome: From Bench to Bedside and Back. *Neurotherapeutics* **2015**, *12*, 584–608. [CrossRef] [PubMed]
12. Kumari, D.; Gazy, I.; Usdin, K. Pharmacological Reactivation of the Silenced FMR1 Gene as a Targeted Therapeutic Approach for Fragile X Syndrome. *Brain Sci.* **2019**, *9*. [CrossRef] [PubMed]
13. Hampson, D.R.; Hooper, A.W.M.; Niibori, Y. The Application of Adeno-Associated Viral Vector Gene Therapy to the Treatment of Fragile X Syndrome. *Brain Sci.* **2019**, *9*. [CrossRef] [PubMed]
14. Abu Diab, M.; Eiges, R. The Contribution of Pluripotent Stem Cell (PSC)-Based Models to the Study of Fragile X Syndrome (FXS). *Brain Sci.* **2019**, *9*. [CrossRef] [PubMed]
15. Yrigollen, C.M.; Davidson, B.L. CRISPR to the Rescue: Advances in Gene Editing for the FMR1 Gene. *Brain Sci.* **2019**, *9*. [CrossRef] [PubMed]
16. Zhao, X.; Gazy, I.; Hayward, B.; Pintado, E.; Hwang, Y.H.; Tassone, F.; Usdin, K. Repeat Instability in the Fragile X-Related Disorders: Lessons from a Mouse Model. *Brain Sci.* **2019**, *9*. [CrossRef]
17. Sare, R.M.; Figueroa, C.; Lemons, A.; Loutaev, I.; Beebe Smith, C. Comparative Behavioral Phenotypes of Fmr1 KO, Fxr2 Het, and Fmr1 KO/Fxr2 Het Mice. *Brain Sci.* **2019**, *9*. [CrossRef] [PubMed]

18. Zafarullah, M.; Tassone, F. Molecular Biomarkers in Fragile X Syndrome. *Brain Sci.* **2019**, *9*. [CrossRef]
19. Pal, R.; Bhattacharya, A. Modelling Protein Synthesis as A Biomarker in Fragile X Syndrome Patient-Derived Cells. *Brain Sci.* **2019**, *9*. [CrossRef]
20. Weber, J.D.; Smith, E.; Berry-Kravis, E.; Cadavid, D.; Hessl, D.; Erickson, C. Voice of People with Fragile X Syndrome and Their Families: Reports from a Survey on Treatment Priorities. *Brain Sci.* **2019**, *9*. [CrossRef]
21. Bartholomay, K.L.; Lee, C.H.; Bruno, J.L.; Lightbody, A.A.; Reiss, A.L. Closing the Gender Gap in Fragile X Syndrome: Review on Females with FXS and Preliminary Research Findings. *Brain Sci.* **2019**, *9*. [CrossRef] [PubMed]
22. Reisinger, D.L.; Shaffer, R.C.; Pedapati, E.V.; Dominick, K.C.; Erickson, C.A. A Pilot Quantitative Evaluation of Early Life Language Development in Fragile X Syndrome. *Brain Sci.* **2019**, *9*. [CrossRef] [PubMed]
23. Schmitt, L.M.; Shaffer, R.C.; Hessl, D.; Erickson, C. Executive Function in Fragile X Syndrome: A Systematic Review. *Brain Sci.* **2019**, *9*. [CrossRef] [PubMed]
24. Okoniewski, K.C.; Wheeler, A.C.; Lee, S.; Boyea, B.; Raspa, M.; Taylor, J.L.; Bailey, D.B., Jr. Early Identification of Fragile X Syndrome through Expanded Newborn Screening. *Brain Sci.* **2019**, *9*. [CrossRef] [PubMed]
25. Erickson, C.A.; Kaufmann, W.E.; Budimirovic, D.B.; Lachiewicz, A.; Haas-Givler, B.; Miller, R.M.; Weber, J.D.; Abbeduto, L.; Hessl, D.; Hagerman, R.J.; et al. Best Practices in Fragile X Syndrome Treatment Development. *Brain Sci.* **2018**, *8*. [CrossRef] [PubMed]
26. Lee, A.W.; Ventola, P.; Budimirovic, D.; Berry-Kravis, E.; Visootsak, J. Clinical Development of Targeted Fragile X Syndrome Treatments: An Industry Perspective. *Brain Sci.* **2018**, *8*. [CrossRef] [PubMed]

Review

Pharmacological Reactivation of the Silenced *FMR1* Gene as a Targeted Therapeutic Approach for Fragile X Syndrome

Daman Kumari *, Inbal Gazy and Karen Usdin

Section on Gene Structure and Disease, Laboratory of Cell and Molecular Biology, National Institute of Diabetes, Digestive and Kidney Diseases, National Institutes of Health, Bethesda, MD 20892, USA; inbal.gazy@nih.gov (I.G.); ku@helix.nih.gov (K.U.)
* Correspondence: damank@niddk.nih.gov; Tel.: +1-301-594-5260

Received: 12 January 2019; Accepted: 8 February 2019; Published: 12 February 2019

Abstract: More than ~200 CGG repeats in the 5′ untranslated region of the *FMR1* gene results in transcriptional silencing and the absence of the *FMR1* encoded protein, FMRP. FMRP is an RNA-binding protein that regulates the transport and translation of a variety of brain mRNAs in an activity-dependent manner. The loss of FMRP causes dysregulation of many neuronal pathways and results in an intellectual disability disorder, fragile X syndrome (FXS). Currently, there is no effective treatment for FXS. In this review, we discuss reactivation of the *FMR1* gene as a potential approach for FXS treatment with an emphasis on the use of small molecules to inhibit the pathways important for gene silencing.

Keywords: fragile X syndrome; gene reactivation; RNA:DNA hybrid; FMRP; histone methylation; DNA methylation; *FMR1*; PRC2

1. Introduction

Fragile X syndrome (FXS, MIM 300624) is the most common form of inherited cognitive disability affecting 1 in 5000 males and 1 in 8000 females [1,2]. In addition to intellectual disability, FXS often presents with a characteristic behavioral and physical phenotype that includes attention deficit, anxiety, and autism spectrum disorders as well as a prominent forehead, long face, and protruding ears [3]. FXS is caused by the loss of function of the fragile X mental retardation protein (FMRP) that is encoded by the fragile X mental retardation 1 (*FMR1*) gene. An unstable CGG repeat tract is present in the 5′ untranslated region (UTR) of the *FMR1* gene. In the general population, the repeat tract has less than 45 repeats [4]. *FMR1* alleles with 55–200 repeats are known as premutations (PM), and those with greater than 200 CGG repeats are referred to as full mutations (FM) [5,6]. Most FM alleles show aberrant DNA methylation and are transcriptionally silenced, resulting in the absence of FMRP and thus FXS [7,8]. A minority of FXS patients who do not carry the FM have deletions or point mutations in critical regions of FMRP that result in a loss of function [9–12]. Some FXS patients have a mixture of PM and FM alleles and/or some proportion of unmethylated FM alleles. These individuals make some FMRP and present with a milder clinical phenotype [13–22].

FMRP is an RNA-binding protein that regulates the transport and translation of many mRNAs in the brain [23–27]. The loss of FMRP results in defects in synaptic plasticity and neuronal development [28,29]. In addition, studies have implicated FMRP in the cellular stress response [30], cancer metastasis [31], the DNA damage response [32,33], pre-mRNA alternative splicing [34], and RNA editing [35,36]. Thus, the loss of FMRP has pleiotropic effects.

There is no cure or effective treatment for FXS. Most available medications provide only symptomatic relief, are not very effective, and can be associated with deleterious side effects.

Two different options for developing an effective treatment for FXS are possible: (i) compensating for the loss of FMRP function by identifying and normalizing the altered pathways, and (ii) restoring FMRP expression either by reactivating the silenced *FMR1* gene or by providing exogenous FMRP using gene therapy or mRNA-based approaches (Figure 1). While preclinical testing of targeted treatment strategies aimed at compensating for the loss of FMRP has been successful in mouse models of FXS (reviewed in [37]), many of the clinical trials based on these studies were unsuccessful (see [38] for a recent review). There are a variety of possible explanations for why this was the case, including heterogeneity in the FXS patient population, the lack of suitable objective outcome measures, and the fact that only a subset of altered pathways were targeted.

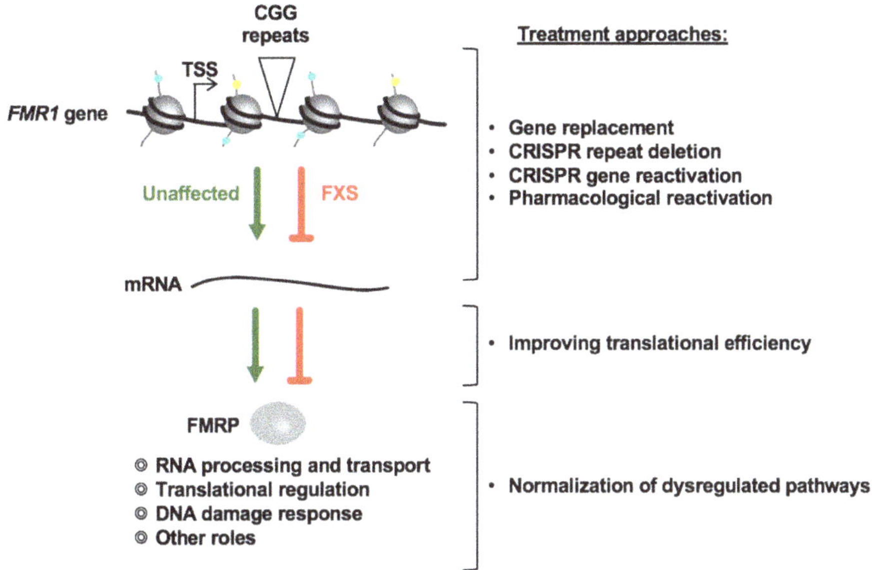

Figure 1. Possible treatment approaches for fragile X syndrome (FXS).

In principle, restoring FMRP expression may be more broadly useful as it targets the root cause of the disease, the absence of FMRP. Different strategies are being pursued for this purpose. Preliminary studies using clustered regularly interspaced short palindromic repeats (CRISPR)/Cas9-mediated gene editing approaches to (i) delete the expanded CGG repeats in FXS patient cells [39,40], (ii) induce DNA demethylation in the *FMR1* promoter region [41], and (iii) target transcriptional activators to the *FMR1* promoter in FXS cells [42] have all been successful in partially reactivating the *FMR1* gene in cell models. Gene therapy approaches are also being pursued to restore FMRP expression. For example, FMRP expression can be achieved in the brains of *Fmr1* knockout (KO) animals using adeno-associated virus (AAV) vectors for gene delivery. Such exogenous expression of FMRP corrects abnormally enhanced hippocampal long-term synaptic depression [43] and reverses some of the abnormal behaviors seen in this mouse model [44]. These approaches are discussed elsewhere in this special issue. In this review we will focus on pharmacological approaches for *FMR1* gene reactivation [45–48]. The use of small molecules for gene reactivation is currently being tested for a number of other disorders including myelodysplatic syndromes [49], Rett Syndrome [50,51], Angelman syndrome [52], frontotemporal dementia [53], and Friedreich ataxia [54]. As a result, the list of small molecules able to reactivate silenced genes that have been approved for use in humans is growing rapidly [55]. The search for small molecules suitable for gene reactivation can be divided into two categories: (i) a rational or candidate approach, in which specific pathways important for silencing are identified and targeted for gene reactivation, and (ii) an unbiased screening approach to identify small molecules that are capable of reactivating the silenced gene in patient cells.

2. Targeting Specific Pathways and Proteins Involved in *FMR1* Gene Silencing in FXS

The rational or candidate approach to reactivating the *FMR1* gene in FXS requires a clear understanding of the underlying silencing mechanism. Despite the fact that it has been more than 25 years since the *FMR1* gene and the causative CGG expansion mutation were identified, the mechanism by which the repeat expansion leads to gene silencing in FXS is still not completely understood. In the following sections, we will review the research that has identified some of the epigenetic modifications present on silenced alleles, some of the proteins important for these modifications, and the various small molecule-based approaches that have been used to date for gene reactivation.

2.1. Epigenetic Marks Associated with the Silenced FMR1 Gene in FXS

The transcriptional activity of a gene is regulated by various epigenetic marks that include DNA methylation and modifications of the N-terminal tails of histone proteins associated with the promoter. In general, transcriptionally active regions are hypomethylated and enriched for acetylated histones. These modifications result in an open chromatin conformation or euchromatin. In contrast, transcriptionally inactive regions often show high levels of CpG methylation of the DNA and are associated with histone modifications that result in compact chromatin (heterochromatin). Heterochromatin is generally hypoacetylated and enriched for H3 and H4 histones methylated at specific lysine residues. Some of these modifications are characteristic of facultative heterochromatin, which is found on developmentally silenced genes, whilst other modifications are more typical of constitutive heterochromatin, which is important for the silencing of repeat elements in the genome.

Most of the knowledge about the contribution of various epigenetic modifications to *FMR1* gene silencing has been obtained from studies done with FXS patient cells. It was observed early on that the FM alleles show increased CpG methylation [7,8]. In vitro methylation of the *FMR1* promoter repressed its activity in transient expression assays [56], perhaps because it abolishes the binding of the transcription factor alpha-Pal/Nrf-1 and reduces binding of upstream stimulatory factor (USF) 1 and USF2 [57]. Moreover, treatment of FXS patient cells with an inhibitor of DNA methyltransferase 1 (DNMT1), 5-azadeoxycytidine (5-aza-dC), leads to gene reactivation [45]. These data, together with the existence of rare individuals with unmethylated FM alleles who are high-functioning, reinforce the importance of DNA hypermethylation and *FMR1* gene silencing for the development of FXS symptoms.

The *FMR1* promoter in FXS patient cells is also associated with a decrease in the levels of active histone marks that include acetylation of histone H3 at lysine 9 (H3K9ac), di-methylation of lysine 4 (H3K4me2), and acetylation of histone H4 at lysine 16 (H4K16ac) [46,47,58]. Moreover, the levels of repressive histone marks are increased on the FM alleles. These include di- and tri-methylation of histone H3 at lysine 9 (H3K9me2, H3K9me3), tri-methylation of lysine 27 (H3K27me3), and tri-methylation of histone H4 at lysine 20 (H4K20me3) [58–60]. Thus, the silenced allele is enriched for histone modifications characteristic of both facultative and constitutive heterochromatin.

2.2. Models for FMR1 Gene Silencing

In addition to the identification of chromatin modifications important for silencing, a knowledge of the timing and sequence of events leading to these modifications may also be important for designing effective strategies for gene reactivation. Early studies of FXS embryonic stem cells (ESCs) suggested that H3K9me2 is a relatively early event in the silencing process, occurring before DNA methylation [61]. This is consistent with the observation that DNA demethylation does not affect the levels of H3K9 methylation in differentiated cells [58,62]. Many regulators of heterochromatin formation bind methylated histone lysines [63] and recruit DNA methyltransferases. For example, methylated H3K9 serves as a binding platform for the recruitment of heterochromatin protein 1 (HP1) that in turn recruits the de novo DNA methyltransferases 3A and 3B [64]. Enhancer of Zeste 2 (EZH2), the catalytic component of polycomb repressive complex 2 (PRC2) that is responsible for H3K27me3, has also been shown to be necessary for DNA methylation of PRC2-target genes [65], although it is not sufficient for methylation at all loci [66]. Since DNA demethylation of FXS alleles results in increased H4K16 acetylation [47], it may be that H4K16 deacetylation occurs downstream of DNA methylation. However, a better understanding of the silencing process is required in order to understand the relationships between all of the epigenetic factors involved.

One model for gene silencing suggests that silencing is initiated by the loss of binding of an insulator protein to a region upstream of the *FMR1* promoter [67]. This in turn is suggested to disrupt the chromatin boundary upstream of the *FMR1* promoter, thus allowing the spread of DNA methylation and repressive histone marks from an upstream heterochromatic zone [60,67]. A recent study has also suggested that CCCTC-binding factor (CTCF) plays a role in maintaining the topologically associated domains (TADs) at disease-associated short tandem repeats like those at the *FMR1* locus and that disruption of the higher-order genome folding alters the enhancer landscape leading to gene silencing [68]. While CTCF binding has been seen in some cells from unaffected individuals and is missing from the same cell type in affected individuals [69,70], demethylation of the *FMR1* promoter in FXS lymphoblastoid cells by 5-aza-dC treatment did not restore CTCF binding. Furthermore, knockdown of CTCF in cells from unaffected individuals and those with unmethylated FMs did not result in the spreading of DNA methylation from the upstream boundary into the *FMR1* promoter [70]. This suggests that CTCF does not act as an insulator at the *FMR1* locus. Whether the loss of the chromatin boundary is relevant to *FMR1* gene silencing is still unclear.

An alternate hypothesis is that silencing is initiated by chromatin changes occurring in the repeat itself. This idea is supported by the observation that the constitutive heterochromatin marks, H3K9me3 and H4K20me3, show a focal distribution on the *FMR1* gene in FXS cells, being enriched in the vicinity of expanded CGG repeats. In contrast, the facultative heterochromatin marks, H3K9me2 and H3K27me3, are more widely distributed, perhaps, due to the propensity of these marks to spread [60]. In this view, the spreading of the facultative heterochromatin from the vicinity of the expanded repeat results in its merging with the upstream heterochromatin zone, and thus giving the impression of a loss of boundary function.

The expanded CGG repeats could trigger heterochromatin formation by processes that are DNA-or RNA-dependent (Figure 2) (reviewed in [71]). The expanded CGG/CCG repeats in the DNA and RNA form unusual structures in vitro that include stem-loop/hairpins, G-tetraplexes/quadruplexes, and R-loops/RNA:DNA hybrids [48,72–86], and there is evidence that such structures are also formed in vivo [84,86]. These secondary structures may in turn recruit chromatin modifiers. For example, it has been suggested that CGG hairpins can directly recruit DNA methyl transferases [73,87]. Sequence-specific factors that bind CGG repeats in the DNA may also play a role in heterochromatization of the *FMR1* locus by directly recruiting chromatin modifiers, as has been reported for Suv39h recruitment to major satellite repeats in mice [88].

An RNA-based mechanism for the initiation of silencing is appealing since it has been shown that the increase in the levels of repressive mark H3K27me3 after 5-aza-dC treatment is dependent on the levels of the *FMR1* transcript and that blocking deposition of this repressive mark is able to significantly delay the re-silencing that happens after 5-aza-dC is withdrawn [48,62]. In addition, the observed similarity between the chromatin marks associated with the silenced *FMR1* gene and *Sat2* repeats [60] suggests that *FMR1* gene silencing might involve a mechanism similar to the RNA-dependent mechanism involved in the formation of pericentromeric heterochromatin (reviewed in [89]). Both RNAi-dependent and RNAi-independent RNA-based models have been proposed for silencing in FXS. RNAi-dependent gene silencing involves the generation of small double stranded (ds) RNAs by Dicer cleavage of larger dsRNAs. The small dsRNAs associate with Argonaute (AGO) proteins which in turn recruit chromatin modifiers to the locus that result in transcriptional silencing. A complex mixture of sense and antisense transcripts have been identified at the *FMR1* locus, and their pairing could provide a substrate for the generation of small dsRNAs [60,69,90]. Another potential source of small dsRNAs could be the hairpins formed by the CGG repeats in the RNA which are known to be substrates for Dicer [80]. Indeed, small RNAs, ~20 nucleotide (nt) in length that are derived from the *FMR1* promoter and CGG repeat region, have been reported in FXS lymphoblastoid cells after treatment with 5-aza-dC. However, similar levels of these small RNAs were also seen in cells from unaffected individuals [60]. A recent study also reported the presence of ~21 nt RNAs containing CGG repeats in association with AGO1 in FXS ESCs. This was suggested to lead to the recruitment of the H3K9 histone methyl transferase, SUV39H, to the *FMR1* locus [91]. However, while members of the AGO protein family have been shown to be important for heterochromatin formation and transcriptional silencing of repetitive sequences in fission yeast [92], *Tetrahymena* [93,94] and *Drosophila* [95], their role in transcriptional gene silencing in mammals in general [96–98] and in the context of FXS in particular [82] is unclear.

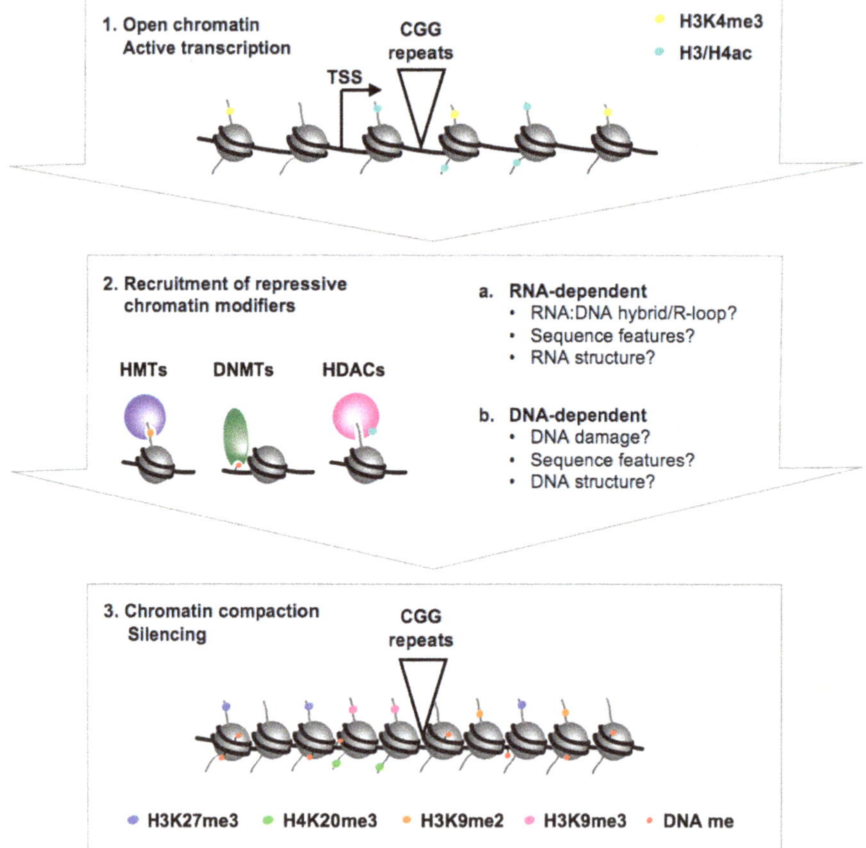

Figure 2. Schematic representation of the recruitment of repressive chromatin modifiers to the *FMR1* locus leading to a heterochromatic silenced state in FXS.

RNAi-independent mechanisms for heterochromatin formation at the *FMR1* locus could involve the recruitment of chromatin modifiers by the *FMR1* mRNA through a mechanism similar to that used by long non-coding RNAs like *HOTAIR* [99], *Air* [100], and *Xist* [101,102]. This would require the interaction of *FMR1* mRNA with the *FMR1* locus either directly or indirectly. The *FMR1* mRNA from reactivated alleles is preferentially enriched in the chromatin fraction and thus supporting such interaction [48]. Moreover, an R-loop is present at the *FMR1* locus [48,82–84,86]. R-loops can potentially recruit chromatin modifiers to a genomic locus, as was proposed for the recruitment of the PRC2 responsible for H3K27me3 deposition at the *RASSF1A* gene [103]. Stable R-loops have also been shown to recruit the G9a histone methyltransferase, which is responsible for H3K9me2, to expanded GAA repeats in the *frataxin* gene in Friedreich ataxia [83]. Furthermore, RNA:DNA hybrid formation has also been implicated in the recruitment of H3K9 trimethylases, Suv39h1 and Suv39h2, to the heterochromatin in mouse ESCs [104]. While an R-loop is present at the *FMR1* locus in unaffected individuals, the R-loop formed on FM alleles would be longer and more stable and thus perhaps more effective at recruitment of epigenetic repressors. PRC2 can directly bind quadruplexes in RNA or unstructured G-rich RNA sequences [105] and has been shown to interact with the *Fmr1* transcript in mouse ESCs [106]. Since the 5′ end of mouse and human *FMR1* transcripts share much sequence similarity, it is possible that PRC2 also directly binds the human *FMR1* transcript. This may represent one way in which R-loops are able to recruit PRC2 to the 5′ end of the *FMR1* gene. In addition, SUV39H1 has also been shown to directly bind RNA in both mice and humans, although it does not show any major preference for particular structures or sequences [104,107–109]. A persistent R-loop could also cause transcription termination. The resulting short nascent RNAs from the 5′ end of the *FMR1* transcript may also have the potential to recruit chromatin modifiers including PRC2 [110].

2.3. Targeting the Activity of Repressive Chromatin Modifiers for Gene Reactivation

In principle, inhibition of the repressive epigenetic modifiers involved in the silencing of FM alleles could result in gene reactivation. For example, inhibition of DNMT1 with 5-aza-dC reactivates the *FMR1* gene in FXS lymphoblastoid, fibroblasts, and induced pluripotent stem cell (iPSC)-derived neural progenitor cells (NPCs). Given that differentiated cells express little, if any, active demethylation factors, DNMT1 inhibitors are thought to be effective primarily in dividing cells where methylation of nascent daughter strands is prevented and existing methylation marks are gradually diluted by successive rounds of replication. In addition to 5-aza-dC, another somewhat less effective DNMT1 inhibitor, 5-azacytidine (5-aza-C), has been shown to partially reactivate the *FMR1* gene in NPCs derived from FXS iPSCs [111,112] and in vitro differentiated neurons [111]. However, given that only ~50% of the cells in the differentiated population were neuronal cells, whether demethylation actually occurred in the neurons is unclear. An effect of DNMT1 inhibitors in non-dividing cells like neurons would be both surprising and important given the need to restore FMRP expression for proper neuronal function. Valproic acid (VPA), a drug used for the treatment of epilepsy and as a mood stabilizer, has been shown to trigger replication-independent active demethylation at other loci [113]. However, treatment of FXS lymphoblastoid cells with VPA had little effect on transcriptional activation and did not induce DNA demethylation of the *FMR1* gene [114]. Another compound that has been suggested to cause DNA demethylation is methotrexate (MTX), an inhibitor of dihydrofolate reductase [115,116]. However, while treatment of FXS fibroblasts with MTX did indeed reactivate the *FMR1* gene, it did not decrease promoter methylation, suggesting that its effect on *FMR1* transcription was independent of local DNA demethylation [117]. There is some evidence to suggest that 5-aza-dC can affect gene expression independently of its effects on DNA demethylation, and it is possible that this accounts for the high efficacy of this compound relative to other DNMT1 inhibitors [118,119].

Treatment of FXS cells with inhibitors of class I, II, and IV histone deacetylases (HDAC) including 4-phenylbutyrate, sodium butyrate (NaB), trichostatin A (TSA), romidepsin, and vorinostat have been shown to be ineffective at reactivating the silenced *FMR1* gene; however, a synergistic effect of 5-aza-dC and some of these inhibitors has been reported [120,121]. Treatment with either NaB or TSA increased total H4 acetylation but did not increase H3 acetylation [58]. In contrast, inhibition of Sirtuin 1(SIRT1), a class III HDAC, is able to reactivate the *FMR1* gene to levels similar to those seen with 5-aza-dC treatment [47] and results in increased acetylation at both H3 lysine 9 and H4 lysine 16. Because SIRT1 inhibitors do not require replication to be effective, this class of inhibitors might be better for the reactivation of FM alleles in non-dividing cells like neurons.

While gene reactivation with 5-aza-dC treatment does not lead to any significant changes in the levels of repressive histone marks H3K9me2, H3K9me3, or H4K20me3 [58,62], it increases the levels of H3K27me3 as well as EZH2, the catalytic component of the PRC2 that is responsible for H3K27me3, at the *FMR1* promoter [62]. EZH2 inhibition by itself does not reactivate the *FMR1* gene in FXS cells; however, EZH2 inhibitors are very effective at preventing the re-silencing of reactivated alleles after 5-aza-dC is withdrawn [48]. This would be consistent with the idea that PRC2 recruitment to the *FMR1* locus likely also occurs prior to DNA methylation. Since DNA methylation is likely to be clonally propagated in differentiated cells [122,123], DNA methylation effectively provides an epigenetic memory of the silenced state in these cells. Thus, inhibition of DNMT1 is likely to be required for inhibitors of early steps in the silencing pathway to be effective at reactivation of silenced alleles.

It is also important to note that the efficacy of all of these compounds varies considerably between different cell lines. Whether this is due to differences in the length of the CGG repeat tract and the extent of DNA methylation or other genetic differences that might impact the drug uptake and efflux is unclear. Furthermore, these compounds are likely to have off-target effects that include altered expression of other repressed genes [124,125] and other effects on cell viability [126,127].

2.4. Targeting the Recruitment of Chromatin Modifiers for Gene Reactivation

While it has been reported that some inhibitors of epigenetic modifying enzymes do not have the global effects on gene expression that one might expect [124,125], in principle, a more gene-specific strategy for the reactivation of *FMR1* would reduce the likelihood of undesirable effects at other loci. Given that inhibition of PRC2 after 5-aza-dC withdrawal prevents the re-silencing of reactivated alleles and that the *FMR1* transcript is involved in PRC2 recruitment and gene silencing, blocking this recruitment might be one way to achieve a more specific effect on *FMR1* expression. This could be accomplished either by preventing the interaction of *FMR1* mRNA with the *FMR1* locus or by blocking the binding of *FMR1* mRNA to PRC2. For example, Compound 1a (9-hydroxy-5, 11-dimethyl-2-[2-(piperidin-1-yl)ethyl]-6H-pyrido[4,3-b]carbazol-2-ium) is a small molecule that stabilizes the hairpins formed by the CGG repeats in the RNA (rCGG) and disrupts protein-binding [128,129]. Treatment of FXS lymphoblastoid cells with Compound 1a reduced H3K27me3 levels and prevented re-silencing of 5-aza-dC-reactivated alleles [48]. However, this treatment had no effect on the formation of RNA:DNA hybrids, suggesting that Compound 1a was acting to prevent re-silencing by blocking the interaction of *FMR1* mRNA with PRC2 [48]. Since Compound 1a also binds other G-rich repeats [130], and such repeat tracts are present at multiple locations in the human genome, it is possible that Compound 1a will have undesirable effects on the expression of other genes. While the extent of off-target effects of Compound 1a requires additional study, the development of more *FMR1*-specific small molecules that are also able to block PRC2 recruitment may be desirable.

3. Unbiased High-Throughput Screens to Identify Compounds that Reverse *FMR1* Gene Silencing

High-throughput screening (HTS) is an approach to accelerate drug discovery that involves testing a large number of potential biological modulators and effectors against a specific target. With respect to identifying compounds that can reactivate the *FMR1* gene in FXS cells, HTS can help identify new biologically active small molecules against known targets, for example, DNA methyltransferases, histone deacetylases, and histone methyltransferases, as well as identify additional targets that might provide new insights into the gene silencing mechanism. This is particularly important as many of the compounds that have been shown to reactivate the silenced *FMR1* gene so far are toxic and may not be suitable for long-term use in humans.

The success of an HTS depends on a robust readout that can be measured using an assay that is reliable and economical. In addition to the detection assay, the cells used in the screen are also important. Most FXS fibroblasts are mosaic for PM and FM alleles as well as their methylation status. Hence, it is difficult to assess if the increase in FMRP levels is due to an increase in translation from transcriptionally active alleles or reactivation of the silenced alleles. Furthermore, primary fibroblasts have a finite life span making it difficult to generate the large number of cells required for HTS. These characteristics make primary human fibroblast cells less desirable for HTS. In contrast, FXS iPSC-derived neural stem cells (NSCs) or NPCs are suitable for HTS because they can be used to rapidly generate a large number of cells. Furthermore, given that FM alleles in FXS iPSCs remain methylated [131,132], it is relatively easy to generate lines with a homogeneous population of silenced FM alleles.

In this respect, HTS designed for identifying drugs that can increase *FMR1* transcription have relied on both the detection of endogenous FMRP in FXS NSCs or NPCs and the use of reporter cell lines. In the following sections we review the library screens that have been done thus far to identify compounds that reactivate the *FMR1* gene.

3.1. HTS Based on Measuring Endogenous FMRP Levels

Two different HTS have been reported for identifying molecules that are able to reactivate the *FMR1* gene [133,134]. For one of these screens, a specific and sensitive time-resolved fluorescence resonance energy transfer (TR-FRET)-based FMRP assay was developed and used to perform HTS in a 1536-well plate format [134]. Given that long CGG repeats negatively affect translation, the authors chose an iPSC line carrying a completely methylated FM allele with relatively short CGG repeats (~300) to generate NSCs and neurons. The NSCs were treated with 5-aza-dC to confirm the production of FMRP by western blot analysis and TR-FRET assay. FXS NSCs and in vitro differentiated neurons were first evaluated in test screens of a LOPAC^{1280} compound library, and two hits were identified, protoporphyrin IX and SB216763. These hit compounds were further validated by the dose response in TR-FRET assay and in a quantitative reverse transcription (qRT)-PCR assay for *FMR1* mRNA. The authors then screened a ~4000 compound FDA-approved library. Four additional compounds (sodium decanehydroxamate, geliomycin, tibrofan, and deserpidine) were identified from this HTS. With the exception of sodium decanehydroxamate, which is known to have HDAC activity, the mode of action of these compounds is unknown. While the compounds identified in this HTS were effective at very high concentrations and thus not likely to be biologically useful, it provides proof of principle of the approach and suggests that better lead compounds might be identified using larger compound libraries.

In another screen, Kaufmann et al. [133] used high-content imaging with FMRP antibodies to screen 50,000 compounds in a 384-well plate format. The cells used were FXS NPCs derived from an iPSC line carrying a single methylated *FMR1* allele with ~480 CGG repeats. The authors also used 5-aza-dC as a positive control compound for the HTS. A total of 2099 compounds were identified that induced a small FMRP increase, and 790 of those were further tested in dose response assays. Only one compound was identified that had a previously known mode of action—a hydroxamate-based HDAC inhibitor, similar to the one identified in the HTS by Kumari and Swaroop et al. [134]. The identity of other hit compounds was not disclosed. The advantage of high-content imaging-based FMRP detection is that it provides information about FMRP levels in single cells. However, careful calibration is required to enable the detection of weak hits.

3.2. HTS Using Knock-In Reporter Cell Lines

While HTS based on assays using antibodies to detect FMRP were successful at identifying a number of hit compounds, the cost of antibodies is a limiting factor. To eliminate the requirement for antibodies, Li et al. [112] inserted the Nano luciferase gene (Nluc) into the endogenous *FMR1* gene locus in FXS iPSCs and control H1 ESCs using CRISPR/Cas9 gene editing. The authors then generated NPCs from the H1 control and FXS *FMR1*-Nluc reporter lines and optimized the HTS in 384-well and 1536-well plates. Both 5-aza-C and 5-aza-dC were used as positive control compounds to screen a

128 epigenetic compound library and ~1134 FDA-approved drug library using NPCs differentiated from the FXS-*FMR1*-Nluc reporter iPSCs. While no new compounds were identified in these screens, the sensitivity and cost-effectiveness of these reporter lines will make it feasible to screen very large compound libraries.

The FMRP-based screening assay and the Nano-luc reporter lines share a common drawback—namely, the negative effect of CGG repeats on translation, which limits the amount of protein that can be detected. Thus, these assays are less sensitive than assays that detect *FMR1* mRNA levels. However, currently, RNA detection assays are not economical, and the generation of new reporter cell lines that allow the effect of compounds on transcription or translation to be distinguished would thus be highly desirable for large library screens.

4. Limitations and Challenges of Using *FMR1* Gene Reactivation as a Treatment Approach for FXS

Restoring FMRP expression in FXS cells by reactivating the endogenous *FMR1* gene is a potentially useful treatment option for FXS. However, many challenges to this approach remain. More work is needed to identify additional proteins important for gene silencing that can be targeted for gene reactivation. Conducting genome-wide CRISPR/siRNA screens for gene reactivation is one way to identify new targets. In addition, better model systems to understand the initiating events leading to gene silencing may allow the identification of novel proteins for pharmacological targeting. Early work in humans had led to the suggestion that *FMR1* gene silencing occurs relatively late in embryonic development at around 10 weeks of gestation [135]. If so, then the FM alleles in ESCs would be expected to be active, thus making these cells useful for understanding early steps in the silencing process. Indeed, a few studies have reported that the *FMR1* gene is actively transcribed in FXS ESCs and undergoes differentiation-induced silencing [61,82,136]. However, many FXS ESC lines already show some DNA methylation, suggesting that differentiation *per se* is not required for gene silencing [137,138]. Moreover, the silenced *FMR1* gene was not reactivated in iPSCs derived from FXS patients [131,132]. Both human ESCs and iPSCs are thought to more closely resemble primed pluripotent stem cells rather than the earlier more naïve state present in the preimplantation embryo, and it may be that *FMR1* gene silencing occurs at an earlier stage of embryonic development. It is therefore possible that naïve FXS iPSCs or ESCs could provide a better cell model for understanding the very earliest events in *FMR1* gene silencing [139].

Additional challenges include the necessity for overcoming the negative effect of the CGG repeats on the translation efficiency of *FMR1* mRNA [140–143] and to reduce the toxicity associated with rCGG expression that is thought to be responsible for fragile X-associated tremor/ataxia syndrome (FXTAS) and fragile X-associated primary ovarian insufficiency (FXPOI) in PM carriers [144–148]. Indeed, some carriers of unmethylated FM alleles have been reported to show symptoms of FXTAS [149–151]. However, there is a wide variability in the expression levels of FMRP in unaffected individuals [152] and FM females who express FMRP in only 50% of their cells usually present with milder intellectual impairment [153,154]. Similarly, males mosaic for PM and FM alleles and those with unmethylated FM alleles can make some FMRP and can be high-functioning. Thus, reactivation that results in expression of even low levels of FMRP may be clinically beneficial. In principle, pharmacological approaches that maximize translation may be useful in reducing the level of *FMR1* gene reactivation required. This would also reduce the risk of developing pathology resulting from rCGG expression. Furthermore, small molecules like Compound 1a that prevent gene re-silencing in a relatively gene-specific way while also reducing the deleterious effects of rCGG may be particularly useful [48,128,129].

Finally, the challenge of if, when, and how a gene reactivation approach could be deployed needs to be considered. For families with a known history of FXS, preimplantation genetic diagnosis might be the preferred option [155]. Most new cases of FXS are diagnosed when the child is already 2–3 years old [156]. While FXS deficits that likely arise during embryonic life are unlikely to be modulated by postnatal gene reactivation, there is evidence from work in *Fmr1* KO mice to suggest that increasing

FMRP production during postnatal life may still have some clinical benefit [157,158]. Whether gene reactivation in utero will be feasible given the potentially detrimental effects of epigenetic modulators in early development remains to be seen.

5. Concluding Remarks

Preliminary studies using cell-based models for FXS have shown that it is possible to reactivate the silenced *FMR1* gene and suggested approaches for gene reactivation that are most likely to be effective. However, we still have a long way to go before this approach is therapeutically useful. A number of new tools are needed, including an animal model that recapitulates the repeat-mediated *FMR1* gene silencing seen in FXS and human cell-based/organoid models that can be used to verify the compounds and approaches that work in the animal models. In addition, the identification of molecular biomarkers, a focus area for FXS, will be useful not only to test the efficacy of treatment strategies based on restoring FMRP expression but also for those aimed at compensating for its loss.

Author Contributions: D.K. and K.U. conceived the idea and D.K., I.G., and K.U. wrote the paper.

Funding: This research and the APC was funded by NIDDK IRP, grant number DK057602.

Acknowledgments: The work described in this manuscript was funded by a grant from the Intramural Program of the NIDDK to K.U. (DK057602), which includes funds for covering the costs to publish in open access.

Conflicts of Interest: The authors declare no conflict of interest.

References

1. Coffee, B.; Keith, K.; Albizua, I.; Malone, T.; Mowrey, J.; Sherman, S.L.; Warren, S.T. Incidence of fragile X syndrome by newborn screening for methylated *FMR1* DNA. *Am. J. Hum. Genet.* **2009**, *85*, 503–514. [CrossRef] [PubMed]
2. Tassone, F.; Iong, K.P.; Tong, T.H.; Lo, J.; Gane, L.W.; Berry-Kravis, E.; Nguyen, D.; Mu, L.Y.; Laffin, J.; Bailey, D.B.; et al. FMR1 CGG allele size and prevalence ascertained through newborn screening in the United States. *Genome Med.* **2012**, *4*, 100. [CrossRef] [PubMed]
3. Kidd, S.A.; Lachiewicz, A.; Barbouth, D.; Blitz, R.K.; Delahunty, C.; McBrien, D.; Visootsak, J.; Berry-Kravis, E. Fragile X syndrome: A review of associated medical problems. *Pediatrics* **2014**, *134*, 995–1005. [CrossRef]
4. Monaghan, K.G.; Lyon, E.; Spector, E.B.; erican College of Medical Genetics and Genomics. ACMG Standards and Guidelines for fragile X testing: A revision to the disease-specific supplements to the Standards and Guidelines for Clinical Genetics Laboratories of the American College of Medical Genetics and Genomics. *Genet. Med.* **2013**, *15*, 575–586. [CrossRef] [PubMed]
5. Fu, Y.H.; Kuhl, D.P.; Pizzuti, A.; Pieretti, M.; Sutcliffe, J.S.; Richards, S.; Verkerk, A.J.; Holden, J.J.; Fenwick, R.G., Jr.; Warren, S.T.; et al. Variation of the CGG repeat at the fragile X site results in genetic instability: Resolution of the Sherman paradox. *Cell* **1991**, *67*, 1047–1058. [CrossRef]
6. Verkerk, A.J.; Pieretti, M.; Sutcliffe, J.S.; Fu, Y.H.; Kuhl, D.P.; Pizzuti, A.; Reiner, O.; Richards, S.; Victoria, M.F.; Zhang, F.P.; et al. Identification of a gene (*FMR-1*) containing a CGG repeat coincident with a breakpoint cluster region exhibiting length variation in fragile X syndrome. *Cell* **1991**, *65*, 905–914. [CrossRef]
7. Pieretti, M.; Zhang, F.P.; Fu, Y.H.; Warren, S.T.; Oostra, B.A.; Caskey, C.T.; Nelson, D.L. Absence of expression of the *FMR-1* gene in fragile X syndrome. *Cell* **1991**, *66*, 817–822. [CrossRef]
8. Sutcliffe, J.S.; Nelson, D.L.; Zhang, F.; Pieretti, M.; Caskey, C.T.; Saxe, D.; Warren, S.T. DNA methylation represses *FMR-1* transcription in fragile X syndrome. *Hum. Mol. Genet.* **1992**, *1*, 397–400. [CrossRef] [PubMed]
9. De Boulle, K.; Verkerk, A.J.; Reyniers, E.; Vits, L.; Hendrickx, J.; Van Roy, B.; Van den Bos, F.; de Graaff, E.; Oostra, B.A.; Willems, P.J. A point mutation in the *FMR-1* gene associated with fragile X mental retardation. *Nat. Genet.* **1993**, *3*, 31–35. [CrossRef]
10. Lugenbeel, K.A.; Peier, A.M.; Carson, N.L.; Chudley, A.E.; Nelson, D.L. Intragenic loss of function mutations demonstrate the primary role of *FMR1* in fragile X syndrome. *Nat. Genet.* **1995**, *10*, 483–485. [CrossRef] [PubMed]

11. Collins, S.C.; Bray, S.M.; Suhl, J.A.; Cutler, D.J.; Coffee, B.; Zwick, M.E.; Warren, S.T. Identification of novel *FMR1* variants by massively parallel sequencing in developmentally delayed males. *Am. J. Med. Genet. A* **2010**, *152A*, 2512–2520. [CrossRef]

12. Suhl, J.A.; Warren, S.T. Single-Nucleotide Mutations in FMR1 Reveal Novel Functions and Regulatory Mechanisms of the Fragile X Syndrome Protein FMRP. *J. Exp. Neurosci.* **2015**, *9*, 35–41. [CrossRef] [PubMed]

13. McConkie-Rosell, A.; Lachiewicz, A.M.; Spiridigliozzi, G.A.; Tarleton, J.; Schoenwald, S.; Phelan, M.C.; Goonewardena, P.; Ding, X.; Brown, W.T. Evidence that methylation of the *FMR-I* locus is responsible for variable phenotypic expression of the fragile X syndrome. *Am. J. Hum. Genet.* **1993**, *53*, 800–809. [PubMed]

14. Merenstein, S.A.; Shyu, V.; Sobesky, W.E.; Staley, L.; Berry-Kravis, E.; Nelson, D.L.; Lugenbeel, K.A.; Taylor, A.K.; Pennington, B.F.; Hagerman, R.J. Fragile X syndrome in a normal IQ male with learning and emotional problems. *J. Am. Acad. Child Adolesc. Psychiatry* **1994**, *33*, 1316–1321. [CrossRef] [PubMed]

15. Rousseau, F.; Robb, L.J.; Rouillard, P.; Der Kaloustian, V.M. No mental retardation in a man with 40% abnormal methylation at the *FMR-1* locus and transmission of sperm cell mutations as premutations. *Hum. Mol. Genet.* **1994**, *3*, 927–930. [CrossRef] [PubMed]

16. Smeets, H.J.; Smits, A.P.; Verheij, C.E.; Theelen, J.P.; Willemsen, R.; van de Burgt, I.; Hoogeveen, A.T.; Oosterwijk, J.C.; Oostra, B.A. Normal phenotype in two brothers with a full *FMR1* mutation. *Hum. Mol. Genet.* **1995**, *4*, 2103–2108. [CrossRef]

17. Lachiewicz, A.M.; Spiridigliozzi, G.A.; McConkie-Rosell, A.; Burgess, D.; Feng, Y.; Warren, S.T.; Tarleton, J. A fragile X male with a broad smear on Southern blot analysis representing 100–500 CGG repeats and no methylation at the EagI site of the *FMR-1* gene. *Am. J. Med. Genet.* **1996**, *64*, 278–282. [CrossRef]

18. Wang, Z.; Taylor, A.K.; Bridge, J.A. *FMR1* fully expanded mutation with minimal methylation in a high functioning fragile X male. *J. Med. Genet.* **1996**, *33*, 376–378. [CrossRef]

19. Tassone, F.; Hagerman, R.J.; Ikle, D.N.; Dyer, P.N.; Lampe, M.; Willemsen, R.; Oostra, B.A.; Taylor, A.K. FMRP expression as a potential prognostic indicator in fragile X syndrome. *Am. J. Med. Genet.* **1999**, *84*, 250–261. [CrossRef]

20. Taylor, A.K.; Tassone, F.; Dyer, P.N.; Hersch, S.M.; Harris, J.B.; Greenough, W.T.; Hagerman, R.J. Tissue heterogeneity of the *FMR1* mutation in a high-functioning male with fragile X syndrome. *Am. J. Med. Genet.* **1999**, *84*, 233–239. [CrossRef]

21. Loesch, D.Z.; Huggins, R.M.; Hagerman, R.J. Phenotypic variation and FMRP levels in fragile X. *Ment. Retard. Dev. Disabil. Res. Rev.* **2004**, *10*, 31–41. [CrossRef] [PubMed]

22. Hatton, D.D.; Sideris, J.; Skinner, M.; Mankowski, J.; Bailey, D.B., Jr.; Roberts, J.; Mirrett, P. Autistic behavior in children with fragile X syndrome: Prevalence, stability, and the impact of FMRP. *Am. J. Med. Genet. A* **2006**, *140A*, 1804–1813. [CrossRef] [PubMed]

23. Eberhart, D.E.; Malter, H.E.; Feng, Y.; Warren, S.T. The fragile X mental retardation protein is a ribonucleoprotein containing both nuclear localization and nuclear export signals. *Hum. Mol. Genet.* **1996**, *5*, 1083–1091. [CrossRef] [PubMed]

24. Brown, V.; Jin, P.; Ceman, S.; Darnell, J.C.; O'Donnell, W.T.; Tenenbaum, S.A.; Jin, X.; Feng, Y.; Wilkinson, K.D.; Keene, J.D.; et al. Microarray identification of FMRP-associated brain mRNAs and altered mRNA translational profiles in fragile X syndrome. *Cell* **2001**, *107*, 477–487. [CrossRef]

25. Kao, D.I.; Aldridge, G.M.; Weiler, I.J.; Greenough, W.T. Altered mRNA transport, docking, and protein translation in neurons lacking fragile X mental retardation protein. *Proc. Natl. Acad. Sci. USA* **2010**, *107*, 15601–15606. [CrossRef] [PubMed]

26. Darnell, J.C.; Van Driesche, S.J.; Zhang, C.; Hung, K.Y.; Mele, A.; Fraser, C.E.; Stone, E.F.; Chen, C.; Fak, J.J.; Chi, S.W.; et al. FMRP stalls ribosomal translocation on mRNAs linked to synaptic function and autism. *Cell* **2011**, *146*, 247–261. [CrossRef]

27. Korb, E.; Herre, M.; Zucker-Scharff, I.; Gresack, J.; Allis, C.D.; Darnell, R.B. Excess Translation of Epigenetic Regulators Contributes to Fragile X Syndrome and Is Alleviated by Brd4 Inhibition. *Cell* **2017**, *170*, 1209.e20–1223.e20. [CrossRef]

28. Bassell, G.J.; Warren, S.T. Fragile X syndrome: Loss of local mRNA regulation alters synaptic development and function. *Neuron* **2008**, *60*, 201–214. [CrossRef]

29. De Rubeis, S.; Bagni, C. Fragile X mental retardation protein control of neuronal mRNA metabolism: Insights into mRNA stability. *Mol. Cell Neurosci.* **2010**, *43*, 43–50. [CrossRef]

30. Jeon, S.J.; Seo, J.E.; Yang, S.I.; Choi, J.W.; Wells, D.; Shin, C.Y.; Ko, K.H. Cellular stress-induced up-regulation of FMRP promotes cell survival by modulating PI3K-Akt phosphorylation cascades. *J. Biomed. Sci.* **2011**, *18*, 17. [CrossRef]

31. Luca, R.; Averna, M.; Zalfa, F.; Vecchi, M.; Bianchi, F.; La Fata, G.; Del Nonno, F.; Nardacci, R.; Bianchi, M.; Nuciforo, P.; et al. The fragile X protein binds mRNAs involved in cancer progression and modulates metastasis formation. *EMBO Mol. Med.* **2013**, *5*, 1523–1536. [CrossRef] [PubMed]

32. Alpatov, R.; Lesch, B.J.; Nakamoto-Kinoshita, M.; Blanco, A.; Chen, S.; Stutzer, A.; Armache, K.J.; Simon, M.D.; Xu, C.; Ali, M.; et al. A chromatin-dependent role of the fragile X mental retardation protein FMRP in the DNA damage response. *Cell* **2014**, *157*, 869–881. [CrossRef] [PubMed]

33. Zhang, W.; Cheng, Y.; Li, Y.; Chen, Z.; Jin, P.; Chen, D. A feed-forward mechanism involving Drosophila fragile X mental retardation protein triggers a replication stress-induced DNA damage response. *Hum. Mol. Genet.* **2014**, *23*, 5188–5196. [CrossRef] [PubMed]

34. Zhou, L.T.; Ye, S.H.; Yang, H.X.; Zhou, Y.T.; Zhao, Q.H.; Sun, W.W.; Gao, M.M.; Yi, Y.H.; Long, Y.S. A novel role of fragile X mental retardation protein in pre-mRNA alternative splicing through RNA-binding protein 14. *Neuroscience* **2017**, *349*, 64–75. [CrossRef] [PubMed]

35. Shamay-Ramot, A.; Khermesh, K.; Porath, H.T.; Barak, M.; Pinto, Y.; Wachtel, C.; Zilberberg, A.; Lerer-Goldshtein, T.; Efroni, S.; Levanon, E.Y.; et al. Fmrp Interacts with Adar and Regulates RNA Editing, Synaptic Density and Locomotor Activity in Zebrafish. *PLoS Genet.* **2015**, *11*, e1005702. [CrossRef] [PubMed]

36. Filippini, A.; Bonini, D.; Lacoux, C.; Pacini, L.; Zingariello, M.; Sancillo, L.; Bosisio, D.; Salvi, V.; Mingardi, J.; La Via, L.; et al. Absence of the Fragile X Mental Retardation Protein results in defects of RNA editing of neuronal mRNAs in mouse. *RNA Biol.* **2017**, *14*, 1580–1591. [CrossRef] [PubMed]

37. Richter, J.D.; Bassell, G.J.; Klann, E. Dysregulation and restoration of translational homeostasis in fragile X syndrome. *Nat. Rev. Neurosci.* **2015**, *16*, 595–605. [CrossRef] [PubMed]

38. Berry-Kravis, E.M.; Lindemann, L.; Jonch, A.E.; Apostol, G.; Bear, M.F.; Carpenter, R.L.; Crawley, J.N.; Curie, A.; Des Portes, V.; Hossain, F.; et al. Drug development for neurodevelopmental disorders: Lessons learned from fragile X syndrome. *Nat. Rev. Drug Discov.* **2018**, *17*, 280–299. [CrossRef] [PubMed]

39. Park, C.Y.; Halevy, T.; Lee, D.R.; Sung, J.J.; Lee, J.S.; Yanuka, O.; Benvenisty, N.; Kim, D.W. Reversion of FMR1 Methylation and Silencing by Editing the Triplet Repeats in Fragile X iPSC-Derived Neurons. *Cell Rep.* **2015**, *13*, 234–241. [CrossRef]

40. Xie, N.; Gong, H.; Suhl, J.A.; Chopra, P.; Wang, T.; Warren, S.T. Reactivation of *FMR1* by CRISPR/Cas9-Mediated Deletion of the Expanded CGG-Repeat of the Fragile X Chromosome. *PLoS ONE* **2016**, *11*, e0165499. [CrossRef]

41. Liu, X.S.; Wu, H.; Krzisch, M.; Wu, X.; Graef, J.; Muffat, J.; Hnisz, D.; Li, C.H.; Yuan, B.; Xu, C.; et al. Rescue of Fragile X Syndrome Neurons by DNA Methylation Editing of the *FMR1* Gene. *Cell* **2018**, *172*, 979–992. [CrossRef] [PubMed]

42. Haenfler, J.M.; Skariah, G.; Rodriguez, C.M.; Monteiro da Rocha, A.; Parent, J.M.; Smith, G.D.; Todd, P.K. Targeted Reactivation of *FMR1* Transcription in Fragile X Syndrome Embryonic Stem Cells. *Front. Mol. Neurosci.* **2018**, *11*, 282. [CrossRef] [PubMed]

43. Zeier, Z.; Kumar, A.; Bodhinathan, K.; Feller, J.A.; Foster, T.C.; Bloom, D.C. Fragile X mental retardation protein replacement restores hippocampal synaptic function in a mouse model of fragile X syndrome. *Gene Ther.* **2009**, *16*, 1122–1129. [CrossRef] [PubMed]

44. Gholizadeh, S.; Arsenault, J.; Xuan, I.C.; Pacey, L.K.; Hampson, D.R. Reduced phenotypic severity following adeno-associated virus-mediated *Fmr1* gene delivery in fragile X mice. *Neuropsychopharmacology* **2014**, *39*, 3100–3111. [CrossRef] [PubMed]

45. Chiurazzi, P.; Pomponi, M.G.; Willemsen, R.; Oostra, B.A.; Neri, G. In vitro reactivation of the *FMR1* gene involved in fragile X syndrome. *Hum. Mol. Genet.* **1998**, *7*, 109–113. [CrossRef] [PubMed]

46. Coffee, B.; Zhang, F.; Warren, S.T.; Reines, D. Acetylated histones are associated with *FMR1* in normal but not fragile X-syndrome cells. *Nat. Genet.* **1999**, *22*, 98–101. [CrossRef] [PubMed]

47. Biacsi, R.; Kumari, D.; Usdin, K. SIRT1 inhibition alleviates gene silencing in Fragile X mental retardation syndrome. *PLoS Genet.* **2008**, *4*, e1000017. [CrossRef] [PubMed]

48. Kumari, D.; Usdin, K. Sustained expression of *FMR1* mRNA from reactivated fragile X syndrome alleles after treatment with small molecules that prevent trimethylation of H3K27. *Hum. Mol. Genet.* **2016**, *25*, 3689–3698. [CrossRef] [PubMed]

49. Issa, J.P. The myelodysplastic syndrome as a prototypical epigenetic disease. *Blood* **2013**, *121*, 3811–3817. [CrossRef] [PubMed]
50. Sripathy, S.; Leko, V.; Adrianse, R.L.; Loe, T.; Foss, E.J.; Dalrymple, E.; Lao, U.; Gatbonton-Schwager, T.; Carter, K.T.; Payer, B.; et al. Screen for reactivation of MeCP2 on the inactive X chromosome identifies the BMP/TGF-beta superfamily as a regulator of XIST expression. *Proc. Natl. Acad. Sci. USA* **2017**, *114*, 1619–1624. [CrossRef] [PubMed]
51. Carrette, L.L.G.; Wang, C.Y.; Wei, C.; Press, W.; Ma, W.; Kelleher, R.J., 3rd; Lee, J.T. A mixed modality approach towards Xi reactivation for Rett syndrome and other X-linked disorders. *Proc. Natl. Acad. Sci. USA* **2018**, *115*, E668–E675. [CrossRef] [PubMed]
52. Huang, H.S.; Allen, J.A.; Mabb, A.M.; King, I.F.; Miriyala, J.; Taylor-Blake, B.; Sciaky, N.; Dutton, J.W., Jr.; Lee, H.M.; Chen, X.; et al. Topoisomerase inhibitors unsilence the dormant allele of Ube3a in neurons. *Nature* **2011**, *481*, 185–189. [CrossRef] [PubMed]
53. She, A.; Kurtser, I.; Reis, S.A.; Hennig, K.; Lai, J.; Lang, A.; Zhao, W.N.; Mazitschek, R.; Dickerson, B.C.; Herz, J.; et al. Selectivity and Kinetic Requirements of HDAC Inhibitors as Progranulin Enhancers for Treating Frontotemporal Dementia. *Cell Chem. Biol.* **2017**, *24*, 892–906. [CrossRef] [PubMed]
54. Soragni, E.; Gottesfeld, J.M. Translating HDAC inhibitors in Friedreich's ataxia. *Expert Opin Orphan Drugs* **2016**, *4*, 961–970. [CrossRef] [PubMed]
55. Singh, A.K.; Halder-Sinha, S.; Clement, J.P.; Kundu, T.K. Epigenetic modulation by small molecule compounds for neurodegenerative disorders. *Pharmacol. Res.* **2018**, *132*, 135–148. [CrossRef] [PubMed]
56. Hwu, W.L.; Lee, Y.M.; Lee, S.C.; Wang, T.R. In vitro DNA methylation inhibits FMR-1 promoter. *Biochem. Biophys. Res. Commun.* **1993**, *193*, 324–329. [CrossRef] [PubMed]
57. Kumari, D.; Usdin, K. Interaction of the transcription factors USF1, USF2, and alpha-Pal/Nrf-1 with the FMR1 promoter. Implications for Fragile X mental retardation syndrome. *J. Biol. Chem.* **2001**, *276*, 4357–4364. [CrossRef] [PubMed]
58. Coffee, B.; Zhang, F.; Ceman, S.; Warren, S.T.; Reines, D. Histone modifications depict an aberrantly heterochromatinized *FMR1* gene in fragile x syndrome. *Am. J. Hum. Genet.* **2002**, *71*, 923–932. [CrossRef]
59. Tabolacci, E.; Moscato, U.; Zalfa, F.; Bagni, C.; Chiurazzi, P.; Neri, G. Epigenetic analysis reveals a euchromatic configuration in the *FMR1* unmethylated full mutations. *Eur. J. Hum. Genet.* **2008**, *16*, 1487–1498. [CrossRef]
60. Kumari, D.; Usdin, K. The distribution of repressive histone modifications on silenced *FMR1* alleles provides clues to the mechanism of gene silencing in fragile X syndrome. *Hum. Mol. Genet.* **2010**, *19*, 4634–4642. [CrossRef]
61. Eiges, R.; Urbach, A.; Malcov, M.; Frumkin, T.; Schwartz, T.; Amit, A.; Yaron, Y.; Eden, A.; Yanuka, O.; Benvenisty, N.; et al. Developmental study of fragile X syndrome using human embryonic stem cells derived from preimplantation genetically diagnosed embryos. *Cell Stem Cell* **2007**, *1*, 568–577. [CrossRef] [PubMed]
62. Kumari, D.; Usdin, K. Polycomb group complexes are recruited to reactivated *FMR1* alleles in Fragile X syndrome in response to *FMR1* transcription. *Hum. Mol. Genet.* **2014**, *23*, 6575–6583. [CrossRef] [PubMed]
63. Daniel, J.A.; Pray-Grant, M.G.; Grant, P.A. Effector proteins for methylated histones: An expanding family. *Cell Cycle* **2005**, *4*, 919–926. [CrossRef]
64. Smallwood, A.; Esteve, P.O.; Pradhan, S.; Carey, M. Functional cooperation between HP1 and DNMT1 mediates gene silencing. *Genes Dev.* **2007**, *21*, 1169–1178. [CrossRef] [PubMed]
65. Vire, E.; Brenner, C.; Deplus, R.; Blanchon, L.; Fraga, M.; Didelot, C.; Morey, L.; Van Eynde, A.; Bernard, D.; Vanderwinden, J.M.; et al. The Polycomb group protein EZH2 directly controls DNA methylation. *Nature* **2006**, *439*, 871–874. [CrossRef] [PubMed]
66. Rush, M.; Appanah, R.; Lee, S.; Lam, L.L.; Goyal, P.; Lorincz, M.C. Targeting of EZH2 to a defined genomic site is sufficient for recruitment of Dnmt3a but not de novo DNA methylation. *Epigenetics* **2009**, *4*, 404–414. [CrossRef] [PubMed]
67. Naumann, A.; Hochstein, N.; Weber, S.; Fanning, E.; Doerfler, W. A distinct DNA-methylation boundary in the 5′-upstream sequence of the *FMR1* promoter binds nuclear proteins and is lost in fragile X syndrome. *Am. J. Hum. Genet.* **2009**, *85*, 606–616. [CrossRef] [PubMed]
68. Sun, J.H.; Zhou, L.; Emerson, D.J.; Phyo, S.A.; Titus, K.R.; Gong, W.; Gilgenast, T.G.; Beagan, J.A.; Davidson, B.L.; Tassone, F.; et al. Disease-Associated Short Tandem Repeats Co-localize with Chromatin Domain Boundaries. *Cell* **2018**, *175*, 224.e15–238.e15. [CrossRef] [PubMed]

69. Ladd, P.D.; Smith, L.E.; Rabaia, N.A.; Moore, J.M.; Georges, S.A.; Hansen, R.S.; Hagerman, R.J.; Tassone, F.; Tapscott, S.J.; Filippova, G.N. An antisense transcript spanning the CGG repeat region of *FMR1* is upregulated in premutation carriers but silenced in full mutation individuals. *Hum. Mol. Genet.* **2007**, *16*, 3174–3187. [CrossRef]

70. Lanni, S.; Goracci, M.; Borrelli, L.; Mancano, G.; Chiurazzi, P.; Moscato, U.; Ferre, F.; Helmer-Citterich, M.; Tabolacci, E.; Neri, G. Role of CTCF protein in regulating *FMR1* locus transcription. *PLoS Genet.* **2013**, *9*, e1003601. [CrossRef]

71. Usdin, K.; Hayward, B.E.; Kumari, D.; Lokanga, R.A.; Sciascia, N.; Zhao, X.N. Repeat-mediated genetic and epigenetic changes at the *FMR1* locus in the Fragile X-related disorders. *Front. Genet.* **2014**, *5*, 226. [CrossRef] [PubMed]

72. Fry, M.; Loeb, L.A. The fragile X syndrome d(CGG)n nucleotide repeats form a stable tetrahelical structure. *Proc. Natl. Acad. Sci. USA* **1994**, *91*, 4950–4954. [CrossRef] [PubMed]

73. Chen, X.; Mariappan, S.V.; Catasti, P.; Ratliff, R.; Moyzis, R.K.; Laayoun, A.; Smith, S.S.; Bradbury, E.M.; Gupta, G. Hairpins are formed by the single DNA strands of the fragile X triplet repeats: Structure and biological implications. *Proc. Natl. Acad. Sci. USA* **1995**, *92*, 5199–5203. [CrossRef] [PubMed]

74. Kettani, A.; Kumar, R.A.; Patel, D.J. Solution structure of a DNA quadruplex containing the fragile X syndrome triplet repeat. *J. Mol. Biol.* **1995**, *254*, 638–656. [CrossRef] [PubMed]

75. Mitas, M.; Yu, A.; Dill, J.; Haworth, I.S. The trinucleotide repeat sequence d(CGG)15 forms a heat-stable hairpin containing Gsyn. Ganti base pairs. *Biochemistry* **1995**, *34*, 12803–12811. [CrossRef] [PubMed]

76. Nadel, Y.; Weisman-Shomer, P.; Fry, M. The fragile X syndrome single strand d(CGG)n nucleotide repeats readily fold back to form unimolecular hairpin structures. *J. Biol. Chem.* **1995**, *270*, 28970–28977. [CrossRef] [PubMed]

77. Usdin, K.; Woodford, K.J. CGG repeats associated with DNA instability and chromosome fragility form structures that block DNA synthesis *in vitro*. *Nucleic Acids Res.* **1995**, *23*, 4202–4209. [CrossRef]

78. Yu, A.; Barron, M.D.; Romero, R.M.; Christy, M.; Gold, B.; Dai, J.; Gray, D.M.; Haworth, I.S.; Mitas, M. At physiological pH, d(CCG)15 forms a hairpin containing protonated cytosines and a distorted helix. *Biochemistry* **1997**, *36*, 3687–3699. [CrossRef]

79. Patel, P.K.; Bhavesh, N.S.; Hosur, R.V. Cation-dependent conformational switches in d-TGGCGGC containing two triplet repeats of Fragile X Syndrome: NMR observations. *Biochem. Biophys. Res. Commun.* **2000**, *278*, 833–838. [CrossRef]

80. Handa, V.; Saha, T.; Usdin, K. The fragile X syndrome repeats form RNA hairpins that do not activate the interferon-inducible protein kinase, PKR, but are cut by Dicer. *Nucleic Acids Res.* **2003**, *31*, 6243–6248. [CrossRef]

81. Zumwalt, M.; Ludwig, A.; Hagerman, P.J.; Dieckmann, T. Secondary structure and dynamics of the r(CGG) repeat in the mRNA of the fragile X mental retardation 1 (*FMR1*) gene. *RNA Biol.* **2007**, *4*, 93–100. [CrossRef] [PubMed]

82. Colak, D.; Zaninovic, N.; Cohen, M.S.; Rosenwaks, Z.; Yang, W.Y.; Gerhardt, J.; Disney, M.D.; Jaffrey, S.R. Promoter-bound trinucleotide repeat mRNA drives epigenetic silencing in fragile X syndrome. *Science* **2014**, *343*, 1002–1005. [CrossRef] [PubMed]

83. Groh, M.; Lufino, M.M.; Wade-Martins, R.; Gromak, N. R-loops associated with triplet repeat expansions promote gene silencing in Friedreich ataxia and fragile X syndrome. *PLoS Genet.* **2014**, *10*, e1004318. [CrossRef] [PubMed]

84. Loomis, E.W.; Sanz, L.A.; Chedin, F.; Hagerman, P.J. Transcription-associated R-loop formation across the human FMR1 CGG-repeat region. *PLoS Genet.* **2014**, *10*, e1004294. [CrossRef] [PubMed]

85. Malgowska, M.; Gudanis, D.; Kierzek, R.; Wyszko, E.; Gabelica, V.; Gdaniec, Z. Distinctive structural motifs of RNA G-quadruplexes composed of AGG, CGG and UGG trinucleotide repeats. *Nucleic Acids Res.* **2014**, *42*, 10196–10207. [CrossRef] [PubMed]

86. Abu Diab, M.; Mor-Shaked, H.; Cohen, E.; Cohen-Hadad, Y.; Ram, O.; Epsztejn-Litman, S.; Eiges, R. The G-rich Repeats in *FMR1* and *C9orf72* Loci Are Hotspots for Local Unpairing of DNA. *Genetics* **2018**, *210*, 1239–1252. [CrossRef] [PubMed]

87. Smith, S.S.; Laayoun, A.; Lingeman, R.G.; Baker, D.J.; Riley, J. Hypermethylation of telomere-like foldbacks at codon 12 of the human c-Ha-ras gene and the trinucleotide repeat of the *FMR-1* gene of fragile X. *J. Mol. Biol.* **1994**, *243*, 143–151. [CrossRef] [PubMed]

88. Bulut-Karslioglu, A.; Perrera, V.; Scaranaro, M.; de la Rosa-Velazquez, I.A.; van de Nobelen, S.; Shukeir, N.; Popow, J.; Gerle, B.; Opravil, S.; Pagani, M.; et al. A transcription factor-based mechanism for mouse heterochromatin formation. *Nat. Struct. Mol. Biol.* **2012**, *19*, 1023–1030. [CrossRef] [PubMed]

89. Saksouk, N.; Simboeck, E.; Dejardin, J. Constitutive heterochromatin formation and transcription in mammals. *Epigenetics Chromatin* **2015**, *8*, 3. [CrossRef] [PubMed]

90. Pastori, C.; Peschansky, V.J.; Barbouth, D.; Mehta, A.; Silva, J.P.; Wahlestedt, C. Comprehensive analysis of the transcriptional landscape of the human *FMR1* gene reveals two new long noncoding RNAs differentially expressed in Fragile X syndrome and Fragile X-associated tremor/ataxia syndrome. *Hum. Genet.* **2014**, *133*, 59–67. [CrossRef] [PubMed]

91. Hecht, M.; Tabib, A.; Kahan, T.; Orlanski, S.; Gropp, M.; Tabach, Y.; Yanuka, O.; Benvenisty, N.; Keshet, I.; Cedar, H. Epigenetic mechanism of *FMR1* inactivation in Fragile X syndrome. *Int. J. Dev. Biol.* **2017**, *61*, 285–292. [CrossRef] [PubMed]

92. Volpe, T.A.; Kidner, C.; Hall, I.M.; Teng, G.; Grewal, S.I.; Martienssen, R.A. Regulation of heterochromatic silencing and histone H3 lysine-9 methylation by RNAi. *Science* **2002**, *297*, 1833–1837. [CrossRef] [PubMed]

93. Mochizuki, K.; Fine, N.A.; Fujisawa, T.; Gorovsky, M.A. Analysis of a piwi-related gene implicates small RNAs in genome rearrangement in tetrahymena. *Cell* **2002**, *110*, 689–699. [CrossRef]

94. Taverna, S.D.; Coyne, R.S.; Allis, C.D. Methylation of histone h3 at lysine 9 targets programmed DNA elimination in tetrahymena. *Cell* **2002**, *110*, 701–711. [CrossRef]

95. Pal-Bhadra, M.; Bhadra, U.; Birchler, J.A. RNAi related mechanisms affect both transcriptional and posttranscriptional transgene silencing in Drosophila. *Mol. Cell* **2002**, *9*, 315–327. [CrossRef]

96. Morris, K.V.; Chan, S.W.; Jacobsen, S.E.; Looney, D.J. Small interfering RNA-induced transcriptional gene silencing in human cells. *Science* **2004**, *305*, 1289–1292. [CrossRef]

97. Kim, D.H.; Villeneuve, L.M.; Morris, K.V.; Rossi, J.J. Argonaute-1 directs siRNA-mediated transcriptional gene silencing in human cells. *Nat. Struct. Mol. Biol.* **2006**, *13*, 793–797. [CrossRef]

98. Sarshad, A.A.; Juan, A.H.; Muler, A.I.C.; Anastasakis, D.G.; Wang, X.; Genzor, P.; Feng, X.; Tsai, P.F.; Sun, H.W.; Haase, A.D.; et al. Argonaute-miRNA Complexes Silence Target mRNAs in the Nucleus of Mammalian Stem Cells. *Mol. Cell* **2018**, *71*, 1040.e8–1050.e8. [CrossRef]

99. Rinn, J.L.; Kertesz, M.; Wang, J.K.; Squazzo, S.L.; Xu, X.; Brugmann, S.A.; Goodnough, L.H.; Helms, J.A.; Farnham, P.J.; Segal, E.; et al. Functional demarcation of active and silent chromatin domains in human HOX loci by noncoding RNAs. *Cell* **2007**, *129*, 1311–1323. [CrossRef]

100. Nagano, T.; Mitchell, J.A.; Sanz, L.A.; Pauler, F.M.; Ferguson-Smith, A.C.; Feil, R.; Fraser, P. The Air noncoding RNA epigenetically silences transcription by targeting G9a to chromatin. *Science* **2008**, *322*, 1717–1720. [CrossRef]

101. Engreitz, J.M.; Pandya-Jones, A.; McDonel, P.; Shishkin, A.; Sirokman, K.; Surka, C.; Kadri, S.; Xing, J.; Goren, A.; Lander, E.S.; et al. The Xist lncRNA exploits three-dimensional genome architecture to spread across the X chromosome. *Science* **2013**, *341*, 1237973. [CrossRef] [PubMed]

102. Simon, M.D.; Pinter, S.F.; Fang, R.; Sarma, K.; Rutenberg-Schoenberg, M.; Bowman, S.K.; Kesner, B.A.; Maier, V.K.; Kingston, R.E.; Lee, J.T. High-resolution Xist binding maps reveal two-step spreading during X-chromosome inactivation. *Nature* **2013**, *504*, 465–469. [CrossRef] [PubMed]

103. Beckedorff, F.C.; Ayupe, A.C.; Crocci-Souza, R.; Amaral, M.S.; Nakaya, H.I.; Soltys, D.T.; Menck, C.F.; Reis, E.M.; Verjovski-Almeida, S. The intronic long noncoding RNA ANRASSF1 recruits PRC2 to the RASSF1A promoter, reducing the expression of RASSF1A and increasing cell proliferation. *PLoS Genet.* **2013**, *9*, e1003705. [CrossRef] [PubMed]

104. Velazquez Camacho, O.; Galan, C.; Swist-Rosowska, K.; Ching, R.; Gamalinda, M.; Karabiber, F.; De La Rosa-Velazquez, I.; Engist, B.; Koschorz, B.; Shukeir, N.; et al. Major satellite repeat RNA stabilize heterochromatin retention of Suv39h enzymes by RNA-nucleosome association and RNA:DNA hybrid formation. *Elife* **2017**, *6*. [CrossRef]

105. Wang, X.; Goodrich, K.J.; Gooding, A.R.; Naeem, H.; Archer, S.; Paucek, R.D.; Youmans, D.T.; Cech, T.R.; Davidovich, C. Targeting of Polycomb Repressive Complex 2 to RNA by Short Repeats of Consecutive Guanines. *Mol. Cell* **2017**, *65*, 1056.e5–1067.e5. [CrossRef]

106. Zhao, J.; Ohsumi, T.K.; Kung, J.T.; Ogawa, Y.; Grau, D.J.; Sarma, K.; Song, J.J.; Kingston, R.E.; Borowsky, M.; Lee, J.T. Genome-wide identification of polycomb-associated RNAs by RIP-seq. *Mol. Cell* **2010**, *40*, 939–953. [CrossRef]

107. Porro, A.; Feuerhahn, S.; Delafontaine, J.; Riethman, H.; Rougemont, J.; Lingner, J. Functional characterization of the TERRA transcriptome at damaged telomeres. *Nat. Commun.* **2014**, *5*, 5379. [CrossRef]

108. Johnson, W.L.; Yewdell, W.T.; Bell, J.C.; McNulty, S.M.; Duda, Z.; O'Neill, R.J.; Sullivan, B.A.; Straight, A.F. RNA-dependent stabilization of SUV39H1 at constitutive heterochromatin. *Elife* **2017**, *6*. [CrossRef]

109. Shirai, A.; Kawaguchi, T.; Shimojo, H.; Muramatsu, D.; Ishida-Yonetani, M.; Nishimura, Y.; Kimura, H.; Nakayama, J.I.; Shinkai, Y. Impact of nucleic acid and methylated H3K9 binding activities of Suv39h1 on its heterochromatin assembly. *Elife* **2017**, *6*. [CrossRef]

110. Kanhere, A.; Viiri, K.; Araujo, C.C.; Rasaiyaah, J.; Bouwman, R.D.; Whyte, W.A.; Pereira, C.F.; Brookes, E.; Walker, K.; Bell, G.W.; et al. Short RNAs are transcribed from repressed polycomb target genes and interact with polycomb repressive complex-2. *Mol. Cell* **2010**, *38*, 675–688. [CrossRef]

111. Bar-Nur, O.; Caspi, I.; Benvenisty, N. Molecular analysis of *FMR1* reactivation in fragile-X induced pluripotent stem cells and their neuronal derivatives. *J. Mol. Cell Biol.* **2012**, *4*, 180–183. [CrossRef] [PubMed]

112. Li, M.; Zhao, H.; Ananiev, G.E.; Musser, M.T.; Ness, K.H.; Maglaque, D.L.; Saha, K.; Bhattacharyya, A.; Zhao, X. Establishment of Reporter Lines for Detecting Fragile X Mental Retardation (*FMR1*) Gene Reactivation in Human Neural Cells. *Stem Cells* **2017**, *35*, 158–169. [CrossRef] [PubMed]

113. Detich, N.; Bovenzi, V.; Szyf, M. Valproate induces replication-independent active DNA demethylation. *J. Biol. Chem.* **2003**, *278*, 27586–27592. [CrossRef] [PubMed]

114. Tabolacci, E.; De Pascalis, I.; Accadia, M.; Terracciano, A.; Moscato, U.; Chiurazzi, P.; Neri, G. Modest reactivation of the mutant *FMR1* gene by valproic acid is accompanied by histone modifications but not DNA demethylation. *Pharmacogenet. Genomics* **2008**, *18*, 738–741. [CrossRef] [PubMed]

115. Nesher, G.; Moore, T.L.; Dorner, R.W. In vitro effects of methotrexate on peripheral blood monocytes: Modulation by folinic acid and S-adenosylmethionine. *Ann. Rheum. Dis.* **1991**, *50*, 637–641. [CrossRef] [PubMed]

116. Kishi, T.; Tanaka, Y.; Ueda, K. Evidence for hypomethylation in two children with acute lymphoblastic leukemia and leukoencephalopathy. *Cancer* **2000**, *89*, 925–931. [CrossRef]

117. Brendel, C.; Mielke, B.; Hillebrand, M.; Gartner, J.; Huppke, P. Methotrexate treatment of FraX fibroblasts results in *FMR1* transcription but not in detectable FMR1 protein levels. *J. Neurodev. Disord.* **2013**, *5*, 23. [CrossRef] [PubMed]

118. Takebayashi, S.; Nakao, M.; Fujita, N.; Sado, T.; Tanaka, M.; Taguchi, H.; Okumura, K. 5-Aza-2'-deoxycytidine induces histone hyperacetylation of mouse centromeric heterochromatin by a mechanism independent of DNA demethylation. *Biochem. Biophys. Res. Commun.* **2001**, *288*, 921–926. [CrossRef] [PubMed]

119. Nguyen, C.T.; Weisenberger, D.J.; Velicescu, M.; Gonzales, F.A.; Lin, J.C.; Liang, G.; Jones, P.A. Histone H3-lysine 9 methylation is associated with aberrant gene silencing in cancer cells and is rapidly reversed by 5-aza-2'-deoxycytidine. *Cancer Res.* **2002**, *62*, 6456–6461. [PubMed]

120. Chiurazzi, P.; Pomponi, M.G.; Pietrobono, R.; Bakker, C.E.; Neri, G.; Oostra, B.A. Synergistic effect of histone hyperacetylation and DNA demethylation in the reactivation of the FMR1 gene. *Hum. Mol. Genet.* **1999**, *8*, 2317–2323. [CrossRef] [PubMed]

121. Dolskiy, A.A.; Pustylnyak, V.O.; Yarushkin, A.A.; Lemskaya, N.A.; Yudkin, D.V. Inhibitors of Histone Deacetylases Are Weak Activators of the FMR1 Gene in Fragile X Syndrome Cell Lines. *Biomed. Res. Int.* **2017**, *2017*, 3582601. [CrossRef] [PubMed]

122. Holliday, R.; Pugh, J.E. DNA modification mechanisms and gene activity during development. *Science* **1975**, *187*, 226–232. [CrossRef] [PubMed]

123. Riggs, A.D. X inactivation, differentiation, and DNA methylation. *Cytogenet. Cell Genet.* **1975**, *14*, 9–25. [CrossRef] [PubMed]

124. Suzuki, H.; Gabrielson, E.; Chen, W.; Anbazhagan, R.; van Engeland, M.; Weijenberg, M.P.; Herman, J.G.; Baylin, S.B. A genomic screen for genes upregulated by demethylation and histone deacetylase inhibition in human colorectal cancer. *Nat. Genet.* **2002**, *31*, 141–149. [CrossRef] [PubMed]

125. Tabolacci, E.; Mancano, G.; Lanni, S.; Palumbo, F.; Goracci, M.; Ferre, F.; Helmer-Citterich, M.; Neri, G. Genome-wide methylation analysis demonstrates that 5-aza-2-deoxycytidine treatment does not cause random DNA demethylation in fragile X syndrome cells. *Epigenetics Chromatin* **2016**, *9*, 12. [CrossRef]

126. Palii, S.S.; Van Emburgh, B.O.; Sankpal, U.T.; Brown, K.D.; Robertson, K.D. DNA methylation inhibitor 5-Aza-2'-deoxycytidine induces reversible genome-wide DNA damage that is distinctly influenced by DNA methyltransferases 1 and 3B. *Mol. Cell Biol.* **2008**, *28*, 752–771. [CrossRef] [PubMed]

127. Orta, M.L.; Calderon-Montano, J.M.; Dominguez, I.; Pastor, N.; Burgos-Moron, E.; Lopez-Lazaro, M.; Cortes, F.; Mateos, S.; Helleday, T. 5-Aza-2′-deoxycytidine causes replication lesions that require Fanconi anemia-dependent homologous recombination for repair. *Nucleic Acids Res.* **2013**, *41*, 5827–5836. [CrossRef]

128. Disney, M.D.; Liu, B.; Yang, W.Y.; Sellier, C.; Tran, T.; Charlet-Berguerand, N.; Childs-Disney, J.L. A small molecule that targets r(CGG)(exp) and improves defects in fragile X-associated tremor ataxia syndrome. *ACS Chem. Biol.* **2012**, *7*, 1711–1718. [CrossRef]

129. Tran, T.; Childs-Disney, J.L.; Liu, B.; Guan, L.; Rzuczek, S.; Disney, M.D. Targeting the r(CGG) repeats that cause FXTAS with modularly assembled small molecules and oligonucleotides. *ACS Chem. Biol.* **2014**, *9*, 904–912. [CrossRef]

130. Su, Z.; Zhang, Y.; Gendron, T.F.; Bauer, P.O.; Chew, J.; Yang, W.Y.; Fostvedt, E.; Jansen-West, K.; Belzil, V.V.; Desaro, P.; et al. Discovery of a biomarker and lead small molecules to target r(GGGGCC)-associated defects in c9FTD/ALS. *Neuron* **2014**, *83*, 1043–1050. [CrossRef]

131. Urbach, A.; Bar-Nur, O.; Daley, G.Q.; Benvenisty, N. Differential modeling of fragile X syndrome by human embryonic stem cells and induced pluripotent stem cells. *Cell Stem Cell* **2010**, *6*, 407–411. [CrossRef] [PubMed]

132. Sheridan, S.D.; Theriault, K.M.; Reis, S.A.; Zhou, F.; Madison, J.M.; Daheron, L.; Loring, J.F.; Haggarty, S.J. Epigenetic characterization of the *FMR1* gene and aberrant neurodevelopment in human induced pluripotent stem cell models of fragile X syndrome. *PLoS ONE* **2011**, *6*, e26203. [CrossRef]

133. Kaufmann, M.; Schuffenhauer, A.; Fruh, I.; Klein, J.; Thiemeyer, A.; Rigo, P.; Gomez-Mancilla, B.; Heidinger-Millot, V.; Bouwmeester, T.; Schopfer, U.; et al. High-Throughput Screening Using iPSC-Derived Neuronal Progenitors to Identify Compounds Counteracting Epigenetic Gene Silencing in Fragile X Syndrome. *J. Biomol. Screen.* **2015**, *20*, 1101–1111. [CrossRef] [PubMed]

134. Kumari, D.; Swaroop, M.; Southall, N.; Huang, W.; Zheng, W.; Usdin, K. High-Throughput Screening to Identify Compounds That Increase Fragile X Mental Retardation Protein Expression in Neural Stem Cells Differentiated From Fragile X Syndrome Patient-Derived Induced Pluripotent Stem Cells. *Stem Cells Transl. Med.* **2015**, *4*, 800–808. [CrossRef] [PubMed]

135. Willemsen, R.; Bontekoe, C.J.; Severijnen, L.A.; Oostra, B.A. Timing of the absence of *FMR1* expression in full mutation chorionic villi. *Hum. Genet.* **2002**, *110*, 601–605. [CrossRef] [PubMed]

136. Telias, M.; Segal, M.; Ben-Yosef, D. Neural differentiation of Fragile X human Embryonic Stem Cells reveals abnormal patterns of development despite successful neurogenesis. *Dev. Biol.* **2013**, *374*, 32–45. [CrossRef] [PubMed]

137. Avitzour, M.; Mor-Shaked, H.; Yanovsky-Dagan, S.; Aharoni, S.; Altarescu, G.; Renbaum, P.; Eldar-Geva, T.; Schonberger, O.; Levy-Lahad, E.; Epsztejn-Litman, S.; et al. *FMR1* epigenetic silencing commonly occurs in undifferentiated fragile X-affected embryonic stem cells. *Stem Cell Rep.* **2014**, *3*, 699–706. [CrossRef]

138. Zhou, Y.; Kumari, D.; Sciascia, N.; Usdin, K. CGG-repeat dynamics and *FMR1* gene silencing in fragile X syndrome stem cells and stem cell-derived neurons. *Mol. Autism.* **2016**, *7*, 42. [CrossRef]

139. Gafni, O.; Weinberger, L.; Mansour, A.A.; Manor, Y.S.; Chomsky, E.; Ben-Yosef, D.; Kalma, Y.; Viukov, S.; Maza, I.; Zviran, A.; et al. Derivation of novel human ground state naive pluripotent stem cells. *Nature* **2013**, *504*, 282–286. [CrossRef]

140. Kenneson, A.; Zhang, F.; Hagedorn, C.H.; Warren, S.T. Reduced FMRP and increased *FMR1* transcription is proportionally associated with CGG repeat number in intermediate-length and premutation carriers. *Hum. Mol. Genet.* **2001**, *10*, 1449–1454. [CrossRef]

141. Primerano, B.; Tassone, F.; Hagerman, R.J.; Hagerman, P.; Amaldi, F.; Bagni, C. Reduced *FMR1* mRNA translation efficiency in fragile X patients with premutations. *RNA* **2002**, *8*, 1482–1488. [PubMed]

142. Brouwer, J.R.; Mientjes, E.J.; Bakker, C.E.; Nieuwenhuizen, I.M.; Severijnen, L.A.; Van der Linde, H.C.; Nelson, D.L.; Oostra, B.A.; Willemsen, R. Elevated *Fmr1* mRNA levels and reduced protein expression in a mouse model with an unmethylated Fragile X full mutation. *Exp. Cell Res.* **2007**, *313*, 244–253. [CrossRef] [PubMed]

143. Entezam, A.; Biacsi, R.; Orrison, B.; Saha, T.; Hoffman, G.E.; Grabczyk, E.; Nussbaum, R.L.; Usdin, K. Regional FMRP deficits and large repeat expansions into the full mutation range in a new Fragile X premutation mouse model. *Gene* **2007**, *395*, 125–134. [CrossRef] [PubMed]

144. Hagerman, R.J.; Leehey, M.; Heinrichs, W.; Tassone, F.; Wilson, R.; Hills, J.; Grigsby, J.; Gage, B.; Hagerman, P.J. Intention tremor, parkinsonism, and generalized brain atrophy in male carriers of fragile X. *Neurology* **2001**, *57*, 127–130. [CrossRef] [PubMed]

145. Hundscheid, R.D.; Braat, D.D.; Kiemeney, L.A.; Smits, A.P.; Thomas, C.M. Increased serum FSH in female fragile X premutation carriers with either regular menstrual cycles or on oral contraceptives. *Hum. Reprod.* **2001**, *16*, 457–462. [CrossRef] [PubMed]

146. Hagerman, R.J.; Leavitt, B.R.; Farzin, F.; Jacquemont, S.; Greco, C.M.; Brunberg, J.A.; Tassone, F.; Hessl, D.; Harris, S.W.; Zhang, L.; et al. Fragile-X-associated tremor/ataxia syndrome (FXTAS) in females with the FMR1 premutation. *Am. J. Hum. Genet.* **2004**, *74*, 1051–1056. [CrossRef] [PubMed]

147. Welt, C.K.; Smith, P.C.; Taylor, A.E. Evidence of early ovarian aging in fragile X premutation carriers. *J. Clin. Endocrinol. Metab.* **2004**, *89*, 4569–4574. [CrossRef]

148. Allen, E.G.; Sullivan, A.K.; Marcus, M.; Small, C.; Dominguez, C.; Epstein, M.P.; Charen, K.; He, W.; Taylor, K.C.; Sherman, S.L. Examination of reproductive aging milestones among women who carry the *FMR1* premutation. *Hum. Reprod.* **2007**, *22*, 2142–2152. [CrossRef]

149. Loesch, D.Z.; Sherwell, S.; Kinsella, G.; Tassone, F.; Taylor, A.; Amor, D.; Sung, S.; Evans, A. Fragile X-associated tremor/ataxia phenotype in a male carrier of unmethylated full mutation in the *FMR1* gene. *Clin. Genet.* **2012**, *82*, 88–92. [CrossRef]

150. Santa Maria, L.; Pugin, A.; Alliende, M.A.; Aliaga, S.; Curotto, B.; Aravena, T.; Tang, H.T.; Mendoza-Morales, G.; Hagerman, R.; Tassone, F. FXTAS in an unmethylated mosaic male with fragile X syndrome from Chile. *Clin. Genet.* **2014**, *86*, 378–382. [CrossRef]

151. Basuta, K.; Schneider, A.; Gane, L.; Polussa, J.; Woodruff, B.; Pretto, D.; Hagerman, R.; Tassone, F. High functioning male with fragile X syndrome and fragile X-associated tremor/ataxia syndrome. *Am. J. Med. Genet. A* **2015**, *167A*, 2154–2161. [CrossRef] [PubMed]

152. LaFauci, G.; Adayev, T.; Kascsak, R.; Kascsak, R.; Nolin, S.; Mehta, P.; Brown, W.T.; Dobkin, C. Fragile X screening by quantification of FMRP in dried blood spots by a Luminex immunoassay. *J. Mol. Diagn.* **2013**, *15*, 508–517. [CrossRef] [PubMed]

153. Taylor, A.K.; Safanda, J.F.; Fall, M.Z.; Quince, C.; Lang, K.A.; Hull, C.E.; Carpenter, I.; Staley, L.W.; Hagerman, R.J. Molecular predictors of cognitive involvement in female carriers of fragile X syndrome. *JAMA* **1994**, *271*, 507–514. [CrossRef] [PubMed]

154. De Vries, B.B.; Wiegers, A.M.; Smits, A.P.; Mohkamsing, S.; Duivenvoorden, H.J.; Fryns, J.P.; Curfs, L.M.; Halley, D.J.; Oostra, B.A.; van den Ouweland, A.M.; et al. Mental status of females with an *FMR1* gene full mutation. *Am. J. Hum. Genet.* **1996**, *58*, 1025–1032. [PubMed]

155. Kieffer, E.; Nicod, J.C.; Gardes, N.; Kastner, C.; Becker, N.; Celebi, C.; Pirrello, O.; Rongieres, C.; Koscinski, I.; Gosset, P.; et al. Improving preimplantation genetic diagnosis for Fragile X syndrome: Two new powerful single-round multiplex indirect and direct tests. *Eur. J. Hum. Genet.* **2016**, *24*, 221–227. [CrossRef] [PubMed]

156. Bailey, D.B., Jr.; Raspa, M.; Bishop, E.; Holiday, D. No change in the age of diagnosis for fragile x syndrome: Findings from a national parent survey. *Pediatrics* **2009**, *124*, 527–533. [CrossRef] [PubMed]

157. Guo, W.; Allan, A.M.; Zong, R.; Zhang, L.; Johnson, E.B.; Schaller, E.G.; Murthy, A.C.; Goggin, S.L.; Eisch, A.J.; Oostra, B.A.; et al. Ablation of Fmrp in adult neural stem cells disrupts hippocampus-dependent learning. *Nat. Med.* **2011**, *17*, 559–565. [CrossRef]

158. Siegel, J.J.; Chitwood, R.A.; Ding, J.M.; Payne, C.; Taylor, W.; Gray, R.; Zemelman, B.V.; Johnston, D. Prefrontal Cortex Dysfunction in Fragile X Mice Depends on the Continued Absence of Fragile X Mental Retardation Protein in the Adult Brain. *J. Neurosci.* **2017**, *37*, 7305–7317. [CrossRef]

Review

The Contribution of Pluripotent Stem Cell (PSC)-Based Models to the Study of Fragile X Syndrome (FXS)

Manar Abu Diab [1,2] **and Rachel Eiges** [1,2,*]

[1]   Stem Cell Research Laboratory, Medical Genetics Institute, Shaare Zedek Medical Center, Jerusalem 91031, Israel; manar@szmc.org.il

[2]   School of Medicine, Hebrew University of Jerusalem, Jerusalem 9112102, Israel

*   Correspondence: rachela@szmc.org.il; Tel.: +972-26666721

Received: 1 January 2019; Accepted: 13 February 2019; Published: 15 February 2019

Abstract: Fragile X syndrome (FXS) is the most common heritable form of cognitive impairment. It results from a deficiency in the fragile X mental retardation protein (FMRP) due to a CGG repeat expansion in the 5′-UTR of the X-linked *FMR1* gene. When CGGs expand beyond 200 copies, they lead to epigenetic gene silencing of the gene. In addition, the greater the allele size, the more likely it will become unstable and exhibit mosaicism for expansion size between and within tissues in affected individuals. The timing and mechanisms of *FMR1* epigenetic gene silencing and repeat instability are far from being understood given the lack of appropriate cellular and animal models that can fully recapitulate the molecular features characteristic of the disease pathogenesis in humans. This review summarizes the data collected to date from mutant human embryonic stem cells, induced pluripotent stem cells, and hybrid fusions, and discusses their contribution to the investigation of FXS, their key limitations, and future prospects.

Keywords: fragile X syndrome; unstable repeat diseases; epigenetic gene silencing; DNA methylation; repeat instability; pluripotent stem cells

---

1. Introduction

Fragile X syndrome (FXS; OMIM#300624) is the most common heritable form of cognitive impairment (1 in 4000 male and 1 in 8000 female births). It is inherited as an X-linked condition and results from a deficiency in the fragile X mental retardation protein (FMRP). Nearly all FXS patients lack FMRP due to a CGG repeat expansion in the 5′-UTR of *FMR1* [1–3]. The CGG repeats are located downstream to a CpG island promoter. Rarely, a normal allele in *FMR1* exceeds its standard length to a medium size ((55 < CGGs ≤ 199), premutation or PM) by the addition of CGGs during parent-to-offspring transmission. PM alleles confer a risk of fragile X-associated tremor/ataxia syndrome (FXTAS) and fragile X-associated primary ovarian insufficiency (FXPOI), both of which are thought to result from a combination of toxic gain-of-function RNA and repeat-associated non-ATG (RAN)-translation mechanism [4–8]. The greater the size of PM in the mother, the more likely it will further expand and transform into an FXS-causing mutation (CGGs > 199, full mutation, FM) in the next generation [9]. Once CGGs increase and reach the FM range, they induce aberrant DNA methylation and other changes from active to repressive histone modifications that are typical of densely packed chromatin [1,10–15]. This results in *FMR1* transcriptional silencing by abolishing promoter activity. In addition, the greater the allele size, the more likely it will become unstable [2,16,17]. Despite intensive research, the timing and mechanism(s) by which *FMR1* becomes epigenetically modified or unstable are at present far from clear. It is still unknown which repressive histone marks [10–15] or chromatin modifying enzymes [18–20] are critical for eliciting or maintaining *FMR1* gene silencing.

Moreover, it is commonly assumed that gene silencing is facilitated by an RNA-dependent mechanism, although this question remains to be resolved [18–21]. An additional concern relates to the timing of *FMR1* gene inactivation, which is still a controversial topic [22,23]. Other unresolved issues have to do with the mechanisms underlying repeat instability, and which may differ among germ line, preimplantation stage embryos, and somatic cells. In addition, it is perplexing as to why CGG instability in FXS is at its peak during early fetal development and how this is typically constrained later in life [24,25]. Furthermore, there is conflicting evidence as to the effects of differentiation and methylation in restricting repeat instability in affected tissues [25,26].

Current mouse models, including humanized mice, fail to fully recapitulate the molecular features that are typically associated with the disease in humans. For example, Knock In (KI) mouse models with CGG expansions greater than 199 repeats fail to show hypermethylation of the *FMR1* promoter and inactivate the gene, as observed in humans [27–29]. One approach to circumventing this difficulty is to force greater expansions in PM-sized mice by artificially inducing mutations into various DNA processing pathways [30–33]. These strains have been found useful for investigating the role of DNA repair proteins in promoting CGG instability. However, why these induced expansions do not elicit epigenetic gene silencing in mice remains unclear. An alternative approach for FXS disease modeling is to utilize human pluripotent stem cell (PSC) lines that naturally harbor the disease-causing mutation [21,34–39]. This review summarizes the data collected to date on the contributions of currently available PSC model systems to investigate the timing and mechanisms governing epigenetics and repeat instability in FXS, their apparent limitations, and future prospects. The contribution of these cell models to a better understanding of the neural phenotype of the disease, including the effect of RNA/protein toxicity by gain-of-function mechanisms contributed by unmethylated FM alleles, and their therapeutic potential is beyond the scope of this manuscript and can be found elsewhere [40–42].

2. Currently Available Pluripotent Stem Cell (PSC) Models for Investigating FXS

Pluripotent stem cells are undifferentiated cells that are capable of differentiating into all three embryonic germ layers and their differentiated derivatives [43]. They are transiently present during embryonic development but can also be maintained as established cell lines. PSC lines can be derived from the inner cell mass of blastocysts (embryonic stem cells (ESCs)), primordial germ cells (embryonic germ (EG) cells), or from tumorigenic derivatives of germinal tissues (embryonic carcinoma (EC) cells). The primary advantage of these cell lines is that they can be maintained in vitro indefinitely without undergoing cell senescence while preserving their wide developmental and self-renewal potentials. One fascinating feature of PSCs is their ability to de-differentiate somatic cells by fusion. This is achieved by resetting the epigenetic marks that distinguish somatic cells from undifferentiated embryonic cells and is exhibited by re-establishment of methylation patterns that are characteristic of early embryonic cells [44].

2.1. Human Embryonic Stem Cells (hESCs)

Of the various pluripotent cell lines, ESCs best resemble early stage embryos. This is because they are derived from 7-day old spare in vitro fertilized (IVF) human embryos. In addition, they can be established directly from genetically diseased embryos (Figure 1A), which are occasionally available for research purposes from preimplantation genetic diagnosis (PGD) procedures [45,46]. The derivation of human ESCs (hESCs) from FXS affected embryos constitutes a powerful tool for disease modeling since FXS hESCs naturally harbor the disease-causing mutation. This obviates the need to clone lengthy repetitive elements and accurately integrate them into the genome. In addition, FXS hESCs are expected to complement currently available mouse models, which may not accurately replicate the epigenetics or repeat instability aspects of the disease (possibly due to the lack of sequence conservation at the borders of the region that become differentially methylated in patients [47]). Moreover, by more closely resembling preimplantation embryos, they are likely to be a better model for investigating the earliest phases of disease pathology as compared with other human-based model systems such as aborted fetuses, adult postmortem brain samples, or disease unrelated tissues [48–50]. Thus far, over a dozen FXS cell lines have been established worldwide and are currently being utilized to address some of the key questions outlined above [42]. However, the accessibility of FXS hESCs is a limiting factor because their derivation is totally dependent on the infrequent availability of PGD derived embryos.

2.2. Pluripotent Hybrid Cells Fusions (PHCFs)

An alternative approach to modeling early events in FXS pathogenesis is to de-differentiate patient cells by somatic cell reprogramming. One method to reprogram somatic cells is by fusing them with a PSC [44] (Figure 1B). This can be done by whole-cell fusion or microcell fusion, if the transfer of a single mutant chromosome is desired. The hybrids (hereafter referred to as PHCFs), although chromosomally unbalanced, are considered pluripotent because they preserve their self-renewal and developmental potential as long as they remain undifferentiated. Hence, they constitute a valuable experimental resource for resetting the epigenetic memory of somatic cells/single chromosomes obtained from patients. In the case of FXS this is particularly instructive, since it may allow for reversal of the pathogenic epigenetic modifications that are induced by the CGG expansion in patients' cells.

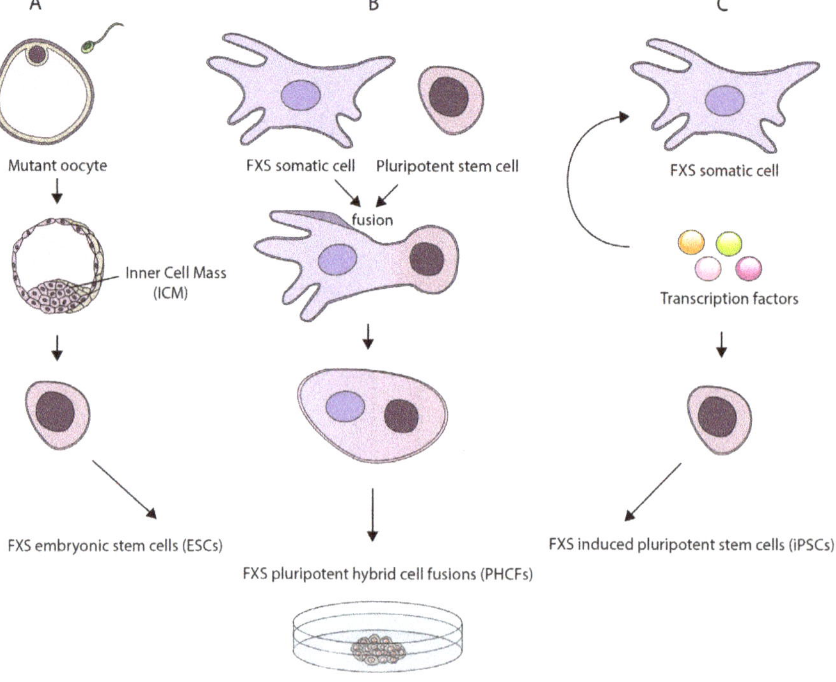

Figure 1. Currently available pluripotent stem cell (PSCs)-based models for investigating the underlying mechanisms for fragile X syndrome (FXS): (**A**) human embryonic stem cell (hESC) lines derived directly from genetically affected FXS embryos following preimplantation genetic diagnosis procedures; (**B**) reprogramming of somatic cells from patients by whole cell/microcell fusion with a pluripotent stem cell, leading to the creation of pluripotent hybrid cell fusions (PHCFs); and (**C**) induced pluripotent cells (iPSCs) derived from patients' somatic cells by over-expression of a defined set of transcription factors.

2.3. Induced Pluripotent Stem Cells (iPSCs)

Another approach for inducing somatic reprogramming of patient cells is to transiently express a small number of transcription factor master regulators [51] (Figure 1C). Quite remarkably, the resulting cell lines, dubbed induced PSCs (iPSCs), closely resemble embryo-derived hESCs in many respects [52,53]. In addition, they are considered superior to PHCFs since they have a normal karyotype and are considered non-tumorigenic. The benefits of iPSC technology are clear. It can be applied, in principle, to any cell type in the body, thus circumventing the obstacle of inaccessibility of spare IVF embryos for hESC line derivation. This technique can be carried out directly on cells obtained from patients, thus enabling any tissue culture laboratory to create pluripotent-based disease models. In addition, the iPSCs derivation procedure typically results in the creation of multiple clones from the same individual (isogenic), that can strengthen the robustness of research findings. Thus far, many iPSCs have been established with the FXS mutation [34,36–39,54] and are used mostly for the study of the neuronal aspects of the disease [55,56]. One limitation of the iPSC system which is also relevant to PHCFs, is that they are derived by clonal selection. As a result, single clones do not accurately reflect the composition of all alleles within individuals, particularly if the primary cells were derived from subjects who are mosaic for methylation or repeat size [57,58].

3. Epigenetics

3.1. The Timing of FMR1 Gene Inactivation in FXS

At the time the gene was discovered, it was assumed that CGG expansion is established and transmitted by the mother to the offspring in its methylated and inactive form. However, later studies on chorionic villus samples (CVS) and fetal tissues appeared to suggest that a developmentally regulated process takes place between 8–16 weeks of gestation [22,24,59–61]. This led to the view that differentiation-dependent factor(s) are necessary to translate the mutation into gene silencing. Nevertheless, attempts to test this model using patient-derived reprogrammed cells (iPSCs and PHCFs) or FXS hESCs yielded conflicting results. For example, when mouse EC/ES cells were used as recipients to reprogram a fragile X chromosome with a heavily methylated and *FMR1*-inactive gene, this frequently led to de-methylation and restoration of *FMR1* activity in the PHCFs [25]. This not only validated the widespread assumption that *FMR1* epigenetic silencing is triggered by differentiation, but also encouraged researchers to posit that somatic cell reprogramming by iPSC technology could also reverse the epigenetic modifications that are acquired as a consequence of the mutation. In fact, when the first FXS hESC line was established it was found to be completely *FMR1* unmethylated and gene active [35]. However, as more FXS hESC lines were characterized, it became apparent that *FMR1* methylation is not restricted to differentiated cells but can be displayed by undifferentiated cells as well [21,34,62]. Specifically, of the 11 FXS hESCs examined to date, the majority present varying levels of *FMR1* methylation, as high as 65% [34].

Clearly, the wide variability in *FMR1* methylation across different FXS hESCs lines runs counter the claim that gene inactivation is initiated by the end of the first trimester and calls for re-evaluation of the timing of *FMR1* hypermethylation in FXS (reviewed by [23]). In addition, it may suggest that the expansion is not evenly methylated or is not long enough in all cells of the preimplantation embryo. This is consistent with the realization that FMRP loss-of-function may not be the sole mechanism contributing to the clinical phenotype in FXS patients, because mosaicism for allele size and methylation within affected tissues [63] may lead to typical features of FXTAS by toxic gain-of-function mechanisms [57]. An alternative explanation for the extent of variability in *FMR1* methylation among the different FXS hESC lines has to do with the pluripotency state of hESCs, which are heterogeneous in that some cells may reflect a less primitive ground state of pluripotency (primed) as compared to the inner cell mass (ICM) cells in the embryo (naïve) [42,64]. It is thus imperative to directly monitor the expansion size and methylation state of *FMR1* in FXS preimplantation embryos to better understand the dynamics of both expansion size and methylation early in development. In addition, it would be worthwhile establishing FXS hESCs with a FM under naïve conditions to examine whether methylation is abolished under these conditions.

Importantly, when patients' skin fibroblasts were reprogrammed to generate FXS iPSCs, *FMR1* was consistently hypermethylated and gene inactive [36–38]. Moreover, when iPSCs were derived from the skin of an atypical individual who carried an unmethylated FM, reprogramming frequently led to *FMR1* hypermethylation (instead of remaining hypomethylated) and transcriptional silencing in the newly established iPSC clones [39]. On the other hand, in a different study, when iPSCs were derived from blood cells from subjects with an unmethylated FM, *FMR1* remained unmethylated and gene active [54]. The inconsistency in the methylation status of the gene following reprogramming may stem from inter-individual heterogeneity among unmethylated FM carriers due to "clonal selection bias" or may simply result from a difference in the type of cells used for reprogramming (fibroblasts vs. peripheral blood mononuclear cells).

3.2. The Role of DNA Methylation in the Silencing Process

When CGGs increase in size and reach the FM range, it results in aberrant DNA methylation in a region that initiates approximately 650-850 nucleotides upstream to the CGGs and extends into intron 1 of the *FMR1* gene [47,65,66]. This disease-associated Differentially Methylated Region, DMR, covers a CpG island (91 CpG sites, GRCh38/hg38 chrX:147,911,574-147,912,682) that overlaps with the *FMR1* promoter and the downstream repeats [47,67]. Upon expansion, the DMR becomes incorrectly methylated presumably by the spread of methylation from the upstream flanking region [47]. This is accompanied by the switch from active (histone acetylations and H3K4me2/3) to repressive (H3K9me2/3, H4K20me3 and H3K27me3) histone modifications [10–15], resulting in heterochromatin induction and transcriptional silencing of the *FMR1* gene.

Little is known about the specific role of DNA methylation or the order of events that lead to *FMR1* epigenetic silencing. This is because methylation is only rarely uncoupled from the repressive histone modifications that are associated with silencing in the cell types examined including PSCs [12]. This makes it hard to decipher the role of each of these epigenetic modifications in this process. It is generally assumed that methylation is a relatively late event in the timeline of gene silencing and is responsible for locking up the inactive state. This is because in rare individuals with an unmethylated FM allele, *FMR1* remains transcriptionally active (nonetheless enriched for H3K9me2/3) [14]. Other evidence to support this supposition comes from the treatment of patients' cells with the demethylation agent 5-aza-dC, which results in partial re-activation of the *FMR1* gene without affecting H3K9me2/3 levels [11,18]. Together, this suggests a mechanism of DNA methylation that is downstream or independent of H3K9me2/3 deposition. With respect to H3K27me3, the results are less consistent. While some FXS PSC lines/clones with a hypermethylated and inactive gene have been significantly enriched for the repressive H3K27me3 mark [68], others were not [69]. The significance of this variation remains to be determined. In addition, it should be mentioned that when *FMR1* was partially re-activated by 5-aza-dC treatment, it led to the recruitment of EZH2, the catalytic subunit of PRC2 (histone methyl transferase that modifies H3K27) and to an increase in H3K27me3 in a manner which is dependent on *FMR1* mRNA expression [18]. The relevance of this to the process of gene silencing, as naturally occurs in patient cells, remains to be determined.

To isolate the functional significance of DNA methylation in FXS, researchers have designed DNA methylation editing tools that exploit the fusion of a catalytically inactive Cas9 with the demethylating enzyme TET1 (dCas9-Tet1) to target methylation at a specific locus in the genome [70]. By designing a single gRNA directed against the CGG repeats, they targeted the dCas9-Tet1 to the *FMR1* locus in multiple FXS patient-derived iPSCs, and efficiently demethylated the repeats [69]. Erasing methylation from the CGGs resulted in hypomethylation of the flanking sequence, increased H3K27 acetylation and H3K4 methylation, and reduced H3K9me2 at the *FMR1* promoter. This unlocked the epigenetic silencing of *FMR1* and restored its activity to nearly normal levels (90%, as compared to 25% after 5-aza-dC treatment [18]) by the recruitment of RNA Pol II to the promoter region. Gene re-activation and demethylation in the FXS iPSC edited clones persisted for at least 2 weeks after inhibition of the dCas9-Tet1 protein in vitro and was sustained in vivo in neurons after transplantation into the mouse brain. This provided the first direct evidence that demethylation of the CGGs is sufficient to re-activate the gene and to switch the region from closed to open chromatin. It would be beneficial if the same programmable dCas9 toolkit approach could be utilized to determine the cause-and-effect relationship between DNA methylation and the other histone marks accompanying the cascade of *FMR1* epigenetic silencing. For example, while there are some hints that H3K9me2/3 is involved in steps preceding DNA hypermethylation, this needs to be experimentally substantiated. By directly tethering the catalytic domain of G9a or Suv39H to the CGGs and depositing H3K9me2/3 in unmethylated FM alleles, it may be possible to uncouple these epigenetic changes to define their causal relationship. Crucially, however, when targeting the CGGs (rather than the flanking sequence) this may lead to off-target effects, given the many CGG repetitive sequences spread throughout the genome [69].

To determine whether CGG expansion is needed at all times to preserve aberrant methylation once silencing is achieved, researchers have taken advantage of XY FXS iPSCs with a heavily methylated expansion to eliminate (as opposed to shorten) the CGGs from *FMR1* with the CRISPR/Cas9 system using gRNAs directed against the flanking sequences [71,72]. By monitoring for changes in *FMR1*, Park et al. demonstrated that excision of the CGGs with a single gRNA completely eliminated aberrant methylation from at least a portion of the *FMR1* promoter (22 CpG sites), switched from H3K9me2 to H3K4me3 enrichments, rescued *FMR1* transcription, and restored protein levels [71]. In a different study Xie et al. were able to reactivate the *FMR1* gene in some of the targeted cells (67% of somatic cell hybrids and only 20% of iPSCs) and increase the mRNA to at least 50% of WT mRNA levels, by efficiently deleting the CGGs using a pair of gRNAs targeting either side of the repeats [72]. However, in the latter study repeat deletion was mostly inefficient or only partly reduced methylation levels to wild type control levels at the promoter (altogether 8 CpG sites). It is hard to reconcile the difference in the methylation changes between the clones. Considering that the precision of the deletion with a single gRNA by Park et al. was not as precise as in the latter report and often led to indels, interpretation calls for caution. Nevertheless, together these studies provide evidence that the repressive marks elicited by the mutation under certain conditions are reversible and need to be established persistently at each DNA replication cycle. Clearly, this should be taken into account when considering the mutation as a potential therapeutic target in non-dividing cells such as affected neurons.

Finally, in a recent study, Haenfler et al. achieved re-activation of a completely methylated *FMR1* gene in an hESC line with approximately 800 repeats by targeting a catalytically inactive Cas9 and transcription activator VAP192 fusion protein directly to the repeats [62]. Using this approach, they induced transcription from the *FMR1* gene despite high levels of methylation in both the promoter and the repeats. This implies that epigenetic silencing in undifferentiated PSCs can be overcome by sustained accessibility of transcription factors to the locus.

3.3. The Significance of DNA Hydroxymethylation at the FMR1 Locus

While the role of DNA methylation in *FMR1* epigenetic silencing has been extensively studied, the contribution of 5-hydroxymethylcytosine (5hmC) to this process is less clear and has not been well characterized in differentiated and undifferentiated iPSC-based models. The epigenetic mark 5hmC is produced by DNA demethylation through oxidation of 5mC by the TET family of deoxygenases [73,74]. It acts as an intermediate in DNA demethylation during the conversion of 5mC into cytosine, and thus is implicated in transcriptional activation of genes. There is evidence to show that TET1 acts in differentiated cells as a maintenance demethylase to prevent aberrant methylation spreading into CpG islands [75–77]. Consistent with this idea, it was hypothesized that repeat expansion elicits *FMR1* epigenetic silencing by impeding TET-mediated demethylation at the otherwise hypomethylated CpG Island. With this in mind, Esanov and colleagues analyzed the levels of 5hmC at the *FMR1* promoter in post-mortem brain samples, primary fibroblasts and immortalized lymphocytes from FXS and control subjects as well as in vitro differentiated neural progenitors (NPCs) from FXS iPSC and hESCs (WCMC-37) [50]. In their study, the *FMR1* promoter was found to be exclusively 5hmC in primary neurons of FXS patients, suggesting that NPCs established from patient-derived iPSCs or FXS hESCs may not reflect the complete repertoire of epigenetic modifications that are typically found in mature neurons in brains of patients. The reason for the discrepancy between the different cell types is unknown however it may be because 5hmC cannot be well maintained in highly proliferating cells [78]. Hence, the significance of this epigenetic mark to gene silencing is uncertain as it may simply reflect changes in other processes that lead to DNA demethylation such as the activity of repair proteins [79].

3.4. The Effect of Differentiation on the Epigenetic Status of the Gene

To date, no simple model can be put forward based on the data collected from FXS PSC-based systems. In some reports, epigenetic modifications were reversed by reprogramming or, conversely, triggered by differentiation [20,21,25]. In other reports, in vitro differentiation, particularly to neurons, had no effect on the methylation status of the gene [54,68]. On the other hand, there is some evidence to suggest that the threshold for silencing lies around 400 repeats in iPSCs derived from individuals with an unmethylated FM, and that this may be the cause of the failure of FXS undifferentiated cells, in general, to methylate the mutant gene [54]. Taking advantage of a selectable reporter system which can be harnessed to identify spontaneous silencing events in *FMR1*, studies have shown that silencing can be induced by increasing the length of the repeats and reversed by contractions [54]. A different study indicated that when the size of the mutation drops below ~400 repeats, but still remains in the FM range, methylation erodes [68]. It would be useful to extend these studies to explore whether PSCs (embryonic or induced) are distinct from other cell types in that they require a greater number of repeats (>199) for methylation to be elicited. In addition, it would be worthwhile exploring whether differentiation can indirectly lead to an increase in methylation levels by selection against unmethylated FMs, as was previously suggested [23,42,68]. In fact, in vitro differentiated iPSC neurons carrying unmethylated FMs presented in addition to FMRP aggregates increased numbers of ubiquitin-positive inclusion bodies as compared to their PM isogenic controls, thus pointing to toxicity of unmethylated FM alleles [54].

3.5. Potential Mechanisms for Epigenetic FMR1 Silencing

Mechanistically, researchers are attempting to address how CGG expansion leads to de novo methylation at the 5′-UTR of *FMR1* using FXS hESCs. FXS hESCs may be particularly useful for this type of study since they carry a FM that is frequently unmethylated [34]. This provides a rare opportunity to uncouple the CGG expansion and the epigenetic marks that are elicited post-fertilization. It is generally accepted that untranslated repeat expansions, including the CGGs in *FMR1*, trigger hypermethylation by a common mechanism that relies on incorrect recruitment of silencing complexes/chromatin modifying enzymes to the otherwise hypomethylated CpG island in which they reside. This may result from binding loss/gain of specific DNA-interacting proteins that normally counteract/promote de novo methylation, respectively.

CTCF and TADs: One model to explain how CGG expansion leads to hypermethylation of the 5′-UTR of *FMR1* suggests that the insulator-binding protein CTCF differentially binds to the *FMR1* locus next to the repeats thereby protecting the region from the spread of adjacent heterochromatin [80]. CTCF is a zinc finger protein that functions as a barrier against the influence of neighboring regulatory elements. It also blocks the spread of inactive chromatin (for a review see [81]) and separates the genome into independent functional domains (called topologically associated domains or TADs), ranging from 10kb to a few megabases. Consistent with functional studies involving other loci, it was initially hypothesized that when the CGGs expand and reach a pathological range, this leads to the loss of CTCF binding in the immediate proximity of the repeats, resulting in the spread of heterochromatin and promoter inactivation.

To explore the potential role of CTCF as an insulating factor in *FMR1*, researchers took advantage of FXS hESCs and examined whether methylation was coupled with the binding loss of CTCF next to the CGGs. However, no enrichments for CTCF could be detected by Chromatin immuno-precipitation (ChIP) analysis in wild type or expanded cells [34]. This is consistent with an earlier study where the knock-down of CTCF in cells from healthy individuals as well as subjects with an unmethylated FM did not induce *FMR1* hypermethylation [66]. Nonetheless, in a recent genome-wide study using Hi-C datasets from wild type hESCs as well as patient tissues and cell lines, Sun et al. bioinformatically identified chromatin folding patterns that are typical of unstable repeat-associated loci, including *FMR1* [82]. Akin to other unstable tandem repeats, the CGGs in *FMR1* co-localize with the boundaries of TADs, which are typically positioned in CpG-island-dense regions and are occupied by CTCF. Upon expansion, when the gene is epigenetically silenced and heavily methylated, the boundaries of the TADs are altered and CTCF occupancy is abolished (100 kb upstream to the repeats).

While the mechanistic relationship between CTCF binding, TAD boundaries, and epigenetic gene silencing remains to be discovered, it is tempting to suggest that the repeat expansion disrupts the TAD structure by interfering with CTCF occupancy by long range interactions. This could ultimately result in hypermethylation and gene silencing in affected cells, possibly by a shift in the location of the gene from a downstream TAD containing active enhancers to an upstream TAD that is poor in active enhancers. Understanding the cause and effect between repeat expansion, boundary disruption and epigenetic silencing is expected to pave the way for research and provide new insights into the mechanisms that may apply to noncoding repeat expansion pathologies as a whole. On the other hand, hypermethylation may be upstream to the changes in TAD boundaries and the loss of CTCF binding. Alternatively, hypermethylation may be unrelated to these events since the majority of disease-causing repeat loci that co-localize with TAD boundaries in the Sun et al. study, were not coupled with aberrant DNA methylation or gene silencing in patients (21 out of 23 examined loci including *HTT*, *ATN1*, *ATXN3*, *ATXN7*, *CSTB*, *JPH3*, *ZIC2* and *CACNA1A*).

RNA-mediated epigenetic silencing: Other models (not necessarily conflicting with insulator-interacting protein binding loss and/or TAD change models), argue for a mechanism that is mediated by RNA [83]. On the basis of accumulating data related to epigenetic gene silencing in other loci, particularly at repetitive elements, it is commonly thought that *FMR1* inactivation is elicited in a way that is RNA-directed. In fact, antisense transcription, siRNAs and R-loop retention have all been put forward as candidates for mediating this process [83]. However, there is no robust experimental evidence to firmly support any of these possibilities. One hypothesized RNA-directed epigenetic silencing mechanism is based on bi-directional transcription, given the identification of two antisense long noncoding RNAs at the 5′-end of *FMR1*; *FMR4* and *ASFMR1* [80,84]. *FMR4*, is a primate-specific noncoding RNA transcript (2.4 kb) that resides immediately upstream and shares a bidirectional promoter with *FMR1*. Like *FMR1*, it is epigenetically silenced in FXS patients and is up-regulated in PM carriers. However, knockdown of *FMR4* did not affect *FMR1* expression, nor vice versa, suggesting that *FMR4* most likely is not involved with epigenetic regulation of the *FMR1* gene [84]. *ASFMR1*, is an overlapping antisense transcript which initiates from intron 2 of *FMR1* and extends past the CGGs. Similar to *FMR4*, *ASFMR1* is upregulated in PM carriers but is silenced in FM individuals. Given the capacity of long double strand RNA (dsRNA) molecules to be processed into small RNA molecules by the RNAi machinery, it was posited that *ASFMR1* may contribute to epigenetic silencing of *FMR1* by forming dsRNA structures together with *FMR1* mRNA [80]. Another hypothesis is that the propensity of CGG-expanded RNAs to fold into hairpin structures provides a favorable substrate for Dicer activity. Regardless of the mechanism (bi-directional transcription or RNA hairpin structures), Dicer-processed short RNA molecules are thought to attract silencing complexes to the region via pairing directly to the DNA or to nascent RNA transcripts (for instance in pericentromeric regions [85]).

By taking advantage of cell fusion technology to reprogram patient cells with mouse Dicer-depleted ESCs (knocked down with 10% residual mRNA levels), Hecht et al. abolished H3K9me3 marking and hence interrupted the silencing process of *FMR1* directed by cell differentiation [20]. Likewise, by fusing patient cells with a mouse ESC that was completely deficient for the histone methyl transferases Suv39h (double mutants for *Suv39h1* and *Suv39h2*) they successfully abolished heterochromatinization and the transcriptional inactivation of the gene. These findings led to the notion that *FMR1* gene silencing is triggered by the recruitment of Suv39h to the repeats via Dicer-processed

CGG-containing small RNAs, resulting in the local induction of heterochromatin (H3K9me3, HP1 and DNA methylation). These authors reported a 2 to 3-fold increase in the number of small RNA molecules with pure CGGs in an FXS hESC with an unmethylated mutation (HEFX), as compared to the wild type control. Their study suggests that *FMR1* transcriptional repression is differentiation-dependent and is mediated by small RNAs working upstream to H3K9me3 heterochromatin induction and DNA hypermethylation. Conversely, in a different study, the knockdown of *Dicer*, *Ago1* and *Ago2* in *FMR1*-expressing FXS hESC lines (WCMC-37 and SI-214) did not prevent epigenetic gene silencing upon differentiation [21], suggesting that gene inactivation may not depend on the RNAi pathway as originally thought. This underscores the need to examine whether over-expression of CGG-pure small RNAs is sufficient to trigger repressive epigenetic modifications in FXS hESCs with an unmethylated FM with/without differentiation.

Other studies exploring the mechanism of *FMR1* gene silencing have pointed to the role of R-loops (three-stranded nucleic structures composed of persistent DNA:RNA hybrids) in the process. For instance, Colak et al., who employed two FXS hESC lines with expansions exceeding 400 CGGs, claimed that hybridization of the mutant mRNA to the promoter and the 5$'$ flanking region of *FMR1* elicits silencing by inducing a change from active (H3K4me2) to repressive (H3K9me2) histone modifications during neural differentiation [21]. Disrupting the interaction of the mRNA with the CGG-repeat portion of the gene only abolished silencing if induced at a critical time point during differentiation. However, careful examination of the methylation status of *FMR1* in the FXS hESC lines in this study indicated that they were methylated to a certain extent to begin with. Clearly, this should be taken into consideration when addressing the issue of how and when aberrant epigenetic modifications are first set in FXS.

A different RNA-based model relies on the observations that R-loops frequently form across the CGGs between nascent RNA and DNA in *FMR1*-expressing cells with wild type, PM and unmethylated FM alleles [19,86–88]. In line with this reasoning, Groh and colleagues suggested that co-transcriptional forming R-loops across the CGGs provide the first trigger for heterochromatinization by directly recruiting the histone methyltransferase G9a and locally depositing H3K9me2 [19]. One assumption is that R-loop stability should be greater in unmethylated FM vs. PM alleles. To confirm this, it would be imperative to stabilize co-transcriptionally formed R-loops in FXS hESCs with an unmethylated FM and monitor for H3K9me2 enrichments. This would attest to a function for R-loops in triggering heterochromatin as proposed in Groh's study.

However, a study suggests that R-loops may play the opposite role *FMR1* silencing. Ginno and colleagues examined the potential role of R-loops as a counteracting mechanism for de novo methylation at CpG island promoters [89]. Using a computational approach, they showed that promoters embedded within CpG islands are particularly enriched for R-loops. They provided experimental evidence that R-loops in CpG island promoters can prevent de novo DNA methylation by interfering with the recruitment of DNMT3b, the most widely expressed DNMT during early embryo development. This may imply that R-loops normally protect the 5$'$-UTR of *FMR1* from de novo methylation by preventing the binding of DNMT3b to the region, and that this is hampered upon expansion. It would be of key interest to design studies to interfere with the formation of R-loops across normal or unmethylated FM alleles, to ascribe a function to R-loops in counteracting de novo methylation at the *FMR1* promoter. A summary of potential factors/mechanism(s) that may be involved in *FMR1* epigenetic silencing, as exhibited by the the currently available PSC-based models, is presented in Table 1.

Table 1. Evidence for the presence (+)/absence (−) of factors that may be involved in *FMR1* epigenetic silencing in various pluripotent stem cell (PSC)-based models of FXS.

	Model System		
	hESCs	**iPSCs**	**PHCFs**
Epigenetic Marks			
CpG methylation	[21] (+/−) [34] (+/−) [62] (+) [68] (+)	[34] (+) [36] (+) [37] (+) [38] (+) [39] (+) [54] (+) [69] (+) [72] (+)	[20] (−) [25] (−) [26] (+) [72] (+)
5-hydroxymethylation	[50] (−)	[50] (−)	
H3K4me3	[34] (+) [68] (−)	[39] (−) [69] (+)	
H3K9me2/3	[21] (−) [34] (+/−) [39] (+) [69] (+)	[39] (+) [69] (+)	[20] (−)
H3K27me3	[68] (+)	[69] (−)	
Modifying Enzymes			
G9a			[20] (+)
Suv39			[20] (+)
DNMT3A			[20] (+)
Insulator Binding Proteins			
CTCF	[34] (−) [82] (+)		
RNA Mediated			
RNAi pathway	[20] (+)		[20] (+)
R-loops	[21] (+) [87] (+)		

4. Repeat Instability

4.1. CGG Instability in FXS PSCs

The mechanism governing CGG instability in FXS is far from understood. Nor do we know when and where precisely repeat instability occurs. There is strong evidence to suggest that FM alleles are initially set by expansions of PM alleles pre-zygotically in the female oocyte [90]. However, large expansions and contractions must also arise in the early embryo since FXS patients typically present somatic mosaicism for expansion length even though FM alleles are stable in terminally differentiated patient cells [24,25,63,91]. Consistent with the concept that instability persists post-fertilization is the data from 2-cell stage PM mouse embryos demonstrating both expansions and a high frequency of large contractions [92]. It remains uncertain whether this equally applies to early cleavage human FXS embryos.

The mechanisms underlying CGG instability in the early embryo may be different from those in the oocyte because in the oocyte (a non-dividing cell) repeat instability must be driven by a mechanism(s) that does not depend on DNA replication (repair or recombination). Whereas, in the early cleavage embryo (high proliferating cells) instability may result from both: DNA-replication-dependent and independent mechanisms. CGG instability in the early embryo [92] most likely involves more than one mechanism. This is because expansions and contractions most likely arise from different mutational events [93].

Given the tight inverse correlation between CGG hypermethylation and instability in patient cells [94], it was assumed that FXS hESCs, which best resemble embryos before/at the time of implantation, would present extensive CGG instability (large expansions and contractions) when the FM is unmethylated [21,34,35]. Indeed, when the FM allele was hypermethylated and transcriptionally inactive in FXS PSCs, mutation size remained unaltered. On the other hand, when the FM allele was hypomethylated, it manifested a high degree of size variation [25,34,54,68]. Nevertheless, as increasingly more FXS hESC/iPSCs cell lines are being established and monitored for expansion size for longer periods, it has become apparent that unmethylated FM alleles rarely expand in culture [54]. In fact, the only clear example of large expansions was reported in iPSCs at very low frequencies and was dependent on selection [54]. In a different study, when single cell clones were monitored for repeat size and methylation in a given FXS hESC line with >400 repeat expansion (WCMC37), unmethylated FM alleles systematically underwent small increases and decreases in repeat numbers reminiscent of microsatellite instability (gain and loss of a few repeat units) [68]. In no case was there evidence for large step-wise expansions such as the ones described in hESC/iPSC models with other repeat-associated mutations [95,96]. One potential explanation for these small length changes may be due to errant DNA replication through the CGGs, an effect that would be compatible with rapidly dividing cells. In addition, studies have shown that upon extended propagation in an initially heterogeneous cell population, there is selection against unmethylated FM alleles, likely due to toxic gain-of-function mechanisms [54,68]. Altogether, the findings in FXS hESCs and iPSCs with unmethylated FM imply that the currently available FXS PSC lines may not be a suitable model for investigating how FM alleles are initially established. Nevertheless, they may be informative for investigating how DNA replication is impaired by CGG expansion in highly proliferating embryonic cells, and for characterizing the DNA structures (that are formed in vivo) which may predispose the locus to instability.

4.2. Potential Mechanisms that May Account for CGG Instability in FXS

DNA replication and repair: Over the years, various models have been suggested to explain how the CGGs become increasingly unstable in FXS cells. One DNA replication-based model relates to the slippage of Okazaki fragments during lagging strand synthesis due to unequal distribution of Gs and Cs (positive GC-skew) between the template (C-rich) and the non-template (G-rich) DNA strands in *FMR1*. It is based on that G-rich single stranded DNA tends to form hairpin and other non-canonical intramolecular structures that are more stable and difficult for the cell to resolve than C-rich repeats (CCG) during DNA replication [97–99]. According to this model, when the CGG-strand provides the template for lagging DNA synthesis, it induces contractions. This is because the replication machinery skips the secondary structures that are frequently formed by strand-slippage in the template DNA. On the other hand, when the CCG-strand serves as the template for lagging strand synthesis, it favors expansions. This is because the newly synthesized CGGs tend to slip and form hairpin structures in the Okazaki fragments that result in the addition of repeats in newly synthesized DNA [100]. Thus, replication direction and ORI usage relative to the CGGs is likely to dictate whether expansions (ORI located 5' to the CGGs) or contractions (ORI located 3' to the CGGs) will be induced. To explain the difference in CGG stability between embryonic (unstable) and adult (stable) tissues, it was postulated that during early embryogenesis there is a switch in ORI usage, leading to a change in the direction of replication across the repeats (origin-shift model).

In line with the replication-based ORI switch model, using a pair of FXS hESCs Gerhardt and colleagues reported a difference in ORI usages between wild type and expanded alleles [100]. By applying an innovative technique for mapping replication origins and measuring fork progression at single molecule resolution (SMARD), they found a switch in the replication direction relative to the repeats combined with fork stalling. Unlike wild type hESCs, they showed that replication initiates mainly from a downstream, rather than an upstream ORI (-53kb), in undifferentiated FXS hESCs, paralleling early embryonic cells. This may indicate that replication across the CGGs can mainly be attributed to lagging strand synthesis and therefore should promote expansions. Gerhardt et al.

suggested that once the cells begin to differentiate, replication patterns change and become more similar to those in normal hESCs, which would imply that replication irregularities are developmentally regulated, and by extension that instability is lost with differentiation. In a different study, a polymorphism in the upstream ORI site was identified and associated with the propensity of the CGGs to expand to FMs, thus providing a potential mechanistic link between CGG instability and the failure to activate an upstream ORI by a cis-acting DNA sequence in undifferentiated FXS embryonic cells [101].

It remains unclear whether the change in ORI usages in mutant undifferentiated cells is the cause or the effect of the expansion, and whether this is mechanistically associated with CGG embryonic instability. This is particularly important given the stabilizing effect of de novo methylation, which apparently does not depend on the differentiation state of the cell.

Another suggested mechanism for CGG instability, besides those that rely on a problem in DNA replication, is incorrect DNA repair. It should be noted that as explained above CGG instability may result from both DNA repair and replication problems, depending on the cell type involved. A driving mechanism for error-prone repair, rather than replication, would be particularly pertinent to non-dividing cells like neurons and oocytes. While evidence in human cells is still lacking, accumulating data in PM mice carrying mutations in various DNA repair pathways suggest that CGG instability is promoted by the incorrect recruitment of mismatch repair proteins, particularly MSH2, MSH3 and MSH6, which leads to the activation of error-prone repair pathways [102]. The data related to repeat instability in hESCs/iPSCs in other loci tend to confirm this reasoning. For example, it was shown that when the mismatch repair (MMR) proteins MSH2, MSH3 and MSH6 are downregulated upon differentiation, repeat instability is considerably reduced in hESC/iPSCs with myotonic dystrophy (DM1; CTG repeats), Huntington's disease (HD; CAG repeats) and Friedrich's ataxia (FRDA; GAA repeats) mutations [95,96]. In addition, knockdown of *MSH2* and *MSH6* in iPSCs with the FRDA mutation impeded repeat expansions [103]. In fact, genome-wide association studies (GWAS) in humans with these conditions support a relationship between MMR proteins and levels of somatic instability [104,105]. It would be imperative to manipulate the levels of these factors in undifferentiated hESCs/iPSCs with an unmethylated FM to confirm the mechanistic link between these factors and CGG instability in FXS early embryos, as suggested in mice [30,32,102].

RNA transcription as a mediator of CGG instability: Regardless of the mechanism (DNA repair or replication), there is a growing consensus that repeat instability is mediated by RNA transcription, potentially through the formation of R-loops. It is assumed that R-loops promote instability by enhancing the formation of unconventional structures such as hairpin and G-quadruplexes by the G-rich unpaired DNA strand in the R-loop. In a recent study, researchers used hESC with wild type and FXS-expanded alleles to finely characterize and precisely map R-loops and single strand DNA displacements across and near the CGG repeats in vivo to better understand the propensity of this locus to become highly unstable [87]. FXS hESCs are particularly useful for this type of study because they often carry unmethylated FM alleles. These authors showed that the CGGs constitute preferential sites (hotspots) for DNA unpairing in normal range alleles. When R-loops are formed, DNA unpairing is more extensive, and is coupled with interruptions of double strand structures by the non-transcribing (G-rich) DNA strand. These interruptions, which were described in an earlier study by Loomis et al. in somatic cells with normal and PM size alleles [86], are likely to reflect unusual structures in the DNA that are hard to process when the CGGs expand significantly. Strikingly, when the *FMR1* gene was hypermethylated and transcriptionally inactive, local unpairing was abolished. This is consistent with the strict inverse correlation between repeat instability and the methylation/transcriptional competence of the gene [87]. It remains unclear whether more complex double strand structures are actually formed in the locus when it is fully expanded and transcribed. Once these structures are identified, they will need to be associated with the induction of DNA damage or replication irregularities to mechanistically link them with CGG instability in FXS.

Three-dimensional DNA structure: Another feature that may relate to instability has to do with the 3D structure of the *FMR1* containing region. In the study by Sun et al. (see Section 3.5) it was suggested that repeat instability occurs as a consequence of TADs boundary disruption [82]. Accordingly, short tandem repeats that are located at the TAD boundaries may be inherently unstable. This may be due to their position at the junctions of physical domains, which interferes with mechanisms that normally suppress excessive expansions. Given that TAD boundaries are hotspots for DNA double strand breaks [106], this may predispose the CGGs to error-prone replication or repair. It would be of great interest to explore whether interfering with the insulator function of the TAD boundary, possibly via disruption of a CTCF binding site, would elicit CGG instability (as in the *Atxn7* transgenic mice model [107]), as suggested by this bioinformatic association study.

5. Conclusions

Despite intensive research, the timing and mechanism(s) by which *FMR1* becomes epigenetically silenced or extensively unstable in FXS patients remains obscure. This is at least partially due to the lack of appropriate animal and cellular models. Given that mutant PSCs most resemble early embryonic FXS development, they may provide a valuable model system to investigate the dynamic changes that take place in diseased cells at stages that are otherwise inaccessible to research.

Based on the data obtained from these cell lines to date, it is becoming increasingly clear that *FMR1* hypermethylation is not limited to or triggered by differentiation, and that the threshold for silencing in undifferentiated PSCs is most likely higher than assumed and is situated around 400 repeats. Clearly the wide variability in *FMR1* methylation among the different FXS hESC lines calls for re-evaluation of the timing of *FMR1* hypermethylation and suggests that the expansion is not evenly methylated or not sufficiently long enough in all the cells in the preimplantation embryo. In terms of the molecular mechanisms involved in this process, it will be important to identify the earliest epigenetic modification that provides the trigger for heterochromatin induction. The role of RNA transcription and CTCF binding far from the repeats, as well as the significance of the topological organization of the locus, need further study. In addition, given the tight association between CGG instability and the epigenetic status of the repeats, it should become possible to validate the mechanistic association between de novo methylation and CGG instability with the advent of locus-specific epigenetic editing tools. The role of ORI switching and the formation of unusual non-canonical DNA structures in the locus should be further characterized and mechanistically linked to the enhancement of CGG instability in FXS PSCs. In addition, the possibility that the de-methylation of FM alleles can naturally occur in post-mitotic neurons should be explored. Finally, attempting to replicate the results in FXS PSCs and reactivate the silenced gene by modifying the chromatin state of the locus in slow/non- dividing cells will be a major challenge as a therapeutic approach. Even if achieved, the problem of toxic RNA/protein will need to be addressed. Clearly, the currently available FXS PSC models, particularly hESCs and iPSCs, will continue to contribute as complementary powerful systems to fill the gaps in our knowledge of the molecular mechanisms that go awry in FXS at very early stages of development.

Author Contributions: M.A.D. and R.E. contributed to the conception, design and writing of the manuscript.

Funding: This research was partly funded by the LEGACY HERITAGE BIOMEDICAL PROGRAM OF THE ISRAEL SCIENCE FOUNDATION [1260/16, R.E.].

Acknowledgments: We thank David Zeevi for critically reading the manuscript and Anna Bialer Tsypin for the illustrations. M.A.D. is supported by the LEGACY HERITAGE BIOMEDICAL PROGRAM OF THE ISRAEL SCIENCE FOUNDATION [1260/16, R.E.].

Conflicts of Interest: The authors declare no conflict of interest. The funding sponsors had no role in the design of the study; in the collection, analyses, or interpretation of data; in the writing of the manuscript, and in the decision to publish the results.

References

1. Oberle, I.; Rousseau, F.; Heitz, D.; Kretz, C.; Devys, D.; Hanauer, A.; Boué, J.; Bertheas, M.; Mandel, J. Instability of a 550-base pair DNA segment and abnormal methylation in fragile X syndrome. *Science* **1991**, *252*, 1097–1102. [CrossRef] [PubMed]
2. Rousseau, F.; Heitz, D.; Biancalana, V.; Blumenfeld, S.; Kretz, C.; Boué, J.; Tommerup, N.; Van Der Hagen, C.; DeLozier-Blanchet, C.; Croquette, M.-F.; et al. Direct diagnosis by DNA analysis of the fragile X syndrome of mental retardation. *N. Engl. J. Med.* **1991**, *325*, 1673–1681. [CrossRef] [PubMed]
3. Verkerk, A.J.; Pieretti, M.; Sutcliffe, J.S.; Fu, Y.-H.; Kuhl, D.P.; Pizzuti, A.; Reiner, O.; Richards, S.; Victoria, M.F.; Zhang, F.; et al. Identification of a gene (*FMR-1*) containing a CGG repeat coincident with a breakpoint cluster region exhibiting length variation in fragile X syndrome. *Cell* **1991**, *65*, 905–914. [CrossRef]
4. Brunberg, J.A.; Jacquemont, S.; Hagerman, R.J.; Berry-Kravis, E.M.; Grigsby, J.; A Leehey, M.; Tassone, F.; Brown, W.T.; Greco, C.M.; Hagerman, P.J. Fragile X premutation carriers: Characteristic MR imaging findings of adult male patients with progressive cerebellar and cognitive dysfunction. *AJNR Am. J. Neuroradiol.* **2002**, *23*, 1757–1766. [PubMed]
5. Jacquemont, S.; Hagerman, R.J.; Leehey, M.; Grigsby, J.; Zhang, L.; Brunberg, J.A.; Greco, C.; Portes, V.D.; Jardini, T.; Levine, R.; et al. Fragile X premutation tremor/ataxia syndrome: Molecular, clinical, and neuroimaging correlates. *Am. J. Hum. Genet.* **2003**, *72*, 869–878. [CrossRef] [PubMed]
6. Iwahashi, C.K.; Yasui, D.H.; An, H.-J.; Greco, C.M.; Tassone, F.; Nannen, K.; Babineau, B.; Lebrilla, C.B.; Hagerman, R.J.; Hagerman, P.J. Protein composition of the intranuclear inclusions of FXTAS. *Brain* **2006**, *129*, 256–271. [CrossRef] [PubMed]
7. Bretherick, K.L.; Fluker, M.R.; Robinson, W.P. *FMR1* repeat sizes in the gray zone and high end of the normal range are associated with premature ovarian failure. *Hum. Genet.* **2005**, *117*, 376–382. [CrossRef]
8. Todd, P.K.; Oh, S.Y.; Krans, A.; He, F.; Sellier, C.; Frazer, M.; Renoux, A.J.; Chen, K.-C.; Scaglione, K.M.; Basrur, V.; et al. CGG repeat-associated translation mediates neurodegeneration in fragile X tremor ataxia syndrome. *Neuron* **2013**, *78*, 440–455. [CrossRef]
9. Nolin, S.L.; Lewis, F.A.; Ye, L.L.; Houck, G.E., Jr.; Glicksman, A.E.; Limprasert, P.; Li, S.Y.; Zhong, N.; Ashley, A.E.; Feingold, E.; et al. Familial transmission of the *FMR1* CGG repeat. *Am. J. Hum. Genet.* **1996**, *59*, 1252–1261.
10. Coffee, B.; Zhang, F.; Warren, S.T.; Reines, D. Acetylated histones are associated with *FMR1* in normal but not fragile X-syndrome cells. *Nat. Genet.* **1999**, *22*, 98–101. [CrossRef]
11. Coffee, B.; Zhang, F.; Ceman, S.; Warren, S.; Reines, D. Histone modifications depict an aberrantly heterochromatinized *FMR1* gene in fragile x syndrome. *Am. J. Hum. Genet.* **2002**, *71*, 923–932. [CrossRef] [PubMed]
12. Tabolacci, E.; Zalfa, F.; Zito, I.; Terracciano, A.; Oostra, B.; Neri, G.; Pietrobono, R.; EMoscato, U.; TBabolaccgni, C.; Chiurazzi, P. Molecular dissection of the events leading to inactivation of the *FMR1* gene. *Hum. Mol. Genet.* **2005**, *14*, 267–277. [CrossRef]
13. Tabolacci, E.; Pietrobono, R.; Moscato, U.; A Oostra, B.; Chiurazzi, P.; Neri, G. Differential epigenetic modifications in the *FMR1* gene of the fragile X syndrome after reactivating pharmacological treatments. *Eur. J. Hum. Genet.* **2005**, *13*, 641–648. [CrossRef] [PubMed]
14. Tabolacci, E.; Moscato, U.; Zalfa, F.; Bagni, C.; Chiurazzi, P.; Neri, G. Epigenetic analysis reveals a euchromatic configuration in the *FMR1* unmethylated full mutations. *Eur. J. Hum. Genet.* **2008**, *16*, 1487–1498. [CrossRef] [PubMed]
15. Kumari, D.; Usdin, K. The distribution of repressive histone modifications on silenced *FMR1* alleles provides clues to the mechanism of gene silencing in fragile X syndrome. *Hum. Mol. Genet.* **2010**, *19*, 4634–4642. [CrossRef] [PubMed]
16. Yu, S.; Pritchard, M.; Kremer, E.; Lynch, M.; Nancarrow, J.; Baker, E.; Holman, K.; Mulley, J.; Warren, S.; Schlessinger, D.; et al. Fragile X genotype characterized by an unstable region of DNA. *Science* **1991**, *252*, 1179–1181. [CrossRef] [PubMed]
17. Kremer, E.J.; Pritchard, M.; Lynch, M.; Yu, S.; Holman, K.; Baker, E.; Warren, S.; Schlessinger, D.; Sutherland, G.; Richards, R. Mapping of DNA instability at the fragile X to a trinucleotide repeat sequence p(CCG)n. *Science* **1991**, *252*, 1711–1714. [CrossRef]

18. Kumari, D.; Usdin, K. Polycomb group complexes are recruited to reactivated *FMR1* alleles in Fragile X syndrome in response to *FMR1* transcription. *Hum. Mol. Genet.* **2014**, *23*, 6575–6583. [CrossRef]
19. Groh, M.; Lufino, M.M.; Wade-Martins, R.; Gromak, N. R-loops associated with triplet repeat expansions promote gene silencing in Friedreich ataxia and fragile X syndrome. *PLoS Genet.* **2014**, *10*, e1004318. [CrossRef]
20. Hecht, M.; Tabib, A.; Kahan, T.; Orlanski, S.; Gropp, M.; Tabach, Y.; Yanuka, O.; Benvenisty, N.; Keshet, I.; Cedar, H. Epigenetic mechanism of *FMR1* inactivation in Fragile X syndrome. *Int. J. Dev. Biol.* **2017**, *61*, 285–292. [CrossRef]
21. Colak, D.; Zaninovic, N.; Cohen, M.S.; Rosenwaks, Z.; Yang, W.-Y.; Gerhardt, J.; Disney, M.D.; Jaffrey, S.R. Promoter-bound trinucleotide repeat mRNA drives epigenetic silencing in fragile X syndrome. *Science* **2014**, *343*, 1002–1005. [CrossRef]
22. Willemsen, R.; Bontekoe, C.J.; Severijnen, L.-A.; Oostra, B.A. Timing of the absence of *FMR1* expression in full mutation chorionic villi. *Hum. Genet.* **2002**, *110*, 601–605. [CrossRef] [PubMed]
23. Mor-Shaked, H.; Eiges, R. Reevaluation of *FMR1* Hypermethylation Timing in Fragile X Syndrome. *Front. Mol. Neurosci.* **2018**, *11*, 31. [CrossRef] [PubMed]
24. Devys, D.; Biancalana, V. Analysis of full fragile X mutations in fetal tissues and monozygotic twins indicate that abnormal methylation and somatic heterogeneity are established early in development. *Am. J. Med. Genet.* **1992**, *43*, 208–216. [CrossRef] [PubMed]
25. Wohrle, D.; Salat, U.; Hameister, H.; Vogel, W.; Steinbach, P. Demethylation, reactivation, and destabilization of human fragile X full-mutation alleles in mouse embryocarcinoma cells. *Am. J. Hum. Genet.* **2001**, *69*, 504–515. [CrossRef] [PubMed]
26. Burman, R.W.; Popovich, B.W.; Jacky, P.B.; Turker, M.S. Fully expanded FMR1 CGG repeats exhibit a length- and differentiation-dependent instability in cell hybrids that is independent of DNA methylation. *Hum. Mol. Genet.* **1999**, *8*, 2293–302. [CrossRef] [PubMed]
27. Bontekoe, C.J.; de Graaff, E.; Nieuwenhuizen, I.M.; Willemsen, R.; A Oostra, B. *FMR1* premutation allele (CGG)81 is stable in mice. *Eur. J. Hum. Genet.* **1997**, *5*, 293–298. [PubMed]
28. Entezam, A.; Biacsi, R.; Orrison, B.; Saha, T.; Hoffman, G.E.; Grabczyk, E.; Nussbaum, R.L.; Usdin, K. Regional FMRP deficits and large repeat expansions into the full mutation range in a new Fragile X premutation mouse model. *Gene* **2007**, *395*, 125–134. [CrossRef]
29. Lavedan, C.; Grabczyk, E.; Usdin, K.; Nussbaum, R.L. Long uninterrupted CGG repeats within the first exon of the human *FMR1* gene are not intrinsically unstable in transgenic mice. *Genomics* **1998**, *50*, 229–240. [CrossRef]
30. Lokanga, R.A.; Zhao, X.N.; Usdin, K. The mismatch repair protein MSH2 is rate limiting for repeat expansion in a fragile X premutation mouse model. *Hum. Mutat.* **2014**, *35*, 129–136. [CrossRef]
31. Lokanga, R.A.; Senejani, A.G.; Sweasy, J.B.; Usdin, K. Heterozygosity for a hypomorphic Polβ mutation reduces the expansion frequency in a mouse model of the Fragile X-related disorders. *PLoS Genet.* **2015**, *11*, e1005181. [CrossRef] [PubMed]
32. Zhao, X.N.; Kumari, D.; Gupta, S.; Wu, D.; Evanitsky, M.; Yang, W.; Usdin, K. Mutsβ generates both expansions and contractions in a mouse model of the Fragile X-associated disorders. *Hum. Mol. Genet.* **2015**, *24*, 7087–7096. [CrossRef] [PubMed]
33. Zhao, X.N.; Usdin, K. The transcription-coupled repair protein ERCC6/CSB also protects against repeat expansion in a mouse model of the fragile X premutation. *Hum. Mutat.* **2015**, *36*, 482–487. [CrossRef] [PubMed]
34. Avitzour, M.; Mor-Shaked, H.; Yanovsky-Dagan, S.; Aharoni, S.; Altarescu, G.; Renbaum, P.; Eldar-Geva, T.; Schonberger, O.; Levy-Lahad, E.; Epsztejn-Litman, S.; et al. *FMR1* epigenetic silencing commonly occurs in undifferentiated fragile X-affected embryonic stem cells. *Stem Cell Rep.* **2014**, *3*, 699–706. [CrossRef] [PubMed]
35. Eiges, R.; Urbach, A.; Malcov, M.; Frumkin, T.; Schwartz, T.; Amit, A.; Yaron, Y.; Eden, A.; Yanuka, O.; Benvenisty, N.; et al. Developmental study of fragile X syndrome using human embryonic stem cells derived from preimplantation genetically diagnosed embryos. *Cell Stem Cell* **2007**, *1*, 568–577. [CrossRef]
36. Urbach, A.; Bar-Nur, O.; Daley, G.Q.; Benvenisty, N. Differential modeling of fragile X syndrome by human embryonic stem cells and induced pluripotent stem cells. *Cell Stem Cell* **2010**, *6*, 407–411. [CrossRef]

37. Alisch, R.S.; Conneely, K.N.; Warren, S.T.; Wang, T.; Chopra, P.; Visootsak, J. Genome-wide analysis validates aberrant methylation in fragile X syndrome is specific to the FMR1 locus. *BMC Med. Genet.* **2013**, *14*, 18. [CrossRef]

38. Sheridan, S.D.; Theriault, K.M.; Reis, S.A.; Zhou, F.; Madison, J.M.; Daheron, L.; Loring, J.F.; Haggarty, S.J. Epigenetic characterization of the *FMR1* gene and aberrant neurodevelopment in human induced pluripotent stem cell models of fragile X syndrome. *PLoS ONE* **2011**, *6*, e26203. [CrossRef]

39. De Esch, C.E.; Ghazvini, M.; Loos, F.; Schelling-Kazaryan, N.; Widagdo, W.; Munshi, S.T.; Van Der Wal, E.; Douben, H.; Gunhanlar, N.; Kushner, S.A.; et al. Epigenetic characterization of the *FMR1* promoter in induced pluripotent stem cells from human fibroblasts carrying an unmethylated full mutation. *Stem Cell Rep.* **2014**, *3*, 548–555. [CrossRef]

40. Bhattacharyya, A.; Zhao, X. Human pluripotent stem cell models of Fragile X syndrome. *Mol. Cell Neurosci.* **2016**, *73*, 43–51. [CrossRef]

41. Telias, M.; Ben-Yosef, D. Neural stem cell replacement: A possible therapy for neurodevelopmental disorders? *Neural. Regen. Res.* **2015**, *10*, 180–182. [CrossRef] [PubMed]

42. Mor-Shaked, H.; Eiges, R. Modeling Fragile X Syndrome Using Human Pluripotent Stem Cells. *Genes* **2016**, *7*, 77. [CrossRef] [PubMed]

43. Eiges, R.; Benvenisty, N. A molecular view on pluripotent stem cells. *FEBS Lett.* **2002**, *529*, 135–141. [CrossRef]

44. Cowan, C.A.; Atienza, J.; Melton, D.A.; Eggan, K. Nuclear reprogramming of somatic cells after fusion with human embryonic stem cells. *Science* **2005**, *309*, 1369–1373. [CrossRef] [PubMed]

45. Verlinsky, Y.; Strelchenko, N.; Kukharenko, V.; Rechitsky, S.; Verlinsky, O.; Galat, V.; Kuliev, A. Human embryonic stem cell lines with genetic disorders. *Reprod. Biomed. Online* **2005**, *10*, 105–110. [CrossRef]

46. Ben-Yosef, D.; Amit, A.; Malcov, M.; Frumkin, T.; Ben-Yehudah, A.; Eldar, I.; Mey-Raz, N.; Azem, F.; Altarescu, G.; Renbaum, P.; et al. Female sex bias in human embryonic stem cell lines. *Stem Cells Dev.* **2012**, *21*, 363–372. [CrossRef] [PubMed]

47. Naumann, A.; Hochstein, N.; Weber, S.; Fanning, E.; Doerfler, W. A distinct DNA-methylation boundary in the 5′-upstream sequence of the FMR1 promoter binds nuclear proteins and is lost in fragile X syndrome. *Am. J. Hum. Genet.* **2009**, *85*, 606–616. [CrossRef]

48. Wöhrle, D.; Hennig, I.; Vogel, W.; Steinbach, P. Mitotic stability of fragile X mutations in differentiated cells indicates early post-conceptional trinucleotide repeat expansion. *Nat. Genet.* **1993**, *4*, 140–142. [CrossRef] [PubMed]

49. Godler, D.E.; Tassone, F.; Loesch, D.Z.; Taylor, A.K.; Gehling, F.; Hagerman, R.J.; Burgess, T.; Ganesamoorthy, D.; Hennerich, D.; Gordon, L.; et al. Methylation of novel markers of fragile X alleles is inversely correlated with FMRP expression and FMR1 activation ratio. *Hum. Mol. Genet.* **2010**, *19*, 1618–1632. [CrossRef] [PubMed]

50. Esanov, R.; Andrade, N.S.; Bennison, S.; Wahlestedt, C.; Zeier, Z. The *FMR1* promoter is selectively hydroxymethylated in primary neurons of fragile X syndrome patients. *Hum. Mol. Genet.* **2016**, *25*, 4870–4880. [CrossRef] [PubMed]

51. Takahashi, K.; Tanabe, K.; Ohnuki, M.; Narita, M.; Ichisaka, T.; Tomoda, K.; Yamanaka, S. Induction of pluripotent stem cells from adult human fibroblasts by defined factors. *Cell* **2007**, *131*, 861–872. [CrossRef] [PubMed]

52. Choi, J.; Lee, S.; Mallard, W.; Clement, K.; Tagliazucchi, G.M.; Lim, H.; Choi, I.Y.; Ferrari, F.; Tsankov, A.M.; Pop, R.; et al. A comparison of genetically matched cell lines reveals the equivalence of human iPSCs and ESCs. *Nat. Biotechnol.* **2015**, *33*, 1173–1181. [CrossRef] [PubMed]

53. Narsinh, K.H.; Plews, J.; Wu, J.C. Comparison of human induced pluripotent and embryonic stem cells: Fraternal or identical twins? *Mol. Ther.* **2011**, *19*, 635–638. [CrossRef] [PubMed]

54. Brykczynska, U.; Pecho-Vrieseling, E.; Thiemeyer, A.; Klein, J.; Fruh, I.; Doll, T.; Manneville, C.; Fuchs, S.; Iazeolla, M.; Beibel, M.; et al. CGG Repeat-Induced FMR1 Silencing Depends on the Expansion Size in Human iPSCs and Neurons Carrying Unmethylated Full Mutations. *Stem Cell Rep.* **2016**, *7*, 1059–1071. [CrossRef] [PubMed]

55. Sunamura, N.; Iwashita, S.; Enomoto, K.; Kadoshima, T.; Isono, F. Loss of the fragile X mental retardation protein causes aberrant differentiation in human neural progenitor cells. *Sci. Rep.* **2018**, *8*, 11585. [CrossRef] [PubMed]

56. Halevy, T.; Czech, C.; Benvenisty, N. Molecular mechanisms regulating the defects in fragile X syndrome neurons derived from human pluripotent stem cells. *Stem Cell Rep.* **2015**, *4*, 37–46. [CrossRef] [PubMed]

57. Hwang, Y.T.; Dudding, T.; Arpone, S.M.; Arpone, M.; Francis, D.; Li, X.; Slater, H.R.; Rogers, C.; Bretherton, L.; du Sart, D.; et al. Molecular Inconsistencies in a Fragile X Male with Early Onset Ataxia. *Genes* **2016**, *7*, 68. [CrossRef] [PubMed]

58. Aliaga, S.M.; Slater, H.R.; Francis, D.; Du Sart, D.; Li, X.; Amor, D.J.; Alliende, A.M.; Maria, L.S.; Faundes, V.; Morales, P.; et al. Identification of Males with Cryptic Fragile X Alleles by Methylation-Specific Quantitative Melt Analysis. *Clin. Chem.* **2016**, *62*, 343–352. [CrossRef] [PubMed]

59. Castellví-Bel, S.; Milà, M.; Soler, A.; Carrió, A.; Sánchez, A.; Villa, M.; Jimenez, M.D.; Estivill, X. Prenatal diagnosis of fragile X syndrome: (CGG)n expansion and methylation of chorionic villus samples. *Prenat. Diagn.* **1995**, *15*, 801–807. [CrossRef] [PubMed]

60. Sutherland, G.R.; Gedeon, A.; Kornman, L.; Donnelly, A.; Byard, R.W.; Mulley, J.C.; Kremer, E.; Lynch, M.; Pritchard, M.; Yu, S.; et al. Prenatal diagnosis of fragile X syndrome by direct detection of the unstable DNA sequence. *N. Engl. J. Med.* **1991**, *325*, 1720–1722. [CrossRef] [PubMed]

61. Suzumori, K.; Yamauchi, M.; Seki, N.; Kondo, I.; Hori, T. Prenatal diagnosis of a hypermethylated full fragile X mutation in chorionic villi of a male fetus. *J. Med. Genet.* **1993**, *30*, 785–787. [CrossRef] [PubMed]

62. Haenfler, J.M.; Skariah, G.; Rodriguez, C.M.; Da Rocha, A.M.; Parent, J.M.; Smith, G.D.; Todd, P.K. Targeted Reactivation of FMR1 Transcription in Fragile X Syndrome Embryonic Stem Cells. *Front. Mol. Neurosci.* **2018**, *11*, 282. [CrossRef] [PubMed]

63. Pretto, D.; Yrigollen, C.M.; Tang, H.-T.; Williamson, J.; Espinal, G.; Iwahashi, C.K.; Durbin-Johnson, B.; Hagerman, R.J.; Hagerman, P.J.; Tassone, F. Clinical and molecular implications of mosaicism in FMR1 full mutations. *Front. Genet.* **2014**, *5*, 318. [CrossRef] [PubMed]

64. Gafni, O.; Weinberger, L.; Mansour, A.A.; Manor, Y.S.; Chomsky, E.; Ben-Yosef, D.; Kalma, Y.; Viukov, S.; Maza, I.; Zviran, A.; et al. Derivation of novel human ground state naive pluripotent stem cells. *Nature* **2013**, *504*, 282–286. [CrossRef] [PubMed]

65. Naumann, A.; Kraus, C.; Hoogeveen, A.; Ramirez, C.M.; Doerfler, W. Stable DNA methylation boundaries and expanded trinucleotide repeats: Role of DNA insertions. *J. Mol. Biol.* **2014**, *426*, 2554–2566. [CrossRef] [PubMed]

66. Lanni, S.; Goracci, M.; Borrelli, L.; Mancano, G.; Chiurazzi, P.; Moscato, U.; Ferre', F.; Helmer-Citterich, M.; Tabolacci, E.; Neri, G. Role of CTCF protein in regulating FMR1 locus transcription. *PLoS Genet.* **2013**, *9*, e1003601. [CrossRef] [PubMed]

67. Godler, D.E.; Slater, H.R.; Bui, Q.M.; Ono, M.; Gehling, F.; Francis, D.; Amor, D.J.; Hopper, J.L.; Hagerman, R.; Loesch, D.Z. FMR1 intron 1 methylation predicts FMRP expression in blood of female carriers of expanded FMR1 alleles. *J. Mol. Diagn.* **2011**, *13*, 528–536. [CrossRef]

68. Zhou, Y.; Kumari, D.; Sciascia, N.; Usdin, K. CGG-repeat dynamics and *FMR1* gene silencing in fragile X syndrome stem cells and stem cell-derived neurons. *Mol. Autism* **2016**, *7*, 42. [CrossRef]

69. Liu, X.S.; Wu, H.; Krzisch, M.; Wu, X.; Graef, J.; Muffat, J.; Hnisz, D.; Li, C.H.; Yuan, B.; Xu, C.; et al. Rescue of Fragile X Syndrome Neurons by DNA Methylation Editing of the FMR1 Gene. *Cell* **2018**, *172*, 979–992. [CrossRef]

70. Liu, X.S.; Wu, H.; Ji, X.; Stelzer, Y.; Wu, X.; Czauderna, S.; Shu, J.; Dadon, D.; Young, R.A.; Jaenisch, R. Editing DNA Methylation in the Mammalian Genome. *Cell* **2016**, *167*, 233–247. [CrossRef]

71. Park, C.Y.; Halevy, T.; Lee, D.R.; Sung, J.J.; Lee, J.S.; Yanuka, O.; Benvenisty, N.; Kim, D.-W. Reversion of FMR1 Methylation and Silencing by Editing the Triplet Repeats in Fragile X iPSC-Derived Neurons. *Cell Rep.* **2015**, *13*, 234–241. [CrossRef] [PubMed]

72. Xie, N.; Gong, H.; Suhl, J.A.; Chopra, P.; Wang, T.; Warren, S.T. Reactivation of *FMR1* by CRISPR/Cas9-Mediated Deletion of the Expanded CGG-Repeat of the Fragile X Chromosome. *PLoS ONE* **2016**, *11*, e0165499. [CrossRef] [PubMed]

73. Ito, S.; D'Alessio, A.C.; Taranova, O.V.; Hong, K.; Sowers, L.C.; Zhang, Y. Role of Tet proteins in 5mC to 5hmC conversion, ES-cell self-renewal and inner cell mass specification. *Nature* **2010**, *466*, 1129–1133. [CrossRef] [PubMed]

74. Tahiliani, M.; Koh, K.P.; Shen, Y.; Pastor, W.A.; Bandukwala, H.; Brudno, Y.; Agarwal, S.; Iyer, L.M.; Liu, D.R.; Aravind, L.; et al. Conversion of 5-methylcytosine to 5-hydroxymethylcytosine in mammalian DNA by MLL partner TET1. *Science* **2009**, *324*, 930–935. [CrossRef] [PubMed]

75. Williams, K.; Christensen, J.; Pedersen, M.T.; Johansen, J.V.; Cloos, P.A.C.; Rappsilber, J.; Helin, K. TET1 and hydroxymethylcytosine in transcription and DNA methylation fidelity. *Nature* **2011**, *473*, 343–348. [CrossRef] [PubMed]

76. Wu, H.; Zhang, Y. Tet1 and 5-hydroxymethylation: A genome-wide view in mouse embryonic stem cells. *Cell Cycle* **2011**, *10*, 2428–2436. [CrossRef] [PubMed]

77. Jin, C.; Lu, Y.; Jelinek, J.; Liang, S.; Estecio, M.R.; Barton, M.C.; Issa, J.-P.J. TET1 is a maintenance DNA demethylase that prevents methylation spreading in differentiated cells. *Nucleic. Acids Res.* **2014**, *42*, 6956–6971. [CrossRef]

78. Pfeifer, G.P.; Kadam, S.; Jin, S.G. 5-hydroxymethylcytosine and its potential roles in development and cancer. *Epigenet. Chromatin.* **2013**, *6*, 10. [CrossRef]

79. Gavin, D.P.; Chase, K.A.; Sharma, R.P. Active DNA demethylation in post-mitotic neurons: A reason for optimism. *Neuropharmacology* **2013**, *75*, 233–245. [CrossRef]

80. Ladd, P.D.; Smith, L.E.; Rabaia, N.A.; Moore, J.M.; Georges, S.A.; Hansen, R.S.; Hagerman, R.J.; Tassone, F.; Tapscott, S.J.; Filippova, G.N. An antisense transcript spanning the CGG repeat region of FMR1 is upregulated in premutation carriers but silenced in full mutation individuals. *Hum. Mol. Genet.* **2007**, *16*, 3174–3187. [CrossRef]

81. Filippova, G.N. Genetics and epigenetics of the multifunctional protein CTCF. *Curr. Top. Dev. Biol.* **2008**, *80*, 337–360. [CrossRef] [PubMed]

82. Sun, J.H.; Zhou, L.; Emerson, D.J.; Phyo, S.A.; Titus, K.R.; Gong, W.; Gilgenast, T.G.; Beagan, J.A.; Davidson, B.L.; Tassone, F.; et al. Disease-Associated Short Tandem Repeats Co-localize with Chromatin Domain Boundaries. *Cell* **2018**, *175*, 224–238. [CrossRef] [PubMed]

83. Usdin, K.; Hayward, B.E.; Kumari, D.; Lokanga, R.A.; Sciascia, N.; Zhao, X.-N. Repeat-mediated genetic and epigenetic changes at the FMR1 locus in the Fragile X-related disorders. *Front. Genet.* **2014**, *5*, 226–10. [CrossRef] [PubMed]

84. Khalil, A.M.; Faghihi, M.A.; Modarresi, F.; Brothers, S.P.; Wahlestedt, C. A novel RNA transcript with antiapoptotic function is silenced in fragile X syndrome. *PLoS ONE* **2008**, *3*, e1486. [CrossRef] [PubMed]

85. Jih, G.; Iglesias, N.; Currie, M.A.; Bhanu, N.V.; Paulo, J.A.; Gygi, S.P.; Garcia, B.A.; Moazed, D. Unique roles for histone H3K9me states in RNAi and heritable silencing of transcription. *Nature* **2017**, *547*, 463–467. [CrossRef] [PubMed]

86. Loomis, E.W.; Sanz, L.A.; Chedin, F.; Hagerman, P.J. Transcription-associated R-loop formation across the human *FMR1* CGG-repeat region. *PLoS Genet.* **2014**, *10*, e1004294. [CrossRef] [PubMed]

87. Abu Diab, M.; Mor-Shaked, H.; Cohen, E.; Cohen-Hadad, Y.; Ram, O.; Epsztejn-Litman, S.; Eiges, R. The G-rich Repeats in FMR1 and C9orf72 Loci Are Hotspots for Local Unpairing of DNA. *Genetics* **2018**, *210*, 1239–1252. [CrossRef] [PubMed]

88. Kumari, D.; Usdin, K. Sustained expression of FMR1 mRNA from reactivated fragile X syndrome alleles after treatment with small molecules that prevent trimethylation of H3K27. *Hum. Mol. Genet.* **2016**, *25*, 3689–3698. [CrossRef]

89. Ginno, P.A.; Lott, P.L.; Christensen, H.C.; Korf, I.; Chédin, F. R-loop formation is a distinctive characteristic of unmethylated human CpG island promoters. *Mol. Cell* **2012**, *45*, 814–825. [CrossRef]

90. Yrigollen, C.M.; Martorell, L.; Durbin-Johnson, B.; Naudó, M.; Genovés, J.; Murgia, A.; Polli, R.; Zhou, L.; Barbouth, D.; Rupchock, A.; et al. AGG interruptions and maternal age affect FMR1 CGG repeat allele stability during transmission. *J. Neurodev. Disord.* **2014**, *6*, 24. [CrossRef]

91. Genç, B.; Müller-Hartmann, H. Methylation mosaicism of 5′-(CGG)(n)-3′ repeats in fragile X, premutation and normal individuals. *Nucleic Acids Res.* **2000**, *28*, 2141–2152. [CrossRef] [PubMed]

92. Zhao, X.N.; Usdin, K. Timing of Expansion of Fragile X Premutation Alleles During Intergenerational Transmission in a Mouse Model of the Fragile X-Related Disorders. *Front. Genet.* **2018**, *9*, 314. [CrossRef] [PubMed]

93. Nolin, S.L.; Glicksman, A.; Ersalesi, N.; Dobkin, C.; Brown, W.T.; Cao, R.; Blatt, E.; Sah, S.; Latham, G.J.; Hadd, A.G. Fragile X full mutation expansions are inhibited by one or more AGG interruptions in premutation carriers. *Genet. Med.* **2015**, *17*, 358–364. [CrossRef] [PubMed]

94. Wöhrle, D.; Schwemmle, S.; Steinbach, P. DNA methylation and triplet repeat stability: New proposals addressing actual questions on the CGG repeat of fragile X syndrome. *Am. J. Med. Genet.* **1996**, *64*, 266–267. [CrossRef] [PubMed]

95. Seriola, A.; Spits, C.; Simard, J.P.; Hilven, P.; Haentjens, P.; E Pearson, C.; Sermon, K. Huntington's and myotonic dystrophy hESCs: Down-regulated trinucleotide repeat instability and mismatch repair machinery expression upon differentiation. *Hum. Mol. Genet.* **2011**, *20*, 176–185. [CrossRef] [PubMed]

96. De Temmerman, N.; Seneca, S.; Van Steirteghem, A.; Haentjens, P.; Van Der Elst, J.; Liebaers, I.; Sermon, K. CTG repeat instability in a human embryonic stem cell line carrying the myotonic dystrophy type 1 mutation. *Mol. Hum. Reprod.* **2008**, *14*, 405–412. [CrossRef] [PubMed]

97. Paiva, A.M.; Sheardy, R.D. Influence of sequence context and length on the structure and stability of triplet repeat DNA oligomers. *Biochemistry* **2004**, *43*, 14218–14227. [CrossRef] [PubMed]

98. Mitas, M.; Yu, A.; Dill, J.; Haworth, I.S. The trinucleotide repeat sequence d(CGG)15 forms a heat-stable hairpin containing Gsyn. Ganti base pairs. *Biochemistry* **1995**, *34*, 12803–12811. [CrossRef]

99. Yu, A.; Barron, M.D.; Romero, R.M.; Christy, M.; Gold, B.; Dai, J.; Gray, D.M.; Haworth, I.S.; Mitas, M. At physiological pH, d(CCG)15 forms a hairpin containing protonated cytosines and a distorted helix. *Biochemistry* **1997**, *36*, 3687–3699. [CrossRef]

100. Gerhardt, J.; Tomishima, M.J.; Zaninovic, N.; Colak, D.; Yan, Z.; Zhan, Q.; Rosenwaks, Z.; Jaffrey, S.R.; Schildkraut, C.L. The DNA replication program is altered at the FMR1 locus in fragile X embryonic stem cells. *Mol. Cell* **2014**, *53*, 19–31. [CrossRef]

101. Gerhardt, J.; Zaninovic, N.; Zhan, Q.; Madireddy, A.; Nolin, S.L.; Ersalesi, N.; Yan, Z.; Rosenwaks, Z.; Schildkraut, C.L. Cis-acting DNA sequence at a replication origin promotes repeat expansion to fragile X full mutation. *J. Cell Biol.* **2014**, *206*, 599–607. [CrossRef] [PubMed]

102. Zhao, X.N.; Lokanga, R.; Allette, K.; Gazy, I.; Wu, D.; Usdin, K. A MutSβ-Dependent Contribution of MutSα to Repeat Expansions in Fragile X Premutation Mice. *PLoS Genet.* **2016**, *12*, e1006190. [CrossRef] [PubMed]

103. Du, J.; Campau, E.; Soragni, E.; Ku, S.; Puckett, J.W.; Dervan, P.B.; Gottesfeld, J.M. Role of mismatch repair enzymes in GAA·TTC triplet-repeat expansion in Friedreich ataxia induced pluripotent stem cells. *J. Biol. Chem.* **2012**, *287*, 29861–29872. [CrossRef] [PubMed]

104. Morales, F.; Vásquez, M.; Santamaría, C.; Cuenca, P.; Corrales, E.; Monckton, D.G. A polymorphism in the MSH3 mismatch repair gene is associated with the levels of somatic instability of the expanded CTG repeat in the blood DNA of myotonic dystrophy type 1 patients. *DNA Repair* **2016**, *40*, 57–66. [CrossRef] [PubMed]

105. Bettencourt, C.; Hensman-Moss, D.; Flower, M.; Wiethoff, S.; Brice, A.; Goizet, C.; Stevanin, G.; Koutsis, G.; Karadima, G.; Panas, M.; et al. DNA repair pathways underlie a common genetic mechanism modulating onset in polyglutamine diseases. *Ann. Neurol.* **2016**, *79*, 983–990. [CrossRef] [PubMed]

106. Canela, A.; Maman, Y.; Jung, S.; Wong, N.; Callen, E.; Day, A.; Kieffer-Kwon, K.-R.; Pekowska, A.; Zhang, H.; Rao, S.S.; et al. Genome Organization Drives Chromosome Fragility. *Cell* **2017**, *170*, 507–521. [CrossRef]

107. Libby, R.T.; Hagerman, K.A.; Pineda, V.V.; Lau, R.; Cho, D.H.; Baccam, S.L.; Axford, M.M.; Moore, J.M.; Sopher, B.L.; Filippova, G.N.; et al. CTCF cis-regulates trinucleotide repeat instability in an epigenetic manner: A novel basis for mutational hot spot determination. *PLoS Genet.* **2008**, *4*, e1000257. [CrossRef]

Review

The Application of Adeno-Associated Viral Vector Gene Therapy to the Treatment of Fragile X Syndrome

David R. Hampson *, Alexander W. M. Hooper and Yosuke Niibori

Department of Pharmaceutical Sciences, Leslie Dan Faculty of Pharmacy, University of Toronto, Toronto, ON M5S 3M2, Canada; alex.hooper@utoronto.ca (A.W.M.H.); yo.niibori@gmail.com (Y.N.)

* Correspondence: d.hampson@utoronto.ca

Received: 14 December 2018; Accepted: 31 January 2019; Published: 2 February 2019

Abstract: Viral vector-mediated gene therapy has grown by leaps and bounds over the past several years. Although the reasons for this progress are varied, a deeper understanding of the basic biology of the viruses, the identification of new and improved versions of viral vectors, and simply the vast experience gained by extensive testing in both animal models of disease and in clinical trials, have been key factors. Several studies have investigated the efficacy of adeno-associated viral (AAV) vectors in the mouse model of fragile X syndrome where AAVs have been used to express fragile X mental retardation protein (FMRP), which is missing or highly reduced in the disorder. These studies have demonstrated a range of efficacies in different tests from full correction, to partial rescue, to no effect. Here we provide a backdrop of recent advances in AAV gene therapy as applied to central nervous system disorders, outline the salient features of the fragile X studies, and discuss several key issues for moving forward. Collectively, the findings to date from the mouse studies on fragile X syndrome, and data from clinical trials testing AAVs in other neurological conditions, indicate that AAV-mediated gene therapy could be a viable strategy for treating fragile X syndrome.

Keywords: adeno-associated virus; autism spectrum disorders; cerebral spinal fluid; fragile X mental retardation protein; neurodevelopmental disorders; viral vector

---

1. Molecular and Clinical Aspects of FXS

Fragile X syndrome (FXS) is a genetic disorder caused by a pathological expansion of a triplet repeat in the 5′ untranslated region of the FMR1 gene. Expansion from the normal 5–55 repeats to 200 or more causes hypermethylation of the gene promoter and shutdown of transcription and translation of the encoded protein, fragile X mental retardation protein (FMRP) [1,2]. The expanded repeat also induces formation of RNA:DNA duplexes that induce epigenetic gene silencing [3]. Although most persons with FXS do not express FMRP, some individuals with the full mutation do produce low amounts of the protein (<10% of "normal levels"). In persons without FXS, the levels of FMRP in human brain [4], and in human blood platelets, a rich source of FMRP [5], vary over a wide range, and interestingly in persons with FXS, FMRP levels have been shown to positively correlate with intelligence scores [5,6]. As delineated below, these findings have implications for FXS gene therapy. Individuals with an intermediate triplet expansion of about 60–199 repeats, called the "premutation", are at risk for developing adult onset fragile X-associated tremor/ataxia syndrome (FXTAS) or premature ovarian insufficiency [7]. Fragile X-associated tremor/ataxia syndrome is a late-onset neurodegenerative condition that manifests in some carriers of the *FMR1* premutation; approximately 40–75% of males and 16–20% of females with a premutation develop FXTAS [8].

FMRP is a master regulator of gene expression in various organs including the brain, testes, and ovaries where it is highly expressed. The *FMR1* gene undergoes alternative splicing to generate at least 16 mRNA isoforms [9]. FMRP is a pleiotropic protein that plays a critical role in both CNS development and oogenesis [10,11], and in cognitive function in the mature brain as demonstrated in a fragile X conditional restoration mouse line [12]. FMRP contains several mRNA binding motifs that are capable of binding hundreds of mRNAs [13,14]. However, the ability to bind hundreds of mRNA substrates has been questioned by some who instead suggest that the protein regulates only a restricted number of key mRNA substrates including diacylglycerol kinase kappa, which is thought to act as a master regulator that controls switching between the diacylglycerol and phosphatidic acid signaling pathways [15].

In addition to its mRNA binding role, FMRP also associates with and regulates other proteins directly. Salient examples include voltage-gated potassium channels which modulate the sodium-dependent action potential in neurons [16–18]. In the case of the Kv1.2 potassium channel, it has been demonstrated using wild-type and *Fmr1* KO mice, that FMRP plays an essential role in cerebellar inhibitory interneurons by both assisting in the trafficking of the channel to nerve terminals, and by facilitating Kv1.2 channel activity. In cerebellar basket cell interneurons, Kv1.2 controls (inhibits) gamma-aminobutyric acid (GABA) release from basket cell terminals [18]. Therefore, the absence of FMRP at the cerebellar basket cell-Purkinje neuron synapses results in elevated GABA release onto Purkinje dendrites causing a reduction in Purkinje neuron activity. In addition to the Fmr1 KO mouse, reduced Purkinje neuron activity has been observed in other mouse models of autism; therefore, we speculated that this property might be a common denominator in mediating some features of the autistic phenotype [18].

The clinical profile of FXS overlaps with that of autism spectrum disorders (ASD). Shared characteristics include impaired communication, sensory hypersensitivity, anxiety, stereotyped or repetitive behaviors, aggression, and cognitive impairment [19]. However, not all persons with FXS meet the clinical diagnostic criteria for ASD; approximately 50% of males and 20% of females with FXS meet the criteria for autism [20]. As seen in various forms of ASD, individuals with FXS are at increased risk for developing epileptic seizures in childhood. It has been estimated that about 12% of males and 6% of females with FXS experience spontaneous seizures during early childhood [19,21]. The seizures are partial complex, generalized tonic-clonic, and/or absence seizures that typically resolve by puberty.

Current pharmacotherapy for FXS still consists exclusively of symptomatic drug treatment. Examples include stimulants, antidepressants, antipsychotics, and valproate that are each somewhat effective in suppressing a subset of symptoms [22]. Over the past decade more than a dozen clinical trials of compounds considered to be potential second generation drug candidates were conducted based on what was thought to be mechanisms more closely linked to the underlying dysfunctional neurochemical pathways in the disorder. The most prominent examples include metabotropic glutamate receptor 5 antagonists and an agonist at the GABA$_B$ receptor. Progression to Phase 2 (mGluR5 antagonists) and Phase 3 (a GABA$_B$ agonist) clinical testing was based on extensive encouraging results from both in vitro tests, and in vivo animal analyses using the fragile X knockout mouse model. However, none of the clinical trials led to new drug approvals due in large part to lack of efficacy [23].

The failure of small molecule drugs in clinical trials to date might reflect, in part, the pleiotropic nature of FMRP. In light of the many varied roles of FMRP it should not be completely surprising that drugs that specifically block or activate individual receptors, enzymes, or other proteins may not be sufficient to provide a deeper and more comprehensive correction of the disorder. *A priori*, a major motivation for pursuing small molecule drugs has been based on the idea that a particular receptor, enzyme, or other type of protein that is over- or under-expressed, or is over- or under-active, induces a major symptom or cluster of symptoms, and that when normalized, will result in major therapeutic improvements. The fundamental issue here stems from the difficulty in determining which of the many potential targets of FMRP, when corrected, will actually result in robust, measurable improvements in physiology, behavior, and health of persons with FXS.

An alternative approach is to try to correct the basic underlying biochemical defect of the disorder-the absent or dramatically reduced levels of FMRP in the brain. Viral vector-mediated gene therapy is one potential avenue for rectifying the fundamental molecular defect in FXS. This approach may also be amenable to other genetic disorders associated with ASDs [24]. The essence of the strategy is simple - to incorporate the coding sequence of FMRP, along with appropriate regulatory elements, into the genome of a recombinant viral vector so as to provide a vehicle for transducing (expressing) the recombinant FMRP "transgene" protein in brain cells [22]. However, as outlined below, achieving this to obtain substantial therapeutic improvement and correction of FXS is complicated and challenging from several perspectives.

2. General Features of Adeno-Associated Viral Vectors Used as Gene Therapy Vehicles

Recombinant adeno-associated viral (AAV) vectors are currently the most widely employed class of viral vectors for gene therapy. Less commonly used vectors include adenovirus vectors and lentivirus-based vectors. The lentiviral vectors have been used successfully in ex vivo treatments, for example in treating X-linked cerebral adrenoleukodystrophy where therapy requires infusion of purified autologous CD34+ cells transduced with a lentiviral vector [25]. Nevertheless, AAV vectors have advantages over other types of vectors for gene therapy. In addition to not causing any known pathology, additional upsides of the AAV class of vectors include high infectivity of cells and tissues, small particle size (about 25 nanometers in diameter) facilitating diffusion through tissues, multiple unmodified natural serotypes and modified synthetic serotypes encoding viral capsid proteins, non-replicating, low (but not zero under some conditions) genomic DNA integration, and relatively low immunogenicity [26].

Limitations of AAV vectors are the restricted room for target DNA insertion into the AAV vector and the presence of pre-existing circulating anti-AAV capsid antibodies in up to about half of the human population [27,28]. Preexisting neutralizing anti-AAV antibodies present in the body prior to gene therapy administration can reduce therapeutic efficacy, while additional antibody and T cell induction after AAV vector treatment can further impair efficacy. Obviously, this also presents a problem for a second injection of AAV in cases where a boost in the transgene expression level is desirable.

In addition to multiple serotypes, recombinant AAV vectors are of two main types; single-stranded vectors (ssAAV) and self-complementary vectors (scAAV). The former possesses a larger insertion capacity of about 4.6 kilobases of DNA, while the latter has a very limited capacity restricted to a maximum of 2.4 kilobases, but is more efficient at expressing the inserted transgene [26]. It should be noted that these size limits must include not only the coding sequence of the desired transgene but also the regulatory elements (e.g., promoter) and other regulatory sequences such as the short but mandatory inverted terminal repeats required in all AAV vectors. In the context of the *FMR1* gene sequences, ssAAV vectors are deployable for all isoforms including the full-length isoform 1 [29,30]. The scAAV vectors may also be amenable for use with the shorter isoforms and possibly isoform 1 depending on the size of the promoter used. Another potential issue with AAVs is the possibility of insertional genotoxicity leading to oncogenesis, particularly in the liver, where AAV9 encapsulated genome capsids have been shown to integrate under certain conditions, for example when a strong ubiquitous promoter is used to drive expression [31]. So far, this phenomenon has only been reported in mice, and whether or not AAV vector incorporation into human genomic DNA occurs, and whether it presents a genotoxic threat in humans remains unknown [32].

Recombinant AAV vectors used for gene therapy are packaged into virus particles and subjected to purification by density gradient centrifugation (e.g., with iodixanol) followed by high performance liquid chromatography with ion exchange chromatography. After binding to an AAV receptor protein on the surface of cells, the viral particle enters the cell and begins to transcribe and translate copies of the encoded transgene, but the virus itself does not replicate. Each AAV serotype has one or more receptor protein(s) that is primarily responsible for mediating viral uptake into cells and tissues [28]. The choice of AAV serotype is important as the available array of virus serotypes have different but partially overlapping tissue and cellular preferences. For CNS applications, AAV9 has been the most widely studied in animal studies as it has an excellent ability to transduce neurons and glia. In addition to AAV9, AAV2 and AAVrh10 have also been used to express proteins in the brain in clinical trials [33]. The development of novel modified AAV serotypes is an active area of research with the goal of discovering improved versions of AAVs that possess higher selectivity and binding to more restricted subsets of cells and tissues.

In addition to relatively low tissue and cell-type selectivity imparted by most of the natural viral serotypes, further restriction of transgene expression is achieved via the use of customized gene promoters. Several CNS specific or selective promoters have been tested in AAV gene therapy experiments in preclinical animal studies. Examples include the synapsin promoter for neuron-selective expression [29], the dlx5/6 promoter for GABAergic inhibitory neurons [34], glia fibrillary acidic protein for expression in astrocytes [35], and myelin basic protein for expression in oligodendroglia [35]. Further refinement of short, cell-type specific gene promoters for use in AAV gene therapy will likely be actively pursued over the next few years.

Another very attractive aspect of AAV vectors for treating CNS disorders is the capacity for long-term expression of the therapeutic protein. Upon administration, recombinant AAV-mediated therapeutic protein expression gradually ramps up over the first few weeks after injection and plateaus about 3–4 weeks post-injection. The epichromosomal presence of the AAV is static and can translate into long-term expression of the desired transgene. In tissues containing dividing cells with cellular turnover, expression levels will dwindle over time. In non-dividing long-living cells like differentiated neurons, recombinant transgenes are typically expressed for years. For example, transgenes expressed from injected AAVs have been shown to persist in the primate brain and maintain a therapeutic effect for up to 10 years [36–38]. This is a crucial aspect of CNS gene therapy considering the invasive nature of the treatment when administered into the CNS via direct parenchymal injections or into the cerebral spinal fluid (CSF).

3. Theoretical Aspects of Treating FXS with Viral Vector-Mediated Gene Therapy

As noted above, developing AAV-FMRP gene therapy for FXS presents a numbers of issues that need to be resolved for moving this potential biological therapeutic drug from the laboratory to clinical trials and beyond. At the top of the list are issues associated with CNS delivery of biologically-based therapeutic drugs such as viral vectors. Because of the pan distribution of FMRP throughout the brain, direct injections into the parenchyma are largely precluded due to lack of sufficient spread of the virus to other brain regions. The logical solution would be to administer the vector systemically, for example, via intravenous injection. Several obvious problems here include the very large quantities of a very expensive vector that need to be injected, much or most of which binds to and is taken up by peripheral organs such as the liver and heart that act as sinks for several types of viral vectors, and the potential complications of virus-induced side effects from treating patients with the large doses that would be required to obtain adequate FMRP expression levels in the CNS [39].

A potential solution is to infuse the vector into the cerebral spinal fluid (CSF) through the spinal canal (intrathecal injection, i.t.), by intra-cisterna magna (i.c.m.) injection near the base of the skull, or by intracerebral ventricular injection (i.c.v.). All three of these routes are applicable to studies in experimental mammals, while i.t. injections have been most the widely used mode for administering drugs into the CSF in humans and non-human primates [40–43]. However, i.c.m. infusions are also feasible and may become more widely used in clinical testing. An important parameter after injection into the CSF is the extent of diffusion of the vector from the point(s) of injection. It has been suggested that the flow of CSF through the so-called "glymphatic system" of the brain may facilitate dispersion of injected AAV particles [29]. For treatment of FXS, the goal is to mimic as closely as possible the natural brain-wide expression of FMRP. As discussed in Arsenault et al., 2016 [30], i.c.v. injection of AAV-FMRP in mice at Postnatal Day 2 gave a wider brain distribution of the vector than injection at Postnatal Day 5. This might be related, in part, to the less mature ependymal lining surrounding the walls of the ventricles in the mouse brain two days after birth compared with a few days later.

Because the very early postnatal rodent brain corresponds approximately to a third trimester human fetus, vector injection into the brain of a rodent immediately after birth is not directly translatable to the situation for human therapeutics. Whether or not the drop off of vector diffusion seen in mice occurs in other mammalian species has not been sufficiently explored to date, but such experiments conducted in other species, including non-human primates, would be of considerable value.

A second issue that must be dealt with is proper dosing with the goal of achieving normal "wild-type" levels of FMRP in as much of the brain as possible. Achieving vector-derived FMRP expression at an adequate level in the brain, and in the appropriate types of cells are important parameters that are discussed further below. Viral vectors differ from small molecule drugs in that they are typically given only once or a few times, and are not easily amenable to dose modulation. This means that the injected dose(s) must be carefully established prior to initiation of treatment because with current vector systems, there is no opportunity for reducing the dosage and level of transgene expression downward. Additional dosing to achieve a higher level of expression can be accomplished by one or more subsequent treatments, but as noted above this comes with increased risk of inducing the immune system to generate anti-capsid antibodies that can neutralize or hinder the efficacy of the viral vector therapeutic effects (see Reference [28] for discussion of this topic).

Cell-type selectivity is another variable that needs to be considered for CNS treatment with viral vectors and represents a strength of this approach. Unlike many small molecule CNS drugs which generally lack cell-type targeting specificity, viral vectors capitalize on the use of selective promoters to direct transgene expression in subtypes of neurons and glia. In the case of FXS, selection of a suitable promoter to use is somewhat complicated by the observation that in mice, early in postnatal brain development, FMRP is expressed in virtually all types of neurons and glia. Over the first four weeks after birth, with several exceptions such as the corpus callosum, glial expression gradually down-regulates so that in the mature brain FMRP expression in glia is very low relative to the moderate to high expression in neurons [44]. Extensive analyses of FMR1 gene regulatory elements have been carried out, and although a minimal FMR1 promoter region that works in cells in vitro has been identified, it is not known if additional regulatory elements might be necessary for proper neuronal regulation of the FMR1 gene in vivo [45–51]. Therefore, due to the size constraints of some viral vectors, and the fact that a compact "mini" version of the Fmr1 promoter has not been identified for use in vivo, other mini-gene promoters have been tested in gene therapy experiments conducted in the Fmr1 KO mouse.

4. Successes and Shortcomings of AAV-Mediated Gene Therapy Studies Conducted in the *Fmr1* Knockout Mouse

To date, studies using viral vector gene therapy in the *Fmr1* KO mouse model of FXS have employed AAV5 and AAV9 vectors and the coding sequence of the full-length *Fmr1* isoform 1. The first published report on the testing of a viral vector for FXS was by Zeier et al. (2009) [52]. This group used an AAV5 vector with a chicken β-actin core promoter containing elements from the cytomegalovirus immediate-early enhancer, the full-length *Fmr1* coding sequence, and a FLAG epitope tag. The virus was injected bilaterally directly into the hippocampus of 5 week-old *Fmr1* KO mice. At the cellular level, *Fmr1* KO mice have been shown to display abnormal synaptic plasticity as manifested by impaired long-term potentiation and enhanced long-term depression. Zeier et al. (2009) conducted brain slice recordings of extracellular postsynaptic field potentials from area CA1 of the hippocampus 3–5 weeks post-injection [52]. *Fmr1* KO mice treated with AAV5-FMRP showed correction of abnormally enhanced long-term depression. This finding is important because altered synaptic plasticity in FXS is thought to contribute to the cognitive deficits in the disorder.

The next published study (Gholizadeh et al., 2014) was conducted by our group which utilized an AAV9 vector administered by i.c.v. injection at Postnatal Day 5 [29]. Here, a series of biochemical and behavioral tests was used to assess the degree of reversal of abnormal endophenotypes at four to eight weeks of age. The AAV vector contained the full-length mouse *Fmr1* isoform 1 under the control of the neuron-specific synapsin promoter. Anatomical and cellular analyses using immunocytochemistry revealed transgene expression in neurons in various forebrain regions with lower expression in more caudal areas of the CNS. The observed rostral-to-caudal gradient was expected based of the location of the injection site in the lateral ventricles. Quantitative western blotting using an anti-FMRP antibody of samples collected at Postnatal Days 30 to 60 showed a range of transgene expression levels across different brain regions, from a high of 71% in the cerebral cortex to a low of 18% in the striatum compared to baseline wild-type expression. Low but detectable levels were also seen in the cerebellum. AAV-FMRP expression was still detected at 7-months post-injection demonstrating the stability of the FMRP transgene.

For behavioral analyses, a range of tasks were employed to judge normalization after AAV-FMRP [29]. In uninjected *Fmr1* KO mice, motor activity is increased relative to wild-type mice, marble burying reflecting repetitive behavior, is increased, social dominance as determined by the tube test is decreased, as is the number of ultrasonic vocalizations. Treatment of KO mice with AAV-FMRP normalized stereotyped behavior and social dominance, whereas no significant changes were observed in motor activity or the frequency of vocalizations. *Fmr1* KOs but not wild-type mice are highly susceptible to sound-induced (audiogenic) seizures. This is a dramatic and robust endophenotype of Fmr1 mice in which typically 50–75% of the tested KO mice undergo seizures [53,54]. Gholizadeh et al., 2014 reported that the incidence of audiogenic seizures in both PBS-injected Fmr1 KO mice and AAV-FMRP-injected mice was significantly elevated compared to PBS-injected wild-type mice, however, the AAV–FMRP group was not different to the PBS-injected Fmr1 group.

To summarize, the Gholizadeh et al. (2014) study injected AAV9-FMRP with a neuron-specific promoter into the CSF of *Fmr1* KO mouse neonates and obtained FMRP expression mainly in forebrain regions; this translated into correction of repetitive behavior and social dominance but not motor hyperactivity, abnormal calling frequency, or increased seizure susceptibility [29].

The third published study to date was also carried out by our group. In the Arsenault et al. (2016) study we again used AAV9 and mouse isoform 1 driven by neuron-specific promoters, but instead of injecting control mice with phosphate-buffered saline, an AAV9 carrying no transgene was used as the negative control (termed AAV empty vector or "AAV-EV" here and "AAV-null" in Arsenault et al., 2016) [30]. This study had several new objectives. First, we compared FMRP transgene dispersion in the brain after i.c.v. injection on Postnatal Days 1, 2, 3, or 5. It was concluded that treatment on Postnatal Day 2 or 3 gave the best results with respect to the health of the mice, spread of the vector within the brain, transgene expression levels, and therapeutic outcome. It should be noted however that this early postnatal age in mice corresponds to roughly the third trimester of a human fetus, and therefore, strictly in terms of the timing, is not directly translatable to a human therapeutic situation (see additional discussion on this topic below).

A second objective was to extend behavioral testing to additional tasks not carried out in the previous Gholizadeh et al. (2014) study [29]. The new tasks included the elevated plus maze, ostensibly measuring "anxiety", and prepulse inhibition together with the acoustic startle response to measure auditory sensory perception. AAV-FMRP corrected the abnormal "reduced anxiety" and the elevated acoustic startle response observed in the control KO mice injected with AAV-EV. Interestingly, on the elevated plus maze task, we (Arsenault et al., 2016) [30], and others have shown that *Fmr1* KO mice display an increase in the number entries and the time spent in the open arms of the maze [55–57]. This phenomenon has also been described for other mouse models of autism spectrum disorders [58,59]. Traditionally, more entries into and more time spent in the open arms are interpreted as a reduction in anxiety. This is perplexing as it is well established that persons with FXS and ASD typically display high anxiety. Arsenault et al. (2016) offered an alternate interpretation whereby rather than measuring anxiety, performance on the elevated plus maze test in *Fmr1* KO mice, and perhaps other lines of mice with autistic features, may instead reflect cognitive impairment [30]. In this interpretation, cognitively impaired mice, unlike wild-type mice, are not cognizant of the potential danger of spending time in open spaces. The inability to appreciate potentially dangerous situations may be a trait characteristic of some and possibly many persons with autism (see References [60,61]).

A third objective of the Arsenault et al. (2016) study was to examine the normalization of FMRP substrates in the CNS after treatment [30]. The highly abundant post-synaptic density protein 95 (PSD-95), a scaffolding protein required for synaptic function and plasticity, is known to be down-regulated in *Fmr1* KO mice. We also ascertained that the transcription modulator MeCP2, the mutation of which causes Rett Syndrome and is associated with some autistic features, is a substrate for FMRP (i.e., MeCP2 mRNA binds to FMRP). We found that MeCP2 protein levels are elevated in the *Fmr1* KO mouse brain. Both PSD-95 and MeCP2 proteins reverted to wild-type levels 4 weeks after injection with AAV-FMRP under the control of the synapsin promoter. Therefore, at the biochemical level, treatment with AAV-FMRP was capable of correcting the expression of key proteins that are regulated by FMRP.

Finally, the fourth major objective was to probe the consequences of under and over-expression of FMRP relative to normal wild-type levels in the CNS. The purpose of this part of the study was to conduct the first dose ranging study of FMRP in *Fmr1* KO mice. We compared five treatment groups injected with one of three test vectors: wild-type (WT) mice treated with AAV-EV (the control baseline group), WT mice treated with the AAV-FMRP vector, and *Fmr1* mice treated with AAV-EV (control baseline group of impaired KO mice), AAV-FMRP (the therapeutic treatment group), or the AAV-FMRP high-expressing vector representing a high over-expression group. This allowed for an assessment of the effects of a wide range of FMRP levels in the brain. Distillation of the results indicated that partial rescue of the *Fmr1* phenotype was observed in mice with forebrain FMRP levels of approximately 35–70% of wild-type, while moderate over-expression of up to approximately 120–140% did not induce behavioral abnormalities. However, massive over-expression of 200–600% of normal wild-type levels induced aberrant behaviors such as hyperactivity and an abnormally reduced startle response. The take-home message here is that modest under-expression appears to produce some therapeutic effects, while modest over-expression of FMRP does not appear to induce pathology. These findings suggest that a fairly wide range of CNS FMRP levels could translate into therapeutic benefits.

5. The Pathway from Preclinical Experimentation to a Clinical Trial

The three studies reviewed above that explored the efficacy of AAV-FMRP therapy in the mouse KO model have collectively demonstrated a general proof-of-principle whereby AAV-FMRP fully or partially corrected abnormalities on several behavioral tasks and in vitro assays conducted at the cellular and biochemical levels. However, considering the high bar likely to be set by regulatory agencies for moving forward with a Phase 1 clinical trial for FXS, additional evidence supporting the efficacy and safety of viral vector-mediated gene therapy would bolster the case for proceeding to a clinical trial. To date, clinical trials for CNS disorders using viral vector-mediated gene transfer were approved by regulatory agencies, in part, because they addressed unmet needs for treating severe degenerative disorders such as Parkinson's and Alzheimer's disease, or neurodevelopmental disorders that cause paralysis and/or death such as spinal muscular atrophy type 1 and giant axon neuropathy. In contrast, FXS and most other forms of autism spectrum disorders do not lead to dementia, severe paralysis, or premature death. Nevertheless, FXS and ASDs are life-long disorders in which the majority of affected individuals display moderate to severe cognitive impairment along with a slew of additional medical problems. Most will need some type of life-long care from an early age. Therefore, the discovery and development of an effective treatment would have enormous benefits not only for those with FXS, but also their families and health care funders.

Optimization of several technical parameters should also be considered in moving forward towards clinical testing. First, further assessment of drug delivery methods could be beneficial. Ultimately, experimental treatment methods in the laboratory should be representative of routes of administration that will be used in patients. Assessing additional drug delivery protocols that could provide more wide-spread diffusion of the AAV vectors after injection into the CSF, especially to more caudal regions of the brain, would likely boost the therapeutic response. Simply administering a larger amount of the virus from a single point-source injection may not be a solution because of the liability of transgene over-expression near the needle injection site. However, injecting at more than one site might provide a work-around. Our preliminary observations in mouse brain suggest that simultaneously injecting the vector into the CSF at more than one site, for example, an i.c.v. injection together with an i.c.m. (or i.t.) injection, may enhance transgene coverage in the brain (e.g., see Figure 1).

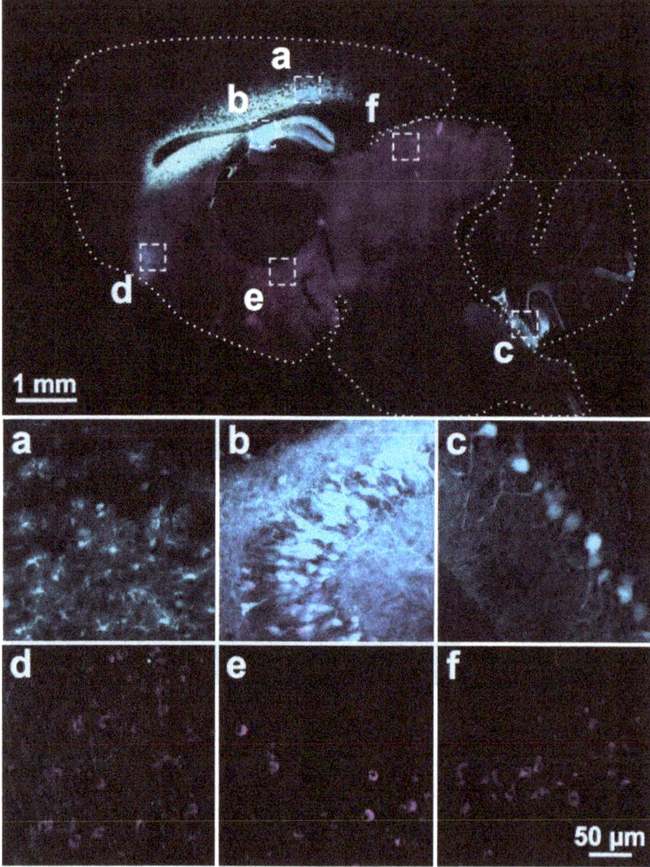

Figure 1. Distribution of AAV serotype 9-mediated transgenes following an i.c.v. and i.c.m. double injection. Two different AAV vectors were injected into a C57BL/6J mouse at Postnatal Day 2. One encoding a cytomegalovirus promoter driving enhanced green fluorescent protein (eGFP; visible in cyan color; dose: 2×10^{10} GC) was injected bilaterally into the lateral ventricles (i.c.v.); the other using a GABAergic neuron promoter driving a myc-tagged sodium channel protein was injected into the cisterna magna (i.c.m., visible as magenta color; dose: 3×10^{10} GC). At Postnatal Day 18, the distributions of the two proteins were examined in sagittal sections of fixed brain tissue using immunocytochemistry and anti-c-myc and anti-green fluorescent protein (GFP) antibodies. GFP immunolabeling was detected in the cerebral cortex (**a**), hippocampus (**b**), and Lobules VII to X of the cerebellum (**c**), while the myc-tagged protein was expressed in soma and eGFP was expressed in axons in the frontal regions of the brain (**d**), hypothalamus, (**e**), and the inferior colliculus (**f**). Thus, the two vectors administered into the CSF at the same time by two different injection routes were largely distributed in distinct brain regions.

The small size of the mouse brain relative to the human brain is a limitation of mouse models. However, previous studies in dogs, pigs, and monkeys have demonstrated successful scale up of AAVs injected into the CSF [62–64]. So far, in human trials testing AAVs in neurological disorders, direct parenchymal injections have been the most widely used, while a few have utilized i.t. or intravenous treatment [33,65]. In the highly successful AAV trial of spinal muscular atrophy type I, the pediatric subjects were injected intravenously with a high dose of an AAV9 therapeutic vector [66]. This is now being followed up by a second Phase 1 trial in which the vector is administered i.t. [32]. Although intravenous AAV administration is attractive from the standpoint of likely attaining a more thorough and uniform coverage of the brain, the downsides include the potential inability to reach a sufficiently high level of transgene expression in the brain to produce a therapeutic effect, the as-yet unclear medical consequences of high systemic dosing with an AAV, and the substantially higher cost of treating patients with large amounts of a highly purified virus.

Second, testing additional isoforms of FMRP could be informative. The human *FMR1* gene on chromosome Xq27.3 is 38 kilobases long and encompasses 17 exons and 16 introns. As outlined above, the three studies on AAV-FMRP published to date have all used the full-length mouse isoform 1 encompassing 4411 nucleotides. This was based on the assumption that the full-length isoform 1 was a major variant. However, this turned out to be incorrect as *FMR1* mRNA isoform expression surveys conducted in human [67,68] and mouse blood cells and brain tissue [69] have indicated that isoform 1 is expressed at lower levels compared to several other splice variants. For example, isoforms lacking exon 12 appear to be among the most abundant [68]. Moreover, deletion of exon 14 has been found to affect the subcellular localization of FMRP [67,70], and in premutation carriers all isoforms are elevated with isoforms 10 and 10b showing the largest increase; the consequences of differential elevations in *Fmr1* isoforms remain unknown [9]. It is important to note that no studies have yet examined the translated FMRP protein isoforms in terms of functional differences or even relative levels of expression in the mammalian brain. This could be important because mRNA levels do not always accurately reflect expression levels of the encoded proteins.

Third, in addition to ongoing work in mice, testing AAV-FMRP distribution and efficacy in additional species could provide further confidence in establishing AAV-FMRP as a viable therapeutic agent. In terms of an alternative disease model, the logical choice would be the *Fmr1* KO rat. Although the *Fmr1* KO mouse has been the most widely used animal model of FXS, it does display somewhat subtle and difficult to replicate endophenotypes on some behavioral tests, especially those tasks probing cognitive function. Although the adult rat brain is only about four times larger than a mouse brain (about 2 g vs. 0.5 g), rats are generally thought to be smarter and capable of performing more difficult cognitive tasks [71]. Deficiencies with other mouse models of CNS disorders have also prompted the neuroscience research community to pursue the development of rat disease models, for example Huntington's Disease, Rett Syndrome, and Parkinson's Disease [72–74].

Because it is much newer on the scene, the *Fmr1* KO rat is relatively uncharacterized compared to the *Fmr1* KO mouse. Nevertheless, the *Fmr1* KO rat has been reported to display impairments or abnormalities on tests measuring sensory perception, communication, cognition, and other functions [75–81]. The first version of the KO rat model was created in collaboration with Autism Speaks in the U.S.A. and underwent initial phenotypic characterization by Paylor and colleagues at Baylor College of Medicine [75]. Juvenile animals exhibited abnormalities in ASD phenotypes including juvenile play, perseverative behaviors, and sensorimotor gating. Further early study of this line showed cortical representation of speech sounds is impaired in *Fmr1* KO rats, despite normal speech discrimination behavior. Evoked potentials and spiking activity in response to speech sounds, noise burst trains, and tones were degraded in primary auditory cortex and the anterior and ventral auditory fields compared to wild-type rats [76].

Abnormalities were also documented in an analysis of neuronal morphology and neurochemistry in the auditory brainstem of *Fmr1* knockout rats [78], and importantly, KO rats have been reported to have deficits in hippocampal-dependent, but not hippocampal-independent memory, indicating that the absence of FMRP causes defects in episodic-like memory [77]. Robust changes were also reported in long-term potentiation, long-term depression, and in hippocampal-dependent memory as assessed using the Morris water maze [80]. Finally, cortical electroencephalographic (EEG) recordings conducted on juvenile *Fmr1* KO rats showed that during quiet rest when activity in wild-type rats became dominated by the inactivated state (3–9 Hz), cortical activity in the KO rats remained activated, resulting in increased high-frequency and reduced low-frequency power during rest [79]. Moreover, firing rate correlations revealed reduced synchronization in *Fmr1* KO rats, particularly between fast-spiking inhibitory interneurons. The findings from this study, together with data from EEG analyses conducted in *Fmr1* KO mice [82], and from clinical EEG studies in subjects with FXS [83], are important because (a) they provide new insight into cortical defects in the disorder, and (b) indicate that EEG may be a very useful tool as an endpoint measurement in clinical trials for FXS. Taken together, these findings suggest that the *Fmr1* KO rat is likely to be a useful addition for testing AAV-FMRP efficacy, particularly in probing cognitive and sensory functions.

6. Conclusions

Gene therapy studies conducted so far have used the mouse model of FXS and demonstrated full or partial correction of selected deficits after early postnatal treatment with AAV-FMRP. Two of the published studies injected AAV-FMRP vectors containing neuron-specific promoters into the CSF of the neonatal mouse brain to drive FMRP transgene expression in neurons [29,30]. The lower levels of expression of FMRP in more caudal regions of the brain, such as the cerebellum and brainstem compared to forebrain regions after ventricular injection, may explain in part, the lack of a more complete correction of the phenotype in the *Fmr1* KO mouse. Whether or not this is an issue for the treatment of human subjects remains unknown. Injecting AAV-FMRP via the intra-cisterna magna route with (Figure 1) or without the simultaneous injection into the ventricles may improve brain coverage and therapeutic efficacy. Although systemic injection (e.g., i.v.) of AAV-FMRP has not been assessed in rodent models of FXS, translating this to human therapy faces the issues of high peripheral organ uptake of the vector, possible elevated immune reactions, and the very high cost of the large dose that would be required for systemic administration. Finally, in terms of the effects of varying levels of FMRP expression, the findings from the Arsenault et al. (2016) study suggest that FMRP transgene expression below wild-type levels may be sufficient to produce at least some therapeutic benefits, and conversely, mild over-expression does not appear to be associated with deleterious consequences [30].

Moving forward, useful experimentation should encompass exploring the therapeutic effects of administering AAV-FMRP at later time points after birth to more closely mimic a clinical trial situation. Additionally, results from the study of *Fmr1* conditional knockout and conditional restoration mice have demonstrated that the selective deletion of FMRP in glia or in the prefrontal cortex after birth results in cellular and behavioral deficits [12,84]. The observation that behavioral deficits induced by the deletion of FMRP in the prefrontal cortex during development could then be reversed in the same line of mice after FMRP expression in the adult cortex suggests that (a) the continuous expression of FMRP is needed in the mature CNS for normal brain function, and (b) that viral vector-mediated production of FMRP initiated in adult or young adults might be capable of rescuing at least some aspects of impaired brain function in FXS.

Additionally, more in-depth characterization of rodent tests that are translatable to a clinical trial, such as prepulse inhibition and EEG, and the effects of AAV-FMRP on normalization of these parameters, could prove essential in eventual approval of this experimental biological therapeutic drug. Finally, encouraging positive results from recent clinical trials using AAVs in severe childhood diseases have generated extremely useful information from several technical standpoints ranging from drug delivery and drug dosage to post-treatment care and follow-up. Overall, the momentum attained in the field of viral vector-mediated gene therapy has now created further incentive to continue to develop this technology for treating not only FXS, but also other neurodevelopmental disorders caused by known gene mutations.

Author Contributions: D.R.H. designed the manuscript and D.R.H., A.W.M.H. and Y.N. wrote the manuscript. All authors contributed and approved the final version.

Funding: This work was supported by funding from the Fragile X Research Foundation of Canada and the Canadian Institutes for Health Research.

Acknowledgments: The authors thank current and previous members of the Hampson Laboratory for various contributions to parts of the work described herein.

Conflicts of Interest: The authors declare no conflict of interest.

References

1. Pieretti, M.; Zhang, F.P.; Fu, Y.H.; Warren, S.T.; Oostra, B.A.; Caskey, C.T.; Nelson, D.L. Absence of expression of the FMR-1 gene in fragile X syndrome. *Cell* **1991**, *66*, 817–822. [CrossRef]
2. Sutcliffe, J.S.; Nelson, D.L.; Zhang, F.; Pieretti, M.; Caskey, C.T.; Saxe, D.; Warren, S.T. DNA methylation represses FMR-1 transcription in fragile X syndrome. *Hum. Mol. Genet.* **1992**, *1*, 397–400. [CrossRef]
3. Colak, D.; Zaninovic, N.; Cohen, M.S.; Rosenwaks, Z.; Yang, W.Y.; Gerhardt, J.; Disney, M.D.; Jaffrey, S.R. Promoter-bound trinucleotide repeat mRNA drives epigenetic silencing in fragile X syndrome. *Science* **2014**, *343*, 1002–1005. [CrossRef]
4. Ludwig, A.L.; Espinal, G.M.; Pretto, D.I.; Jamal, A.L.; Arque, G.; Tassone, F.; Berman, R.F.; Hagerman, P.J. CNS expression of murine fragile X protein (FMRP) as a function of CGG-repeat size. *Hum. Mol. Genet.* **2014**, *23*, 3228–3238. [CrossRef] [PubMed]
5. Lessard, M.; Chouiali, A.; Drouin, R.; Sebire, G.; Corbin, F. Quantitative measurement of FMRP in blood platelets as a new screening test for fragile X syndrome. *Clin. Genet.* **2012**, *82*, 472–477. [CrossRef] [PubMed]
6. Pretto, D.; Yrigollen, C.M.; Tang, H.T.; Williamson, J.; Espinal, G.; Iwahashi, C.K.; Durbin-Johnson, B.; Hagerman, R.J.; Hagerman, P.J.; Tassone, F. Clinical and molecular implications of mosaicism in FMR1 full mutations. *Front. Genet.* **2014**, *5*, 318. [CrossRef] [PubMed]
7. Kong, H.E.; Zhao, J.; Xu, S.; Jin, P.; Jin, Y. Fragile X-Associated Tremor/Ataxia Syndrome: From Molecular Pathogenesis to Development of Therapeutics. *Front. Cell. Neurosci.* **2017**, *11*, 128. [CrossRef] [PubMed]
8. Hagerman, R.J.; Hagerman, P. Fragile X-associated tremor/ataxia syndrome—Features, mechanisms and management. *Nat. Rev. Neurol.* **2016**, *12*, 403–412. [CrossRef]
9. Pretto, D.I.; Eid, J.S.; Yrigollen, C.M.; Tang, H.T.; Loomis, E.W.; Raske, C.; Durbin-Johnson, B.; Hagerman, P.J.; Tassone, F. Differential increases of specific FMR1 mRNA isoforms in premutation carriers. *J. Med. Genet.* **2015**, *52*, 42–52. [CrossRef]
10. Doll, C.A.; Broadie, K. Neuron class-specific requirements for Fragile X Mental Retardation Protein in critical period development of calcium signaling in learning and memory circuitry. *Neurobiol. Dis.* **2016**, *89*, 76–87. [CrossRef]
11. Greenblatt, E.J.; Spradling, A.C. Fragile X mental retardation 1 gene enhances the translation of large autism-related proteins. *Science* **2018**, *361*, 709–712. [CrossRef]
12. Siegel, J.J.; Chitwood, R.A.; Ding, J.M.; Payne, C.; Taylor, W.; Gray, R.; Zemelman, B.V.; Johnston, D. Prefrontal Cortex Dysfunction in Fragile X Mice Depends on the Continued Absence of Fragile X Mental Retardation Protein in the Adult Brain. *J. Neurosci.* **2017**, *37*, 7305–7317. [CrossRef] [PubMed]
13. Darnell, J.C.; Van Driesche, S.J.; Zhang, C.; Hung, K.Y.; Mele, A.; Fraser, C.E.; Stone, E.F.; Chen, C.; Fak, J.J.; Chi, S.W.; et al. FMRP stalls ribosomal translocation on mRNAs linked to synaptic function and autism. *Cell* **2011**, *146*, 247–261. [CrossRef] [PubMed]
14. Ascano, M., Jr.; Mukherjee, N.; Bandaru, P.; Miller, J.B.; Nusbaum, J.D.; Corcoran, D.L.; Langlois, C.; Munschauer, M.; Dewell, S.; Hafner, M.; et al. FMRP targets distinct mRNA sequence elements to regulate protein expression. *Nature* **2012**, *492*, 382–386. [CrossRef]
15. Tabet, R.; Moutin, E.; Becker, J.A.; Heintz, D.; Fouillen, L.; Flatter, E.; Krezel, W.; Alunni, V.; Koebel, P.; Dembele, D.; et al. Fragile X Mental Retardation Protein (FMRP) controls diacylglycerol kinase activity in neurons. *Proc. Natl. Acad. Sci. USA* **2016**, *113*, E3619–E3628. [CrossRef]
16. Zhang, Y.; Brown, M.R.; Hyland, C.; Chen, Y.; Kronengold, J.; Fleming, M.R.; Kohn, A.B.; Moroz, L.L.; Kaczmarek, L.K. Regulation of neuronal excitability by interaction of fragile X mental retardation protein with slack potassium channels. *J. Neurosci.* **2012**, *32*, 15318–15327. [CrossRef] [PubMed]

17. Deng, P.Y.; Rotman, Z.; Blundon, J.A.; Cho, Y.; Cui, J.; Cavalli, V.; Zakharenko, S.S.; Klyachko, V.A. FMRP regulates neurotransmitter release and synaptic information transmission by modulating action potential duration via BK channels. *Neuron* **2013**, *77*, 696–711. [CrossRef]

18. Yang, Y.M.; Arsenault, J.; Bah, A.; Krzeminski, M.; Fekete, A.; Chao, O.Y.; Pacey, L.K.; Wang, A.; Forman-Kay, J.; Hampson, D.R.; et al. Identification of a molecular locus for normalizing dysregulated GABA release from interneurons in the Fragile X brain. *Mol. Psychiatry* **2018**. [CrossRef]

19. Ciaccio, C.; Fontana, L.; Milani, D.; Tabano, S.; Miozzo, M.; Esposito, S. Fragile X syndrome: A review of clinical and molecular diagnoses. *Ital. J. Pediatrics* **2017**, *43*, 39. [CrossRef]

20. Kaufmann, W.E.; Kidd, S.A.; Andrews, H.F.; Budimirovic, D.B.; Esler, A.; Haas-Givler, B.; Stackhouse, T.; Riley, C.; Peacock, G.; Sherman, S.L.; et al. Autism Spectrum Disorder in Fragile X Syndrome: Cooccurring Conditions and Current Treatment. *Pediatrics* **2017**, *139*, S194–S206. [CrossRef]

21. Hagerman, R.J.; Berry-Kravis, E.; Hazlett, H.C.; Bailey, D.B., Jr.; Moine, H.; Kooy, R.F.; Tassone, F.; Gantois, I.; Sonenberg, N.; Mandel, J.L.; et al. Fragile X syndrome. *Nat. Rev. Dis. Primers* **2017**, *3*, 17065. [CrossRef]

22. Hampson, D.R.; Gholizadeh, S.; Pacey, L.K. Pathways to drug development for autism spectrum disorders. *Clin. Pharmacol. Ther.* **2012**, *91*, 189–200. [CrossRef] [PubMed]

23. Berry-Kravis, E.M.; Lindemann, L.; Jonch, A.E.; Apostol, G.; Bear, M.F.; Carpenter, R.L.; Crawley, J.N.; Curie, A.; Des Portes, V.; Hossain, F.; et al. Drug development for neurodevelopmental disorders: Lessons learned from fragile X syndrome. *Nat. Rev. Drug Discov.* **2018**, *17*, 280–299. [CrossRef] [PubMed]

24. Benger, M.; Kinali, M.; Mazarakis, N.D. Autism spectrum disorder: Prospects for treatment using gene therapy. *Mol. Autism* **2018**, *9*, 39. [CrossRef]

25. Eichler, F.; Duncan, C.; Musolino, P.L.; Orchard, P.J.; De Oliveira, S.; Thrasher, A.J.; Armant, M.; Dansereau, C.; Lund, T.C.; Miller, W.P.; et al. Hematopoietic Stem-Cell Gene Therapy for Cerebral Adrenoleukodystrophy. *N. Engl. J. Med.* **2017**, *377*, 1630–1638. [CrossRef]

26. Hastie, E.; Samulski, R.J. Recombinant adeno-associated virus vectors in the treatment of rare diseases. *Expert Opin. Orphan Drugs* **2015**, *3*, 675–689. [CrossRef] [PubMed]

27. Calcedo, R.; Wilson, J.M. Humoral Immune Response to AAV. *Front. Immunol.* **2013**, *4*, 341. [CrossRef]

28. Lykken, E.A.; Shyng, C.; Edwards, R.J.; Rozenberg, A.; Gray, S.J. Recent progress and considerations for AAV gene therapies targeting the central nervous system. *J. Neurodev. Disord.* **2018**, *10*, 16. [CrossRef]

29. Gholizadeh, S.; Arsenault, J.; Xuan, I.C.; Pacey, L.K.; Hampson, D.R. Reduced phenotypic severity following adeno-associated virus-mediated Fmr1 gene delivery in fragile X mice. *Neuropsychopharmacology* **2014**, *39*, 3100–3111. [CrossRef]

30. Arsenault, J.; Gholizadeh, S.; Niibori, Y.; Pacey, L.K.; Halder, S.K.; Koxhioni, E.; Konno, A.; Hirai, H.; Hampson, D.R. FMRP Expression Levels in Mouse Central Nervous System Neurons Determine Behavioral Phenotype. *Hum. Gene Ther.* **2016**, *27*, 982–996. [CrossRef]

31. Chandler, R.J.; Sands, M.S.; Venditti, C.P. Recombinant Adeno-Associated Viral Integration and Genotoxicity: Insights from Animal Models. *Hum. Gene Ther.* **2017**, *28*, 314–322. [CrossRef] [PubMed]

32. Sumner, C.J.; Crawford, T.O. Two breakthrough gene-targeted treatments for spinal muscular atrophy: Challenges remain. *J. Clin. Investig.* **2018**, *128*, 3219–3227. [CrossRef] [PubMed]

33. Hocquemiller, M.; Giersch, L.; Audrain, M.; Parker, S.; Cartier, N. Adeno-Associated Virus-Based Gene Therapy for CNS Diseases. *Hum. Gene Ther.* **2016**, *27*, 478–496. [CrossRef] [PubMed]

34. Dimidschstein, J.; Chen, Q.; Tremblay, R.; Rogers, S.L.; Saldi, G.A.; Guo, L.; Xu, Q.; Liu, R.; Lu, C.; Chu, J.; et al. A viral strategy for targeting and manipulating interneurons across vertebrate species. *Nat. Neurosci.* **2016**, *19*, 1743–1749. [CrossRef]

35. Von Jonquieres, G.; Mersmann, N.; Klugmann, C.B.; Harasta, A.E.; Lutz, B.; Teahan, O.; Housley, G.D.; Frohlich, D.; Kramer-Albers, E.M.; Klugmann, M. Glial promoter selectivity following AAV-delivery to the immature brain. *PLoS ONE* **2013**, *8*, e65646. [CrossRef]

36. Hadaczek, P.; Eberling, J.L.; Pivirotto, P.; Bringas, J.; Forsayeth, J.; Bankiewicz, K.S. Eight years of clinical improvement in MPTP-lesioned primates after gene therapy with AAV2-hAADC. *Mol. Ther. J. Am. Soc. Gene Ther.* **2010**, *18*, 1458–1461. [CrossRef] [PubMed]

37. Leone, P.; Shera, D.; McPhee, S.W.; Francis, J.S.; Kolodny, E.H.; Bilaniuk, L.T.; Wang, D.J.; Assadi, M.; Goldfarb, O.; Goldman, H.W.; et al. Long-term follow-up after gene therapy for canavan disease. *Sci. Transl. Med.* **2012**, *4*, 165ra163. [CrossRef]

38. Sehara, Y.; Fujimoto, K.I.; Ikeguchi, K.; Katakai, Y.; Ono, F.; Takino, N.; Ito, M.; Ozawa, K.; Muramatsu, S.I. Persistent Expression of Dopamine-Synthesizing Enzymes 15 Years After Gene Transfer in a Primate Model of Parkinson's Disease. *Hum. Gene Therapy Clin. Dev.* **2017**, *28*, 74–79. [CrossRef]

39. Zincarelli, C.; Soltys, S.; Rengo, G.; Rabinowitz, J.E. Analysis of AAV serotypes 1-9 mediated gene expression and tropism in mice after systemic injection. *Mol. Ther. J. Am. Soc. Gene Ther.* **2008**, *16*, 1073–1080. [CrossRef] [PubMed]

40. Bailey, R.M.; Armao, D.; Nagabhushan Kalburgi, S.; Gray, S.J. Development of Intrathecal AAV9 Gene Therapy for Giant Axonal Neuropathy. *Mol. Therapy. Methods Clin. Dev.* **2018**, *9*, 160–171. [CrossRef]

41. Hinderer, C.; Bell, P.; Katz, N.; Vite, C.H.; Louboutin, J.P.; Bote, E.; Yu, H.; Zhu, Y.; Casal, M.L.; Bagel, J.; et al. Evaluation of Intrathecal Routes of Administration for Adeno-Associated Viral Vectors in Large Animals. *Hum. Gene Ther.* **2018**, *29*, 15–24. [CrossRef] [PubMed]

42. Castle, M.J.; Cheng, Y.; Asokan, A.; Tuszynski, M.H. Physical positioning markedly enhances brain transduction after intrathecal AAV9 infusion. *Sci. Adv.* **2018**, *4*, eaau9859. [CrossRef]

43. Hardcastle, N.; Boulis, N.M.; Federici, T. AAV gene delivery to the spinal cord: Serotypes, methods, candidate diseases, and clinical trials. *Expert Opin. Biol. Ther.* **2018**, *18*, 293–307. [CrossRef]

44. Gholizadeh, S.; Halder, S.K.; Hampson, D.R. Expression of fragile X mental retardation protein in neurons and glia of the developing and adult mouse brain. *Brain Res.* **2015**, *1596*, 22–30. [CrossRef]

45. Schwemmle, S.; de Graaff, E.; Deissler, H.; Glaser, D.; Wohrle, D.; Kennerknecht, I.; Just, W.; Oostra, B.A.; Doerfler, W.; Vogel, W.; et al. Characterization of FMR1 promoter elements by in vivo-footprinting analysis. *Am. J. Hum. Genet.* **1997**, *60*, 1354–1362. [CrossRef] [PubMed]

46. Drouin, R.; Angers, M.; Dallaire, N.; Rose, T.M.; Khandjian, E.W.; Rousseau, F. Structural and functional characterization of the human FMR1 promoter reveals similarities with the hnRNP-A2 promoter region. *Hum. Mol. Genet.* **1997**, *6*, 2051–2060. [CrossRef]

47. Carrillo, C.; Cisneros, B.; Montanez, C. Sp1 and AP2 transcription factors are required for the human fragile mental retardation promoter activity in SK-N-SH neuronal cells. *Neurosci. Lett.* **1999**, *276*, 149–152. [CrossRef]

48. Kumari, D.; Usdin, K. Interaction of the transcription factors USF1, USF2, and alpha -Pal/Nrf-1 with the FMR1 promoter. Implications for Fragile X mental retardation syndrome. *J. Biol. Chem.* **2001**, *276*, 4357–4364. [CrossRef]

49. Smith, K.T.; Coffee, B.; Reines, D. Occupancy and synergistic activation of the FMR1 promoter by Nrf-1 and Sp1 in vivo. *Hum. Mol. Genet.* **2004**, *13*, 1611–1621. [CrossRef] [PubMed]

50. Gheldof, N.; Tabuchi, T.M.; Dekker, J. The active FMR1 promoter is associated with a large domain of altered chromatin conformation with embedded local histone modifications. *Proc. Natl. Acad. Sci. USA* **2006**, *103*, 12463–12468. [CrossRef]

51. Gray, S.J.; Gerhardt, J.; Doerfler, W.; Small, L.E.; Fanning, E. An origin of DNA replication in the promoter region of the human fragile X mental retardation (FMR1) gene. *Mol. Cell. Biol.* **2007**, *27*, 426–437. [CrossRef] [PubMed]

52. Zeier, Z.; Kumar, A.; Bodhinathan, K.; Feller, J.A.; Foster, T.C.; Bloom, D.C. Fragile X mental retardation protein replacement restores hippocampal synaptic function in a mouse model of fragile X syndrome. *Gene Ther.* **2009**, *16*, 1122–1129. [CrossRef] [PubMed]

53. Pacey, L.K.; Heximer, S.P.; Hampson, D.R. Increased GABA(B) receptor-mediated signaling reduces the susceptibility of fragile X knockout mice to audiogenic seizures. *Mol. Pharmacol.* **2009**, *76*, 18–24. [CrossRef] [PubMed]

54. Pacey, L.K.; Tharmalingam, S.; Hampson, D.R. Subchronic administration and combination metabotropic glutamate and GABAB receptor drug therapy in fragile X syndrome. *J. Pharmacol. Exp. Ther.* **2011**, *338*, 897–905. [CrossRef] [PubMed]

55. Peier, A.M.; McIlwain, K.L.; Kenneson, A.; Warren, S.T.; Paylor, R.; Nelson, D.L. (Over)correction of FMR1 deficiency with YAC transgenics: Behavioral and physical features. *Hum. Mol. Genet.* **2000**, *9*, 1145–1159. [CrossRef] [PubMed]

56. Liu, Z.H.; Smith, C.B. Dissociation of social and nonsocial anxiety in a mouse model of fragile X syndrome. *Neurosci. Lett.* **2009**, *454*, 62–66. [CrossRef] [PubMed]

57. Qin, M.; Xia, Z.; Huang, T.; Smith, C.B. Effects of chronic immobilization stress on anxiety-like behavior and basolateral amygdala morphology in Fmr1 knockout mice. *Neuroscience* **2011**, *194*, 282–290. [CrossRef]

58. Filliol, D.; Ghozland, S.; Chluba, J.; Martin, M.; Matthes, H.W.; Simonin, F.; Befort, K.; Gaveriaux-Ruff, C.; Dierich, A.; LeMeur, M.; et al. Mice deficient for delta- and mu-opioid receptors exhibit opposing alterations of emotional responses. *Nat. Genet.* **2000**, *25*, 195–200. [CrossRef]

59. Becker, J.A.; Clesse, D.; Spiegelhalter, C.; Schwab, Y.; Le Merrer, J.; Kieffer, B.L. Autistic-like syndrome in mu opioid receptor null mice is relieved by facilitated mGluR4 activity. *Neuropsychopharmacology* **2014**, *39*, 2049–2060. [CrossRef]

60. Zurcher, N.R.; Rogier, O.; Boshyan, J.; Hippolyte, L.; Russo, B.; Gillberg, N.; Helles, A.; Ruest, T.; Lemonnier, E.; Gillberg, C.; et al. Perception of social cues of danger in autism spectrum disorders. *PLoS ONE* **2013**, *8*, e81206. [CrossRef]

61. Carlile, K.A.; DeBar, R.M.; Reeve, S.A.; Reeve, K.F.; Meyer, L.S. Teaching help-seeking when lost to individuals with autism spectrum disorder. *J. Appl. Behav. Anal.* **2018**, *51*, 191–206. [CrossRef] [PubMed]

62. Haurigot, V.; Marco, S.; Ribera, A.; Garcia, M.; Ruzo, A.; Villacampa, P.; Ayuso, E.; Anor, S.; Andaluz, A.; Pineda, M.; et al. Whole body correction of mucopolysaccharidosis IIIA by intracerebrospinal fluid gene therapy. *J. Clin. Investig.* **2013**. [CrossRef] [PubMed]

63. Donsante, A.; McEachin, Z.; Riley, J.; Leung, C.H.; Kanz, L.; O'Connor, D.M.; Boulis, N.M. Intracerebroventricular delivery of self-complementary adeno-associated virus serotype 9 to the adult rat brain. *Gene Ther.* **2016**, *23*, 401–407. [CrossRef] [PubMed]

64. Naidoo, J.; Stanek, L.M.; Ohno, K.; Trewman, S.; Samaranch, L.; Hadaczek, P.; O'Riordan, C.; Sullivan, J.; San Sebastian, W.; Bringas, J.R.; et al. Extensive Transduction and Enhanced Spread of a Modified AAV2 Capsid in the Non-human Primate CNS. *Mol. Ther. J. Am. Soc. Gene Ther.* **2018**, *26*, 2418–2430. [CrossRef] [PubMed]

65. Piguet, F.; Alves, S.; Cartier, N. Clinical Gene Therapy for Neurodegenerative Diseases: Past, Present, and Future. *Hum. Gene Ther.* **2017**, *28*, 988–1003. [CrossRef] [PubMed]

66. Mendell, J.R.; Al-Zaidy, S.; Shell, R.; Arnold, W.D.; Rodino-Klapac, L.R.; Prior, T.W.; Lowes, L.; Alfano, L.; Berry, K.; Church, K.; et al. Single-Dose Gene-Replacement Therapy for Spinal Muscular Atrophy. *N. Engl. J. Med.* **2017**, *377*, 1713–1722. [CrossRef] [PubMed]

67. Sittler, A.; Devys, D.; Weber, C.; Mandel, J.L. Alternative splicing of exon 14 determines nuclear or cytoplasmic localisation of fmr1 protein isoforms. *Hum. Mol. Genet.* **1996**, *5*, 95–102. [CrossRef] [PubMed]

68. Fu, X.; Zheng, D.; Liao, J.; Li, Q.; Lin, Y.; Zhang, D.; Yan, A.; Lan, F. Alternatively spliced products lacking exon 12 dominate the expression of fragile X mental retardation 1 gene in human tissues. *Mol. Med. Rep.* **2015**, *12*, 1957–1962. [CrossRef]

69. Brackett, D.M.; Qing, F.; Amieux, P.S.; Sellers, D.L.; Horner, P.J.; Morris, D.R. FMR1 transcript isoforms: Association with polyribosomes; regional and developmental expression in mouse brain. *PLoS ONE* **2013**, *8*, e58296. [CrossRef]

70. Dury, A.Y.; El Fatimy, R.; Tremblay, S.; Rose, T.M.; Cote, J.; De Koninck, P.; Khandjian, E.W. Nuclear Fragile X Mental Retardation Protein is localized to Cajal bodies. *Plos Genet.* **2013**, *9*, e1003890. [CrossRef]

71. Ellenbroek, B.; Youn, J. Rodent models in neuroscience research: Is it a rat race? *Dis. Models Mech.* **2016**, *9*, 1079–1087. [CrossRef] [PubMed]

72. Carreira, J.C.; Jahanshahi, A.; Zeef, D.; Kocabicak, E.; Vlamings, R.; von Horsten, S.; Temel, Y. Transgenic Rat Models of Huntington's Disease. *Curr. Top. Behav. Neurosci.* **2015**, *22*, 135–147. [CrossRef] [PubMed]

73. Patterson, K.C.; Hawkins, V.E.; Arps, K.M.; Mulkey, D.K.; Olsen, M.L. MeCP2 deficiency results in robust Rett-like behavioural and motor deficits in male and female rats. *Hum. Mol. Genet.* **2016**, *25*, 5514–5515. [CrossRef] [PubMed]

74. Creed, R.B.; Goldberg, M.S. New Developments in Genetic rat models of Parkinson's Disease. *Mov. Disord.* **2018**, *33*, 717–729. [CrossRef] [PubMed]

75. Hamilton, S.M.; Green, J.R.; Veeraragavan, S.; Yuva, L.; McCoy, A.; Wu, Y.; Warren, J.; Little, L.; Ji, D.; Cui, X.; et al. Fmr1 and Nlgn3 knockout rats: Novel tools for investigating autism spectrum disorders. *Behav. Neurosci.* **2014**, *128*, 103–109. [CrossRef]

76. Engineer, C.T.; Centanni, T.M.; Im, K.W.; Rahebi, K.C.; Buell, E.P.; Kilgard, M.P. Degraded speech sound processing in a rat model of fragile X syndrome. *Brain Res.* **2014**, *1564*, 72–84. [CrossRef]

77. Till, S.M.; Asiminas, A.; Jackson, A.D.; Katsanevaki, D.; Barnes, S.A.; Osterweil, E.K.; Bear, M.F.; Chattarji, S.; Wood, E.R.; Wyllie, D.J.; et al. Conserved hippocampal cellular pathophysiology but distinct behavioural deficits in a new rat model of FXS. *Hum. Mol. Genet.* **2015**, *24*, 5977–5984. [CrossRef]

78. Ruby, K.; Falvey, K.; Kulesza, R.J. Abnormal neuronal morphology and neurochemistry in the auditory brainstem of Fmr1 knockout rats. *Neuroscience* **2015**, *303*, 285–298. [CrossRef]
79. Berzhanskaya, J.; Phillips, M.A.; Shen, J.; Colonnese, M.T. Sensory hypo-excitability in a rat model of fetal development in Fragile X Syndrome. *Sci. Rep.* **2016**, *6*, 30769. [CrossRef]
80. Tian, Y.; Yang, C.; Shang, S.; Cai, Y.; Deng, X.; Zhang, J.; Shao, F.; Zhu, D.; Liu, Y.; Chen, G.; et al. Loss of FMRP Impaired Hippocampal Long-Term Plasticity and Spatial Learning in Rats. *Front. Mol. Neurosci.* **2017**, *10*, 269. [CrossRef]
81. Saxena, K.; Webster, J.; Hallas-Potts, A.; Mackenzie, R.; Spooner, P.A.; Thomson, D.; Kind, P.; Chatterji, S.; Morris, R.G.M. Experiential contributions to social dominance in a rat model of fragile-X syndrome. *Proc. Biol. Sci.* **2018**, *285*. [CrossRef]
82. Lovelace, J.W.; Wen, T.H.; Reinhard, S.; Hsu, M.S.; Sidhu, H.; Ethell, I.M.; Binder, D.K.; Razak, K.A. Matrix metalloproteinase-9 deletion rescues auditory evoked potential habituation deficit in a mouse model of Fragile X Syndrome. *Neurobiol. Dis.* **2016**, *89*, 126–135. [CrossRef] [PubMed]
83. Wang, J.; Ethridge, L.E.; Mosconi, M.W.; White, S.P.; Binder, D.K.; Pedapati, E.V.; Erickson, C.A.; Byerly, M.J.; Sweeney, J.A. A resting EEG study of neocortical hyperexcitability and altered functional connectivity in fragile X syndrome. *J. Neurodev. Disord.* **2017**, *9*, 11. [CrossRef] [PubMed]
84. Higashimori, H.; Schin, C.S.; Chiang, M.S.; Morel, L.; Shoneye, T.A.; Nelson, D.L.; Yang, Y. Selective Deletion of Astroglial FMRP Dysregulates Glutamate Transporter GLT1 and Contributes to Fragile X Syndrome Phenotypes In Vivo. *J. Neurosci.* **2016**, *36*, 7079–7094. [CrossRef] [PubMed]

Review

CRISPR to the Rescue: Advances in Gene Editing for the *FMR1* Gene

Carolyn M. Yrigollen [1] and Beverly L. Davidson [1,2,*]

[1]   The Raymond G. Perelman Center of Cellular and Molecular Therapeutics, Children's Hospital of Philadelphia, Philadelphia, PA 19104, USA; yrigollenc@email.chop.edu

[2]   Department of Pathology and Laboratory Medicine, University of Pennsylvania, Philadelphia, PA 19104, USA

*   Correspondence: davidsonbl@email.chop.edu; Tel.: +1-267-426-0929

Received: 28 November 2018; Accepted: 15 January 2019; Published: 21 January 2019

Abstract: Gene-editing using Clustered Regularly Interspaced Short Palindromic Repeats (CRISPR) is promising as a potential therapeutic strategy for many genetic disorders. CRISPR-based therapies are already being assessed in clinical trials, and evaluation of this technology in Fragile X syndrome has been performed by a number of groups. The findings from these studies and the advancement of CRISPR-based technologies are insightful as the field continues towards treatments and cures of Fragile X-Associated Disorders (FXADs). In this review, we summarize reports using CRISPR-editing strategies to target Fragile X syndrome (FXS) molecular dysregulation, and highlight how differences in FXS and Fragile X-associated Tremor/Ataxia Syndrome (FXTAS) might alter treatment strategies for each syndrome. We discuss the various modifications and evolutions of the CRISPR toolkit that expand its therapeutic potential, and other considerations for moving these strategies from bench to bedside. The rapidly growing field of CRISPR therapeutics is providing a myriad of approaches to target a gene, pathway, or transcript for modification. As cures for FXADs have remained elusive, CRISPR opens new avenues to pursue.

Keywords: Fragile X syndrome 1; Fragile X-associated Tremor/Ataxia Syndrome 2; CRISPR 3; Trinucleotide Repeat 4; Gene editing

---

1. Introduction

The expansion of the CGG trinucleotide repeat within the 5′ untranslated region (UTR) of the Fragile X Mental Retardation 1 (*FMR1*) gene is the predominant cause of Fragile X syndrome (FXS), and the only known cause of Fragile X-associated Tremor/Ataxia Syndrome (FXTAS) and Fragile X-associated Primary Ovarian Insufficiency (FXPOI) [1]. The trinucleotide repeat is normally between 5 and 44 CGG repeats in length and interspersed with up to 4 AGG interruptions.

Premutation carriers have 55–200 CGG repeats which lead to misregulation of the *FMR1* gene in several ways. Carriers of the premutation allele have elevated *FMR1* mRNA levels but also have a reduction in the translational efficiency of the *FMR1* encoded protein FMRP. *FMR1* transcripts harboring premutation length CGG repeats can give rise to Repeat-Associated Non-ATG (RAN) homopolypeptides (FMRpolyA, FMRpolyG, and FMRpolyR), some of these RAN translation products with homopolymeric amino acid tracts have been shown to be toxic [2–4]. Additionally, RNA toxicity can result from the long CGG tracts with *FMR1* transcripts forming stable hairpin structures and binding with proteins. This aberrant protein-RNA interaction results in protein sequestration and mislocalization, impairing normal cellular processes and leading to the formation of intranuclear inclusions [5–7]. Premutation carriers are at risk of developing the neurodegenerative disorder FXTAS. The age of onset for FXTAS is typically greater than 55 years of age with core features including intention tremor, gait ataxia, executive dysfunction, and neuropathy. Post mortem evaluation of FXTAS patients has identified the presence of intranuclear inclusions throughout the brain [8]. Women have a lower risk of developing FXTAS but are at risk for developing the reproductive disorder FXPOI.

Individuals with more than 200 CGG repeats, categorized as a full mutation allele, are diagnosed with FXS when this mutation becomes hypermethylated leading to a loss of FMRP. Women with FXS are often less severely affected than men; women have a protective second copy of the *FMR1* gene that is expressed in cells when the full mutation resides on the inactive X chromosome. FXS is a neurodevelopmental disorder when the full mutation allele is aberrantly methylated and transcriptionally silenced. *FMR1* epigenetic changes occur early during embryonic development and cause intellectual disability, facial dysmorphia, macroorchidism, and hyperextensible joints, which are diagnosed early in childhood [9].

Both FXS and FXTAS are neurological disorders with no disease modifying therapies. Clustered Regularly Interspaced Short Palindromic Repeats (CRISPR) is a relatively new gene editing technology which has quickly proven effective in correcting a number of pathological mutations in model systems and is advancing to clinical trials [10]. In this review, the current state of CRISPR as a tool to treat Fragile X-Associated Disorders (FXADs) will be presented.

2. CRISPR

The utility of CRISPR as a genome editor in eukaryotic cells was first reported in 2013 [11–13]; since these publications, the technology has quickly evolved and now offers a multitude of modified CRISPR associated (Cas) enzymes capable of an array of genetic modifications [14,15]. CRISPR, which was discovered as an adaptive immune system for bacteria [16], uses Cas enzymes complexed to RNA to identify invading virus and phage DNA. The "memory" mechanism, which occurred during a previous invasion with a similar species, resulted in short DNA sequences from the invaders being stored between short palindromic repeats. When these sequences are expressed into RNA, they are resolved into a structured RNA fragment that is loaded onto the Cas nuclease. The approximately 20 bp RNA sequence that corresponds to the DNA target is accessible to genomic DNA as the nuclease scans the genetic code. When the guideRNA complements with a DNA strand the nuclease changes conformation to a catalytically opened state that cleaves the foreign DNA [17]. To use CRISPR as a genome editor, the endogenous system has been adapted by using synthesized guideRNAs that direct the Cas nucleases to the genomic region of interest. After DNA cleavage, the double stranded breaks at the target site are repaired by one of the host cells' DNA damage response mechanisms, non-homologous end joining (NHEJ) or homology directed repair (HDR) [12].

Cas9 from *Streptococcus pyogenes* (SpCas9) was the first CRISPR enzyme shown to edit eukaryotic DNA using synthetic guideRNAs (Figure 1a). It was successfully adapted for use in eukaryotic cells because cleavage could be achieved using only Cas9 and two short RNA molecules. Since these first reports, novel CRISPR based tools have been developed with a range of functions and advantages [14]. Modified Cas9 enzymes can now be completely deactivated from cleaving DNA while retaining their binding activity (dCas9; Figure 1d) [18,19]. When coupled to repressive proteins, the guideRNAs guide the Cas9 to the appropriate site to repress transcription (dCas9-HDAC; dCas9KRAB; Figure 1e) [20]. Variants have also been made that generate single stranded breaks (Cas9 nickases; Figure 1b) [21], increase transcription (dCas9-Suntag, dCas9-p300, dCas9-VP64, dCas9-SAM; Figure 1g–i) [20,22–24], label DNA (dCas9-GFP; Figure 1f), or directly convert cytosine nucleotides to thymine, or guanine to adenine (Base Editors; Figure 1k) [25,26]. Modifications to SpCas9 and other newly characterized Cas enzymes have also improved the specificity of the nuclease to a target sequence (High Fidelity Cas9) [27–29], altered the required protospacer adjacent motif (PAM) motifs (Figure 1j) [30,31], and demonstrated RNA targeting capability (C2c2; Figure 1l) [32].

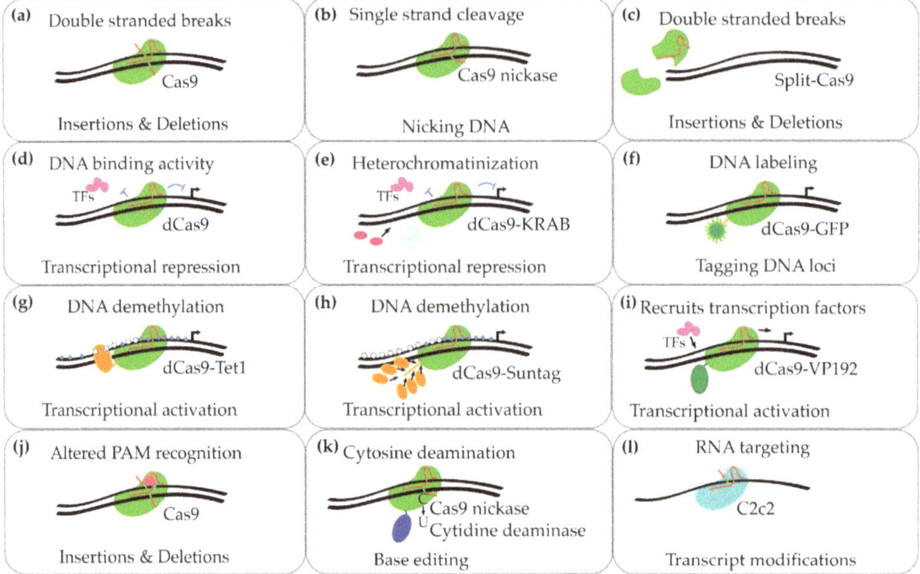

Figure 1. Overview of Clustered Regularly Interspersed Short Palindromic Repeats (CRISPR) technologies with therapeutic utility. (**a**) Wildtype SpCas9 is directed to the target site for editing through guideRNAs possessing sequences complementary to the DNA region of interest. When bound to the target site, the active Cas9 induces a double stranded DNA break. When repaired, this can induce the formation of indels. (**b**) Mutations that inactivate either the RuvC or HNH domain in Cas9 (Cas9 nickase) impacts nuclease such that cleavage in only one strand can occur; the other DNA strand remains intact. (**c**) Expressing split Cas9 that complexes into a functional nuclease when complementing the target DNA sequence. (**d**) Mutations in both RuvC and HNH domains deactivate Cas9; the 'dead' Cas9 retains the ability to bind target sequences but is incapable of generating single or double stranded breaks. When Cas9 is directed to genomic regions near start transcription start sites, expression can be inhibited because normal transcription factor binding sites are blocked. (**e**) Fusing a Krueppel-associated box (KRAB) domain onto dCas9 recruits chromatin remodeling factors that elicit heterochromatinization of the target locus, further reducing transcription (**f**) GFP fused to Cas9 is used as a molecular beacon to monitor specific regions of the genome in vitro and in vivo. (**g–i**) Fusing demethylases or transcriptional activators onto dCas9 to upregulate target genes. Suntag and VP192 use multiple copies of the activating domains to induce higher upregulation. (**j**) SpCas9 nuclease mutations to alter PAM recognition motifs. Wildtype Cas9 recognizes NGG, and mutation variants recognize NGCG, NGAG, NGAN, and NGNG. (**k**) Cytosine deaminase fused onto Cas9 nickase for conversion of cytosine to thymine without inducing double stranded breaks. (**l**) Cas enzymes that target RNA include C2c2, renamed Cas13a.

3. Recently Reported CRISPR-Based Therapies

The first study published using CRISPR to edit *FMR1* targeted the full mutation allele in human induced pluripotent stem cells (iPSCs; Table 1). These cells, harboring an epigenetically silenced full mutation, were electroporated with plasmid DNA encoding SpCas9 and a guideRNA designed to target 47 base pairs upstream of the start of the trinucleotide repeat [33]. Complete deletion of the CGG repeats were observed in iPSCs. These cells, clonally expanded post-editing had a reported 2–3% editing efficiency. The edited iPSCs were shown to reactivate *FMR1* expression to levels similar to the control cells with normal CGG repeat alleles. The authors further reported a loss of methylation at the promoter following CRISPR mediated deletion of the full mutation CGG repeats, and sustained *FMR1* expression was present after reprogramming the iPSCs into neuronal cells [33]. The edited cells that were differentiated into mature neurons stained positive for FMRP compared to a lack of FMRP positive cells in the unedited parental lines. Differences in genes expressed in edited and unedited neurons showed a reduction in three glutamate receptor genes, *GRIA1*, *GRIN2B*, and *GRIN3A* following deletion of the CGG repeat, consistent with restoration of FMRP expression.

Xie and colleagues [34] used a similar strategy to delete the CGG repeats from HEK 293 cells and iPSCs using nucleofection of CRISPR plasmids with guideRNAs targeting 40 bps upstream and 35 bps downstream of the trinucleotide repeat (Table 1). With dual guides, the editing efficiency increased to 20% in iPSCs. Importantly, this study replicated that a silenced full mutation could be reactivated by deletion of the CGG repeats in iPSCs, and this transcriptional activation was stable for a prolonged time in culture (50 days post reactivation). However, it was reported by Xie et al. that not all of the clonal lines reactivated following CRISPR editing. Evaluation of epigenetic status showed that reactivated clones had decreased methylation levels at the CpG sites at the promoter region and adjacent to the CGG repeat locus, while clones that lacked reactivation showed similar methylation levels as unedited iPSC clones. This variability in epigenetic modifications and *FMR1* expression in CGG deleted cells was hypothesized to be the result of incomplete DNA remethylation following DNA replication. As such, more actively replicating cell types would be expected to undergo more efficient *FMR1* reactivation following CRISPR editing of the CGG repeat, while nondividing cell types will be less prone to this demethylation. While promising, neither study showed direct editing in differentiated cells, and further investigation into the reactivation capabilities of nondividing neurons is warranted.

CRISPR has also been used to epigenetically modify the *FMR1* full mutation outright [35]. In the first study of this kind, catalytically deactivated SpCas9 (dCas9) was fused to Tet1, an enzyme that induces demethylation of cytosines to create a methylation eraser (dCas9-Tet1). When dCas9-Tet1 and guideRNAs designed to target the CGG repeats were introduced into iPSCs with lentivirus, epigenetic changes occurred at the *FMR1* locus. As expected, gene expression occurred without sequence modifications of the trinucleotide repeat. Edited FXS iPSCs showed *FMR1* expression levels that were 90% of that measured in a control human embryonic stem cell. There was also DNA hypomethylation of the CpG island adjacent to the repeats, and histone modifications including H3 lysine 27 acetylation, H3 lysine 4 trimethylation, and a decrease of H3 lysine 9 trimethylation at the promoter of *FMR1*. The histone modifications reactivated transcription, and this was maintained for over 35 days in culture. When the edited iPSCs were derived into neurons *FMR1* remained transcriptionally active and the electrophysiological hyperactive firing rate phenotypes were rescued. Gene expression changes were described for 41 identified off target genes to be less than 4-fold upregulated in the edited neurons, while one gene *RGPD1* had a 9-fold increase in gene expression. In contrast, *FMR1* was reported to have a 481-fold increase in gene expression following epigenetic editing. When engrafted into mouse brains, the edited cells expressed FMRP for 3 months post-transplantation. This is the first in vivo analysis of ex vivo edited, transplanted neuron. These findings demonstrate the utility of targeting epigenetic-editors to the CGG repeat locus. A critical question in using gene editing technologies for FXS is whether a constitutively active dCas9-Tet1 is necessary for long term reactivation at the *FMR1* locus. Epigenetic editing of iPSC derived neurons was shown to be less efficient than in iPSCs, *FMR1* mRNA levels were restored to 45% that of control neurons and a 20% decrease in CpG methylation at the *FMR1* promoter was achieved post editing. These differences could have been the technical limitations of isolating the edited neurons or differences in mechanisms to demethylation between cell types. The authors also reported a rescue of hyperactivity in the edited FXS neurons compared to their unedited counterparts.

Table 1. Summary of Fragile X syndrome (FXS) CRISPR gene editing studies.

Study	Cells or Tissue	Host	Delivery	Target Sequence	Nuclease Used	Outcome
Park et al., *Cell Reports* 2015	iPSC	Human	Electroporation	47 bp upstream of CGG repeat	SpCas9	Deletion of CGG repeats in 2–3% clonally expanded cells; reactivation of the *FMR1* gene; sustained *FMR1* reactivation upon differentiation of iPSCs into neurons.
Xie et al., *PLoS ONE* 2016	iPSC	Human	Nucleofection	40 bp upstream and 35 bp downstream of CGG repeat	SpCas9	Deletion of CGG repeats in 20% of cells; variability in reactivaton among edited clones. Reactivation of *FMR1* persisted 50 days in culture.
Liu et al., *Cell* 2018	iPSC	Human	Lenti virus	CGG repeats	dCas9-Tet1	Reactivation of *FMR1* and demethylation of the CGG repeat locus in iPSCs and derived neurons. Lower reactivation efficiency in neurons. Edited iPSCs were reprogrammed into neurons and engrafted into mouse brain; half of neurons actively expressed FMRP 3 months post transplantation.
Haenfler et al., *Frontiers of Molecular Neuroscience* 2018	ESC and neuronal progenitor cells	Human	Lipid mediated transfection	Promoter and CGG repeats	dCas9-VP192	Reactivation of *FMR1* but low FMRP expression with CGG targeting. Higher reactivation efficiency with CGG targeting guideRNAs than promoter targeting.
Lee et al., *Nature Biomedical Engineering* 2018	Striatum	Mouse	Gold nanoparticles	*Grm5*	Cas9 and Cpf1	Reduction of *Grm5* expression; phenotypic rescue of marble burying and jumping behaviors.

In another approach, Haenfler et al. [36] fused dCas9 to multiple VP16 transcriptional activator domains (dCas9-VP192) to drive expression of *FMR1* without altering the genetic sequence (Table 1). Human embryonic stem cells harboring a silenced full mutation allele were transfected in vitro with dCas9-VP192 and guideRNAs targeting either the *FMR1* promoter region or the ~800 CGG trinucleotide repeats. Cells transfected with the transcriptional activator targeting the CGG repeats showed robust transcriptional reactivation but only modest gains in FMRP. The CGG targeting guideRNAs were reported to induce higher transcriptional activity than guideRNAs that targeted the promoter region. The ability of the CGG targeting guideRNAs to target within the trinucleotide repeats multiple times might explain the stronger reactivation of *FMR1*, as more VP192 activating domains are recruited to the gene. They also transfected neuronal progenitor cells derived from the fully methylated hESC with the CGG targeting dCas9-VP192 and reactivated *FMR1* and increased transcriptional expression. Again, there was a limited increase in FMRP. This work highlights a key limitation of reactivating a full mutation; long CGG repeats have been shown to decrease the translational efficiency of *FMR1* [37]. The molecular implications of expressing long CGG repeat tracts is also an area that needs further investigation, as 55–200 repeats can give rise to FXTAS and individuals with an unmethylated full mutation have been reported to develop FXTAS symptoms [38,39]. Thus it will be important to understand if expressing more than 200 repeats will increase the risk of developing FXTAS or increase the severity of clinical pathology.

The first in vivo gene-editing in animal models of FXS was recently reported by Lee et al. This strategy assessed both Cas9 and Cpf1 for their efficiency to knockout gene expression in vivo in mice. The nucleases were delivered to specific regions of the brain using gold-nanoparticles with editing in neurons, astrocytes, and microglia shown to decrease reporter gene expression. In an alternative gene editing strategy to targeting *Fmr1*, the metabotropic glutamate receptor (mGluR5) encoding gene *Grm5* was targeted for knockdown with Cas9 in the striatum of *Fmr1* knockout mice (Table 1) [40]. *Grm5* was targeted because of the over-activation of mGluR5-dependent signaling present in FXS and other autism spectrum disorders. The efficiency of editing *Grm5* was measured to be 14.6%, resulting in a 40–50% reduction in the encoded protein. Repression in mGluR5-dependent signaling resulted in a rescue of a marble burying phenotype in $Fmr1^{-/y}$ mice as well as jumping behaviors, both considered measures of hyperactivity and stereotypy. While this study demonstrated in vivo gene-editing in an animal model of FXS, it did not target the *FMR1* locus directly. However, the work did establish a potential strategy for treating autism. Although the authors used CRISPR editing to treat FXS-like phenotypes in mice, they did so by inducing a loss of function mutation in a second gene rather than correct the initial null mutation in *Fmr1*; the authors suggest that while clinical trials that used drugs to reduce mGluR5 activity were disappointing, using CRISPR to target the overactivation may be more effective. Of note, the delivery strategy resulted in a focused treatment area, which is necessary when knocking out an important regulator of plasticity. However, it has limitations if multiple regions or a large area of the brain must be edited for therapeutic effect, an important consideration in translating from mice to humans. Notably, gold nanoparticles have been shown to have cellular toxicity at high or repeated doses [41,42].

Other trinucleotide repeat disorders are also being targeted with CRISPR-based therapies [43]. One example is the Huntingtin gene (*HTT*), which has been corrected in vitro using patient derived fibroblast cells by targeting only the expanded allele for deletion of the CAG repeat and knocked down expression. This strategy was also used in vivo in a mouse model of the expanded CAG repeat, with similar decreased expression of the gene and encoded protein as was seen in cell lines [44]. An in vivo mouse model of Friedreich Ataxia showed partial increased expression of the frataxin (*FXN*) gene when GAA trinucleotide repeats were deleted from intron 1 with Cas9 [45].

The modifications, fusions, and evolutions of CRISPR for therapeutics opens the field to numerous strategies for using gene-editing to treat and cure FXADs. For example, RNA targeting nucleases can be used to target the CGG repeat in FXTAS and off-target activity can be decreased using split Cas9 or Cas9 nickases that require two guideRNAs to complement the genomic sequence near each other [46,47].

The SpCas9 nuclease has also been genetically modified to decrease off-target activity while maintaining high editing activity at the target loci [28,29]. These designer nucleases can improve the safety of these therapies by minimizing unwanted DNA damage. Cas9 enzymes with alternative PAM motifs are also available, and these make more of the genome accessible for CRISPR modification [31].

4. Further Considerations for Gene-Editing for FXTAS and FXS

4.1. Delivery

Delivering CRISPR-based therapeutics into the brain and specifically neuronal cells is essential for FXS and FXTAS therapy. Several strategies to deliver gene therapies to the brain, some with success in model systems, are described below. Also see a more comprehensive review by Cwetsch, Pinto [48].

Transducing neurons with Adeno Associated Virus (AAV) is the most common method to deliver CRISPR editors in vivo. AAVs have high transduction efficiency and low immunogenicity profiles compared to many other viral vectors, contributing to their broad use in gene therapy applications [49]. AAVs can package approximately 4.8 kb of foreign DNA sequence, which is only 600 bp larger than the coding sequence of SpCas9. This leaves minimal room for cassettes to drive guideRNAs and Cas9 expression. Strategies to bypass this limitation include dual vector systems where the nuclease and guideRNAs are packaged into separate AAV vectors [50] or splitting Cas9 into two fragments that dimerize in the presence of a small molecule (Figure 1c) [51]. Alternative enzymes that are encoded by sequences smaller than the 4.2 kB sequence for SpCas9 include the 3.2 kB *Staphylococcus aureus* Cas9, which is small enough to allow for cis expression of one or two guideRNAs [30,52].

AAV capsids vary in their tropism, defined as their ability to bind to and be internalized by specific cell types. AAV serotype 9 has been used to deliver an *FMR1* transgene into *Fmr1* knockout neonate mice via intracerebroventricular injection [53]. These experiments showed neuronal expression of FMRP in several brain regions including the cortex, hippocampus, and striatum. The cortex and hippocampus had the highest transduction efficiency as measured by the number of FMRP-positive cells for a given area. However, FMRP was not detected beyond the midbrain and cerebellum. Together with AAV9, AAV1, AAV2, AAV5 and AAV8 have been reported to express transgenes throughout the brain [54,55]. Developing genetically engineered AAV serotype variants that increase overall transduction efficiency or specificity is an active area of research [56,57].

Lentiviruses have also been engineered to deliver CRISPR components into cells. Unlike AAVs, that remain mostly episomal; lentiviruses integrate into the host genome, a mechanism that can increase the risk of insertional mutagenesis. An advantage of lentiviruses for gene editing is that they have a packaging capacity of approximately 9 kb. Lentiviruses have been used to deliver CGG targeting CRISPR components into iPSCs in vitro for gene editing as discussed above [35].

Lipid nanoparticles (LNPs) are a non-viral strategy to deliver CRISPR components in vivo. In the liver, high editing efficiency was shown after a single intravenous injection of CRISPR-mRNA containing LNPs [42], or LNPs containing plasmids encoding Cas9 and guideRNAs [58,59]. Currently, efficient editing in the brain using LNP-based systems have not been shown. However, *Fmr1* knockout mice have been focally transfected with gold nanoparticles, as described above [40].

The main advantage of non-viral delivery is that the Cas nuclease is present and functional for a limited time. In recent studies comparing plasmid expression systems or RNPs of Cas9/guide RNAs, Behlke and colleagues showed that transient Cas9 exposure minimizes off target activity [60]. Additionally, transiently expressing Cas and guideRNAs provides the availability of the full CRISPR toolkit for editing new targets without interference from a previous treatment.

The route of CRISPR delivery into the brain has to be considered. Some modalities will allow for widespread delivery but require larger dosages to be administered (i.e., intraventricular injection) while others can provide for targeting a precise region (i.e., intracerebellar or cerebral injection). Localized injections will minimize the amount of therapeutic administered while having a high coverage of cells in the targeted area. Such a delivery strategy will also narrow the functional range of the therapy. This is unlikely to be useful for FXS and FXTAS therapy where pathology occurs in many brain regions [61,62]. FXS is associated with increased white matter and gray matter volume in the thalamus, frontal and temporal lobe, cerebellum and caudate by magnetic resonance imaging (MRI). Histopathological analysis of brains from individuals with FXS identified abnormalities in the hippocampus and cerebellar vermis [63]. Patients with FXTAS often have increased T2 intensity by MRI in the middle cerebellar peduncles (MCPs), this is referred to as the MCP sign and occurs in approximately 60% of diagnosed cases of FXTAS. Other hallmarks of the disorder include cerebellar and cerebral atrophy and a thinning of the corpus callosum. Hippocampal and amygdala structural differences are less obvious and not consistently seen across studies. Balancing precision and spread of the therapeutic components will be important for successful clinical results, and will be heavily dependent on the delivery vehicle [64].

Another important consideration is when to deliver the gene-editors. The timeframe for treatment will be different depending on whether FXS or FXTAS is being treated. While FXS symptoms occur early during development and early interventions show the greatest outcomes [65], FXTAS is a late onset and incompletely penetrant disorder. Premutation carriers may never develop FXTAS or require medical intervention, but we are yet unclear whether features of FXTAS can be improved post onset and what treatment windows will be beneficial.

4.2. Off-Target Editing

The main concerns of researchers developing CRISPR therapies are the off-target and undesired editing effects permanently incorporated into patients' somatic cells. Off target editing has been studied in vitro and in vivo using a variety of genome-wide protocols (i.e., Genome-wide, Unbiased Identification of DSBs Enabled by Sequencing [GUIDE-seq] [66]; High-Throughput Genome-Wide Translocation Sequencing [HTGTS] [67]; Integrative-Deficient Lentiviral Vectors [IDLV] [68]; Digenome-sequencing [Digenome-seq] [69,70]; Circularization for In vitro Reporting of Cleavage Effects by sequencing [CIRCLE-seq] [51]; Selective enrichment and Identification of Tagged genomic DNA Ends by sequencing [SITE-seq] [71]; Breaks Labeling, Enrichment on Streptavidin, and Sequencing/Breaks Labeling In Situ and Sequencing [BLESS/BLISS] [30,72,73], and others [74]). These off-target detection screens have demonstrated that prediction of sites prone to erroneous cleavage by CRISPR is challenging and impacted by the guideRNA sequence, tissue or cell type, the specific nuclease used, and the method of delivery. Off-target activity is balanced with on-target editing efficiency. A high on-target, low off-target ratio is of course the holy grail of therapeutic editing. To reduce off-target binding of guide RNAs, most current guideRNA design tools use algorithms that allow minimizing the number of mismatches of off-target sequences in the host genome to the on-target sequence. Some algorithms also consider chromatin accessibility and guideRNA stability in vivo, although this information changes with cell type [75–77].

Characterizing and reducing off-target effects are critical to move any gene-editing therapy into the clinic. None of the recently published *FMR1* editing studies have surveyed off-target editing using one of the genome wide unbiased methods [51,66,68,69,72,73], with predicted off-targets shown to have little overlap with off targets identified using these approaches. Off-target loci have been shown to occur with as many as 6 mismatches to the 20 nucleotide guideRNA sequence [78], and are significantly reduced with transient Cas9 expression, either by delivery of mRNA encoding the nuclease and guideRNAs, or delivery of the complexed RNPs [79]. Self-destroying CRISPR (KamiCas9) systems has also been shown to limit the time Cas9 is present in treated cells, reducing off-target accumulation [80].

Editing systems must be empirically tested for each gene, and possibly for each cell type, of interest. In general, multiple guideRNAs are tested and their on- and off-target editing efficiencies evaluated. A consideration of off-target effects is whether a biological consequence (e.g., truncated protein) would result from misdirected editing and how well such an outcome is tolerated by the cell type.

CRISPR-editing of the *FMR1* gene may erroneously repair the on-target sequence resulting in large deletions, incorporation of plasmid or viral DNA sequences, or fusion of *FMR1* with other genes at other loci. A non-biased genetic analysis of repaired sequences will quantify which mistakes are prone to occur and at what frequency. Large deletions extending beyond the start codon, or upstream of the transcriptional start site, would disrupt *FMR1* expression or function. Incorporation of foreign DNA sequences may dysregulate *FMR1* reducing therapeutic rescue post editing [81,82]. Fusions of genes after editing could create novel, immunogenic or toxic proteins or protein chimeras that could alter the function of the cell or even be tumorigenic [83,84]. While these possibilities seem dire, careful preclinical testing in relevant model systems would identify problematic editing components such that alternative approaches could be developed and tested.

One of the unique considerations of gene-editing for FXADs therapy is the presence of functional and dysfunctional CGG repeats in females. Females are genetically protected by the presence of a normal CGG repeat allele in a portion of their cells [85,86]. It is important to avoid editing the normal allele-containing cells, while targeting premutation or full mutation expressing cells. Liu et al suggest that preferential editing at the expanded allele occurs because the CGG repeats provide more targetable sequence compared to the few CGG repeat in the normal allele or at other genomic loci Liu, Wu [35]. While possible, this requires more thorough investigation using single cell RNA seq analysis of cells after CRISPR/Cas9 editing of a mixture of cells harboring normal and expanded repeat sequences.

4.3. CRISPR Limitations

There are several additional limitations when implementing the CRISPR technology as a therapeutic strategy. The stringency of CRISPR nucleases to a PAM sequence can prevent researchers from targeting a specific genomic locus because there are no available PAM sites within that region, this results in changing the gene-editing strategy to conform to the currently accessible sequences. However, CRISPR tools are being developed to broaden the sites that Cas9 and other nucleases recognize [31]. The incomplete editing of CRISPR in somatic cells can result in mosaicism of desired and off target modifications in vivo. These mosaic modifications can result in unpredictable outcomes [87]. As the field continues to progress, we will have an improved perspective on how often these sporadic events occur and the severity they have in vivo.

4.4. Ethical Considerations

With the rapid progress of CRISPR in the biomedical field, keeping pace with the ethical issues that arise is important. Like clinical trials that are currently underway for new pharmaceuticals and gene therapies, gene-editing has the potential for unforeseen outcomes and side effects. Because the therapy is a permanent modification of the genome, these unexpected effects can carry a greater burden to the participant. Treatment of an embryo, child, or patient incapable of understanding the risks associated with the therapy mean informed consent cannot be obtained. In a disorder as variable as FXS, balancing the potential benefit of treating a patient during early development with the ability to assess the severity of the individual's symptoms later in life comes into question. Mosaic off-target mutations can result in unpredictable side effects to the therapy in a single participant. Additionally, as access to gene-editing therapies becomes available, equal access to these technologies across society will be essential to not disenfranchise sectors of our society [88]. The ethical issues that arise as genome editing in humans continues to advance are important and more each come with serious consequences can result from how these issues are resolved. As we move forward with developing genome editing therapies that have the potential to greatly improve the quality of life of individuals within the Fragile X community, it is important to also move forward in our understanding of these issues and decide how to address these ethical considerations.

5. Final Remarks

The recent successes that have been achieved using CRISPR for genetic and epigenetic editing and the advances being made to deliver, control, and modify Cas nucleases are driving rapid development of editing-based gene therapies for disease treatment. The continued developments in the field, even for other disorders, will allow us to optimize CRISPR technology for FXADs with regards to efficacy and safety. Finally, CRISPR/Cas9 technologies may also provide tools to develop an in vivo model of the *FMR1* epigenetically silenced full mutation, which to date does not exist. Thus, the technological advancement of CRISPR/Cas9 has infiltrated many sectors of biomedical research and provides one of the more promising approaches for FXS and FXTAS therapy.

Author Contributions: C.M.Y. and B.L.D. wrote the paper.

Funding: This research received no external funding.

Acknowledgments: This work was supported by: the Burroughs Wellcome Fund Postdoctoral Enrichment Program and the University of Pennsylvania Academic Diversity Fellowship and the Children's Hospital of Philadelphia Research Institute.

Conflicts of Interest: B.L.D. is a founder of Spark Therapeutics and Talee Bio, and serves on the Scientific Advisory Board of Sarepta Therapeutics, Prevail Therapeutics, Intellia Therapeutics and Homology Medicines.

References

1. Oostra, B.A.; Willemsen, R. Fmr1: A gene with three faces. *Biochim. Biophys. Acta* **2009**, *1790*, 467–477. [CrossRef] [PubMed]
2. Hukema, R.K.; Buijsen, R.A.; Schonewille, M.; Raske, C.; Severijnen, L.A.; Nieuwenhuizen-Bakker, I.; Verhagen, R.F.; van Dessel, L.; Maas, A.; Charlet-Berguerand, N.; et al. Reversibility of neuropathology and motor deficits in an inducible mouse model for fxtas. *Hum. Mol. Genet.* **2015**, *24*, 4948–4957. [CrossRef]
3. Sellier, C.; Buijsen, R.A.M.; He, F.; Natla, S.; Jung, L.; Tropel, P.; Gaucherot, A.; Jacobs, H.; Meziane, H.; Vincent, A.; et al. Translation of expanded cgg repeats into fmrpolyg is pathogenic and may contribute to fragile x tremor ataxia syndrome. *Neuron* **2017**, *93*, 331–347. [CrossRef] [PubMed]
4. Todd, P.K.; Oh, S.Y.; Krans, A.; He, F.; Sellier, C.; Frazer, M.; Renoux, A.J.; Chen, K.C.; Scaglione, K.M.; Basrur, V.; et al. Cgg repeat-associated translation mediates neurodegeneration in fragile x tremor ataxia syndrome. *Neuron* **2013**, *78*, 440–455. [CrossRef]
5. Muslimov, I.A.; Patel, M.V.; Rose, A.; Tiedge, H. Spatial code recognition in neuronal rna targeting: Role of rna-hnrnp a2 interactions. *J. Cell Biol.* **2011**, *194*, 441–457. [CrossRef]
6. Morriss, G.R.; Cooper, T.A. Protein sequestration as a normal function of long noncoding rnas and a pathogenic mechanism of rnas containing nucleotide repeat expansions. *Hum. Genet.* **2017**, *136*, 1247–1263. [CrossRef] [PubMed]
7. Sellier, C.; Rau, F.; Liu, Y.; Tassone, F.; Hukema, R.K.; Gattoni, R.; Schneider, A.; Richard, S.; Willemsen, R.; Elliott, D.J.; et al. Sam68 sequestration and partial loss of function are associated with splicing alterations in fxtas patients. *EMBO J.* **2010**, *29*, 1248–1261. [CrossRef] [PubMed]
8. Hagerman, P.J.; Hagerman, R.J. Fragile x-associated tremor/ataxia syndrome. *Ann. N. Y. Acad. Sci.* **2015**, *1338*, 58–70. [CrossRef]
9. Hagerman, R.J.; Berry-Kravis, E.; Hazlett, H.C.; Bailey, D.B., Jr.; Moine, H.; Kooy, R.F.; Tassone, F.; Gantois, I.; Sonenberg, N.; Mandel, J.L.; et al. Fragile x syndrome. *Nat. Rev. Dis. Primers* **2017**, *3*, 17065. [CrossRef]
10. Bak, R.O.; Gomez-Ospina, N.; Porteus, M.H. Gene editing on center stage. *Trends Genet.* **2018**, *34*, 600–611. [CrossRef]
11. Cong, L.; Ran, F.A.; Cox, D.; Lin, S.; Barretto, R.; Habib, N.; Hsu, P.D.; Wu, X.; Jiang, W.; Marraffini, L.A.; et al. Multiplex genome engineering using crispr/cas systems. *Science* **2013**, *339*, 819–823. [CrossRef] [PubMed]
12. Jinek, M.; East, A.; Cheng, A.; Lin, S.; Ma, E.; Doudna, J. Rna-programmed genome editing in human cells. *Elife* **2013**, *2*, e00471. [CrossRef] [PubMed]
13. Mali, P.; Yang, L.; Esvelt, K.M.; Aach, J.; Guell, M.; DiCarlo, J.E.; Norville, J.E.; Church, G.M. Rna-guided human genome engineering via cas9. *Science* **2013**, *339*, 823–826. [CrossRef] [PubMed]
14. Adli, M. The crispr tool kit for genome editing and beyond. *Nat. Commun.* **2018**, *9*, 1911. [CrossRef]

15. Hsu, P.D.; Lander, E.S.; Zhang, F. Development and applications of crispr-cas9 for genome engineering. *Cell* **2014**, *157*, 1262–1278. [CrossRef] [PubMed]

16. Barrangou, R.; Fremaux, C.; Deveau, H.; Richards, M.; Boyaval, P.; Moineau, S.; Romero, D.A.; Horvath, P. Crispr provides acquired resistance against viruses in prokaryotes. *Science* **2007**, *315*, 1709–1712. [CrossRef] [PubMed]

17. Jiang, F.; Doudna, J.A. Crispr-cas9 structures and mechanisms. *Annu. Rev. Biophys.* **2017**, *46*, 505–529. [CrossRef]

18. Larson, M.H.; Gilbert, L.A.; Wang, X.; Lim, W.A.; Weissman, J.S.; Qi, L.S. Crispr interference (crispri) for sequence-specific control of gene expression. *Nat. Protoc.* **2013**, *8*, 2180–2196. [CrossRef]

19. Qi, L.S.; Larson, M.H.; Gilbert, L.A.; Doudna, J.A.; Weissman, J.S.; Arkin, A.P.; Lim, W.A. Repurposing crispr as an rna-guided platform for sequence-specific control of gene expression. *Cell* **2013**, *152*, 1173–1183. [CrossRef]

20. Gilbert, L.A.; Larson, M.H.; Morsut, L.; Liu, Z.; Brar, G.A.; Torres, S.E.; Stern-Ginossar, N.; Brandman, O.; Whitehead, E.H.; Doudna, J.A.; et al. Crispr-mediated modular rna-guided regulation of transcription in eukaryotes. *Cell* **2013**, *154*, 442–451. [CrossRef]

21. Shen, B.; Zhang, W.; Zhang, J.; Zhou, J.; Wang, J.; Chen, L.; Wang, L.; Hodgkins, A.; Iyer, V.; Huang, X.; et al. Efficient genome modification by crispr-cas9 nickase with minimal off-target effects. *Nat. Methods* **2014**, *11*, 399–402. [CrossRef]

22. Tanenbaum, M.E.; Gilbert, L.A.; Qi, L.S.; Weissman, J.S.; Vale, R.D. A protein-tagging system for signal amplification in gene expression and fluorescence imaging. *Cell* **2014**, *159*, 635–646. [CrossRef] [PubMed]

23. Konermann, S.; Brigham, M.D.; Trevino, A.E.; Joung, J.; Abudayyeh, O.O.; Barcena, C.; Hsu, P.D.; Habib, N.; Gootenberg, J.S.; Nishimasu, H.; et al. Genome-scale transcriptional activation by an engineered crispr-cas9 complex. *Nature* **2015**, *517*, 583–588. [CrossRef]

24. Hilton, I.B.; D'Ippolito, A.M.; Vockley, C.M.; Thakore, P.I.; Crawford, G.E.; Reddy, T.E.; Gersbach, C.A. Epigenome editing by a crispr-cas9-based acetyltransferase activates genes from promoters and enhancers. *Nat. Biotechnol.* **2015**, *33*, 510–517. [CrossRef]

25. Kim, Y.B.; Komor, A.C.; Levy, J.M.; Packer, M.S.; Zhao, K.T.; Liu, D.R. Increasing the genome-targeting scope and precision of base editing with engineered cas9-cytidine deaminase fusions. *Nat. Biotechnol.* **2017**, *35*, 371–376. [CrossRef] [PubMed]

26. Komor, A.C.; Kim, Y.B.; Packer, M.S.; Zuris, J.A.; Liu, D.R. Programmable editing of a target base in genomic DNA without double-stranded DNA cleavage. *Nature* **2016**, *533*, 420–424. [CrossRef] [PubMed]

27. Vakulskas, C.A.; Dever, D.P.; Rettig, G.R.; Turk, R.; Jacobi, A.M.; Collingwood, M.A.; Bode, N.M.; McNeill, M.S.; Yan, S.; Camarena, J.; et al. A high-fidelity cas9 mutant delivered as a ribonucleoprotein complex enables efficient gene editing in human hematopoietic stem and progenitor cells. *Nat. Med.* **2018**, *24*, 1216–1224. [CrossRef] [PubMed]

28. Slaymaker, I.M.; Gao, L.; Zetsche, B.; Scott, D.A.; Yan, W.X.; Zhang, F. Rationally engineered cas9 nucleases with improved specificity. *Science* **2016**, *351*, 84–88. [CrossRef] [PubMed]

29. Kleinstiver, B.P.; Pattanayak, V.; Prew, M.S.; Tsai, S.Q.; Nguyen, N.T.; Zheng, Z.; Joung, J.K. High-fidelity crispr-cas9 nucleases with no detectable genome-wide off-target effects. *Nature* **2016**, *529*, 490–495. [CrossRef] [PubMed]

30. Ran, F.A.; Cong, L.; Yan, W.X.; Scott, D.A.; Gootenberg, J.S.; Kriz, A.J.; Zetsche, B.; Shalem, O.; Wu, X.; Makarova, K.S.; et al. In vivo genome editing using staphylococcus aureus cas9. *Nature* **2015**, *520*, 186–191. [CrossRef]

31. Kleinstiver, B.P.; Prew, M.S.; Tsai, S.Q.; Topkar, V.V.; Nguyen, N.T.; Zheng, Z.; Gonzales, A.P.; Li, Z.; Peterson, R.T.; Yeh, J.R.; et al. Engineered crispr-cas9 nucleases with altered pam specificities. *Nature* **2015**, *523*, 481–485. [CrossRef]

32. Cox, D.B.T.; Gootenberg, J.S.; Abudayyeh, O.O.; Franklin, B.; Kellner, M.J.; Joung, J.; Zhang, F. Rna editing with crispr-cas13. *Science* **2017**, *358*, 1019–1027. [CrossRef]

33. Park, C.Y.; Halevy, T.; Lee, D.R.; Sung, J.J.; Lee, J.S.; Yanuka, O.; Benvenisty, N.; Kim, D.W. Reversion of fmr1 methylation and silencing by editing the triplet repeats in fragile x ipsc-derived neurons. *Cell Rep.* **2015**, *13*, 234–241. [CrossRef] [PubMed]

34. Xie, N.; Gong, H.; Suhl, J.A.; Chopra, P.; Wang, T.; Warren, S.T. Reactivation of fmr1 by crispr/cas9-mediated deletion of the expanded cgg-repeat of the fragile x chromosome. *PLoS ONE* **2016**, *11*, e0165499. [CrossRef] [PubMed]

35. Liu, X.S.; Wu, H.; Krzisch, M.; Wu, X.; Graef, J.; Muffat, J.; Hnisz, D.; Li, C.H.; Yuan, B.; Xu, C.; et al. Rescue of fragile x syndrome neurons by DNA methylation editing of the fmr1 gene. *Cell* **2018**, *172*, 979–992 e976. [CrossRef]

36. Haenfler, J.M.; Skariah, G.; Rodriguez, C.M.; Monteiro da Rocha, A.; Parent, J.M.; Smith, G.D.; Todd, P.K. Targeted reactivation of fmr1 transcription in fragile x syndrome embryonic stem cells. *Front. Mol. Neurosci.* **2018**, *11*, 282. [CrossRef] [PubMed]

37. Peprah, E.; He, W.; Allen, E.; Oliver, T.; Boyne, A.; Sherman, S.L. Examination of fmr1 transcript and protein levels among 74 premutation carriers. *J. Hum. Genet.* **2010**, *55*, 66–68. [CrossRef] [PubMed]

38. Loesch, D.Z.; Sherwell, S.; Kinsella, G.; Tassone, F.; Taylor, A.; Amor, D.; Sung, S.; Evans, A. Fragile x-associated tremor/ataxia phenotype in a male carrier of unmethylated full mutation in the fmr1 gene. *Clin. Genet.* **2012**, *82*, 88–92. [CrossRef]

39. Santa Maria, L.; Pugin, A.; Alliende, M.A.; Aliaga, S.; Curotto, B.; Aravena, T.; Tang, H.T.; Mendoza-Morales, G.; Hagerman, R.; Tassone, F. Fxtas in an unmethylated mosaic male with fragile x syndrome from chile. *Clin. Genet.* **2014**, *86*, 378–382. [CrossRef] [PubMed]

40. Lee, B.; Lee, K.; Panda, S.; Gonzales-Rojas, R.; Chong, A.; Bugay, V.; Park, H.M.; Brenner, R.; Murthy, N.; Lee, H.Y. Nanoparticle delivery of crispr into the brain rescues a mouse model of fragile x syndrome from exaggerated repetitive behaviours. *Nat. Biomed. Eng.* **2018**, *2*, 497–507. [CrossRef]

41. Ferreira, G.K.; Cardoso, E.; Vuolo, F.S.; Galant, L.S.; Michels, M.; Goncalves, C.L.; Rezin, G.T.; Dal-Pizzol, F.; Benavides, R.; Alonso-Nunez, G.; et al. Effect of acute and long-term administration of gold nanoparticles on biochemical parameters in rat brain. *Mater. Sci. Eng. C Mater. Biol. Appl.* **2017**, *79*, 748–755. [CrossRef] [PubMed]

42. Finn, J.D.; Smith, A.R.; Patel, M.C.; Shaw, L.; Youniss, M.R.; van Heteren, J.; Dirstine, T.; Ciullo, C.; Lescarbeau, R.; Seitzer, J.; et al. A single administration of crispr/cas9 lipid nanoparticles achieves robust and persistent in vivo genome editing. *Cell Rep.* **2018**, *22*, 2227–2235. [CrossRef] [PubMed]

43. Simpson, B.P.; Davidson, B.L. Crispr-cas gene editing for neurological disease. In *Nervous System Drug Delivery: Principles and Practice*; Lonser, R., Sarntinoranont, M., Bankiewicz, K., Eds.; Academic Press/Elsevier: San Diego, CA, USA, 2019; in press.

44. Monteys, A.M.; Ebanks, S.A.; Keiser, M.S.; Davidson, B.L. Crispr/cas9 editing of the mutant huntingtin allele in vitro and in vivo. *Mol. Ther* **2017**, *25*, 12–23. [CrossRef] [PubMed]

45. Ouellet, D.L.; Cherif, K.; Rousseau, J.; Tremblay, J.P. Deletion of the gaa repeats from the human frataxin gene using the crispr-cas9 system in yg8r-derived cells and mouse models of friedreich ataxia. *Gene Ther.* **2017**, *24*, 265–274. [CrossRef] [PubMed]

46. Tsai, S.Q.; Wyvekens, N.; Khayter, C.; Foden, J.A.; Thapar, V.; Reyon, D.; Goodwin, M.J.; Aryee, M.J.; Joung, J.K. Dimeric crispr rna-guided foki nucleases for highly specific genome editing. *Nat. Biotechnol.* **2014**, *32*, 569–576. [CrossRef] [PubMed]

47. Dai, X.; Chen, X.; Fang, Q.; Li, J.; Bai, Z. Inducible crispr genome-editing tool: Classifications and future trends. *Crit. Rev. Biotechnol.* **2018**, *38*, 573–586. [CrossRef] [PubMed]

48. Cwetsch, A.W.; Pinto, B.; Savardi, A.; Cancedda, L. In vivo methods for acute modulation of gene expression in the central nervous system. *Prog. Neurobiol.* **2018**, *168*, 69–85. [CrossRef]

49. Naso, M.F.; Tomkowicz, B.; Perry, W.L., 3rd; Strohl, W.R. Adeno-associated virus (aav) as a vector for gene therapy. *BioDrugs* **2017**, *31*, 317–334. [CrossRef]

50. Nelson, C.E.; Hakim, C.H.; Ousterout, D.G.; Thakore, P.I.; Moreb, E.A.; Castellanos Rivera, R.M.; Madhavan, S.; Pan, X.; Ran, F.A.; Yan, W.X.; et al. In vivo genome editing improves muscle function in a mouse model of duchenne muscular dystrophy. *Science* **2016**, *351*, 403–407. [CrossRef]

51. Tsai, S.Q.; Nguyen, N.T.; Malagon-Lopez, J.; Topkar, V.V.; Aryee, M.J.; Joung, J.K. Circle-seq: A highly sensitive in vitro screen for genome-wide crispr-cas9 nuclease off-targets. *Nat. Methods* **2017**, *14*, 607–614. [CrossRef]

52. Friedland, A.E.; Baral, R.; Singhal, P.; Loveluck, K.; Shen, S.; Sanchez, M.; Marco, E.; Gotta, G.M.; Maeder, M.L.; Kennedy, E.M.; et al. Characterization of staphylococcus aureus cas9: A smaller cas9 for all-in-one

adeno-associated virus delivery and paired nickase applications. *Genome Biol.* **2015**, *16*, 257. [CrossRef] [PubMed]

53. Gholizadeh, S.; Arsenault, J.; Xuan, I.C.; Pacey, L.K.; Hampson, D.R. Reduced phenotypic severity following adeno-associated virus-mediated fmr1 gene delivery in fragile x mice. *Neuropsychopharmacology* **2014**, *39*, 3100–3111. [CrossRef] [PubMed]

54. Aschauer, D.F.; Kreuz, S.; Rumpel, S. Analysis of transduction efficiency, tropism and axonal transport of aav serotypes 1, 2, 5, 6, 8 and 9 in the mouse brain. *PLoS ONE* **2013**, *8*, e76310. [CrossRef] [PubMed]

55. Markakis, E.A.; Vives, K.P.; Bober, J.; Leichtle, S.; Leranth, C.; Beecham, J.; Elsworth, J.D.; Roth, R.H.; Samulski, R.J.; Redmond, D.E., Jr. Comparative transduction efficiency of aav vector serotypes 1-6 in the substantia nigra and striatum of the primate brain. *Mol. Ther.* **2010**, *18*, 588–593. [CrossRef] [PubMed]

56. Bartel, M.A.; Weinstein, J.R.; Schaffer, D.V. Directed evolution of novel adeno-associated viruses for therapeutic gene delivery. *Gene Ther.* **2012**, *19*, 694–700. [CrossRef] [PubMed]

57. Rincon, M.Y.; de Vin, F.; Duque, S.I.; Fripont, S.; Castaldo, S.A.; Bouhuijzen-Wenger, J.; Holt, M.G. Widespread transduction of astrocytes and neurons in the mouse central nervous system after systemic delivery of a self-complementary aav-php.B vector. *Gene Ther.* **2018**, *25*, 83–92. [CrossRef]

58. Jiang, C.; Mei, M.; Li, B.; Zhu, X.; Zu, W.; Tian, Y.; Wang, Q.; Guo, Y.; Dong, Y.; Tan, X. A non-viral crispr/cas9 delivery system for therapeutically targeting hbv DNA and pcsk9 in vivo. *Cell Res.* **2017**, *27*, 440–443. [CrossRef]

59. Miller, J.B.; Zhang, S.; Kos, P.; Xiong, H.; Zhou, K.; Perelman, S.S.; Zhu, H.; Siegwart, D.J. Non-viral crispr/cas gene editing in vitro and in vivo enabled by synthetic nanoparticle co-delivery of cas9 mrna and sgrna. *Angew. Chem. Int. Ed. Engl.* **2017**, *56*, 1059–1063. [CrossRef]

60. Kim, S.; Kim, D.; Cho, S.W.; Kim, J.; Kim, J.S. Highly efficient rna-guided genome editing in human cells via delivery of purified cas9 ribonucleoproteins. *Genome Res.* **2014**, *24*, 1012–1019. [CrossRef]

61. Brown, S.S.; Stanfield, A.C. Fragile x premutation carriers: A systematic review of neuroimaging findings. *J. Neurol. Sci.* **2015**, *352*, 19–28. [CrossRef]

62. Romano, D.; Nicolau, M.; Quintin, E.M.; Mazaika, P.K.; Lightbody, A.A.; Cody Hazlett, H.; Piven, J.; Carlsson, G.; Reiss, A.L. Topological methods reveal high and low functioning neuro-phenotypes within fragile x syndrome. *Hum. Brain Mapp.* **2014**, *35*, 4904–4915. [CrossRef]

63. Greco, C.M.; Navarro, C.S.; Hunsaker, M.R.; Maezawa, I.; Shuler, J.F.; Tassone, F.; Delany, M.; Au, J.W.; Berman, R.F.; Jin, L.W.; et al. Neuropathologic features in the hippocampus and cerebellum of three older men with fragile x syndrome. *Mol. Autism* **2011**, *2*, 2. [CrossRef]

64. Ramanathan, S.; Archunan, G.; Sivakumar, M.; Tamil Selvan, S.; Fred, A.L.; Kumar, S.; Gulyas, B.; Padmanabhan, P. Theranostic applications of nanoparticles in neurodegenerative disorders. *Int. J. Nanomed.* **2018**, *13*, 5561–5576. [CrossRef]

65. Winarni, T.I.; Schneider, A.; Borodyanskara, M.; Hagerman, R.J. Early intervention combined with targeted treatment promotes cognitive and behavioral improvements in young children with fragile x syndrome. *Case Rep. Genet.* **2012**, *2012*, 280813. [CrossRef]

66. Tsai, S.Q.; Zheng, Z.; Nguyen, N.T.; Liebers, M.; Topkar, V.V.; Thapar, V.; Wyvekens, N.; Khayter, C.; Iafrate, A.J.; Le, L.P.; et al. Guide-seq enables genome-wide profiling of off-target cleavage by crispr-cas nucleases. *Nat. Biotechnol.* **2015**, *33*, 187–197. [CrossRef]

67. Frock, R.L.; Hu, J.; Meyers, R.M.; Ho, Y.J.; Kii, E.; Alt, F.W. Genome-wide detection of DNA double-stranded breaks induced by engineered nucleases. *Nat. Biotechnol.* **2015**, *33*, 179–186. [CrossRef]

68. Wang, X.; Wang, Y.; Wu, X.; Wang, J.; Wang, Y.; Qiu, Z.; Chang, T.; Huang, H.; Lin, R.J.; Yee, J.K. Unbiased detection of off-target cleavage by crispr-cas9 and talens using integrase-defective lentiviral vectors. *Nat. Biotechnol* **2015**, *33*, 175–178. [CrossRef]

69. Kim, D.; Bae, S.; Park, J.; Kim, E.; Kim, S.; Yu, H.R.; Hwang, J.; Kim, J.I.; Kim, J.S. Digenome-seq: Genome-wide profiling of crispr-cas9 off-target effects in human cells. *Nat. Methods* **2015**, *12*, 237–243. [CrossRef]

70. Kim, D.; Kim, S.; Kim, S.; Park, J.; Kim, J.S. Genome-wide target specificities of crispr-cas9 nucleases revealed by multiplex digenome-seq. *Genome Res.* **2016**, *26*, 406–415. [CrossRef]

71. Cameron, P.; Fuller, C.K.; Donohoue, P.D.; Jones, B.N.; Thompson, M.S.; Carter, M.M.; Gradia, S.; Vidal, B.; Garner, E.; Slorach, E.M.; et al. Mapping the genomic landscape of crispr-cas9 cleavage. *Nat. Methods* **2017**, *14*, 600–606. [CrossRef]

72. Crosetto, N.; Mitra, A.; Silva, M.J.; Bienko, M.; Dojer, N.; Wang, Q.; Karaca, E.; Chiarle, R.; Skrzypczak, M.; Ginalski, K.; et al. Nucleotide-resolution DNA double-strand break mapping by next-generation sequencing. *Nat. Methods* **2013**, *10*, 361–365. [CrossRef]

73. Yan, W.X.; Mirzazadeh, R.; Garnerone, S.; Scott, D.; Schneider, M.W.; Kallas, T.; Custodio, J.; Wernersson, E.; Li, Y.; Gao, L.; et al. Bliss is a versatile and quantitative method for genome-wide profiling of DNA double-strand breaks. *Nat. Commun.* **2017**, *8*, 15058. [CrossRef]

74. Zischewski, J.; Fischer, R.; Bortesi, L. Detection of on-target and off-target mutations generated by crispr/cas9 and other sequence-specific nucleases. *Biotechnol. Adv.* **2017**, *35*, 95–104. [CrossRef]

75. Singh, R.; Kuscu, C.; Quinlan, A.; Qi, Y.; Adli, M. Cas9-chromatin binding information enables more accurate crispr off-target prediction. *Nucleic Acids Res.* **2015**, *43*, e118. [CrossRef]

76. Moreno-Mateos, M.A.; Vejnar, C.E.; Beaudoin, J.D.; Fernandez, J.P.; Mis, E.K.; Khokha, M.K.; Giraldez, A.J. Crisprscan: Designing highly efficient sgrnas for crispr-cas9 targeting in vivo. *Nat. Methods* **2015**, *12*, 982–988. [CrossRef]

77. Cui, Y.; Xu, J.; Cheng, M.; Liao, X.; Peng, S. Review of crispr/cas9 sgrna design tools. *Interdiscip Sci.* **2018**, *10*, 455–465. [CrossRef]

78. Haeussler, M.; Schonig, K.; Eckert, H.; Eschstruth, A.; Mianne, J.; Renaud, J.B.; Schneider-Maunoury, S.; Shkumatava, A.; Teboul, L.; Kent, J.; et al. Evaluation of off-target and on-target scoring algorithms and integration into the guide rna selection tool crispor. *Genome Biol.* **2016**, *17*, 148. [CrossRef]

79. Hendel, A.; Bak, R.O.; Clark, J.T.; Kennedy, A.B.; Ryan, D.E.; Roy, S.; Steinfeld, I.; Lunstad, B.D.; Kaiser, R.J.; Wilkens, A.B.; et al. Chemically modified guide rnas enhance crispr-cas genome editing in human primary cells. *Nat. Biotechnol.* **2015**, *33*, 985–989. [CrossRef]

80. Merienne, N.; Vachey, G.; de Longprez, L.; Meunier, C.; Zimmer, V.; Perriard, G.; Canales, M.; Mathias, A.; Herrgott, L.; Beltraminelli, T.; et al. The self-inactivating kamicas9 system for the editing of cns disease genes. *Cell Rep.* **2017**, *20*, 2980–2991. [CrossRef]

81. Zelensky, A.N.; Schimmel, J.; Kool, H.; Kanaar, R.; Tijsterman, M. Inactivation of pol theta and c-nhej eliminates off-target integration of exogenous DNA. *Nat. Commun.* **2017**, *8*, 66. [CrossRef]

82. Hendel, A.; Kildebeck, E.J.; Fine, E.J.; Clark, J.; Punjya, N.; Sebastiano, V.; Bao, G.; Porteus, M.H. Quantifying genome-editing outcomes at endogenous loci with smrt sequencing. *Cell Rep.* **2014**, *7*, 293–305. [CrossRef] [PubMed]

83. Torres, R.; Martin, M.C.; Garcia, A.; Cigudosa, J.C.; Ramirez, J.C.; Rodriguez-Perales, S. Engineering human tumour-associated chromosomal translocations with the rna-guided crispr-cas9 system. *Nat. Commun.* **2014**, *5*, 3964. [CrossRef] [PubMed]

84. Hendel, A.; Fine, E.J.; Bao, G.; Porteus, M.H. Quantifying on- and off-target genome editing. *Trends Biotechnol.* **2015**, *33*, 132–140. [CrossRef] [PubMed]

85. Pretto, D.; Yrigollen, C.M.; Tang, H.T.; Williamson, J.; Espinal, G.; Iwahashi, C.K.; Durbin-Johnson, B.; Hagerman, R.J.; Hagerman, P.J.; Tassone, F. Clinical and molecular implications of mosaicism in fmr1 full mutations. *Front. Genet.* **2014**, *5*, 318. [CrossRef] [PubMed]

86. Pretto, D.I.; Mendoza-Morales, G.; Lo, J.; Cao, R.; Hadd, A.; Latham, G.J.; Durbin-Johnson, B.; Hagerman, R.; Tassone, F. Cgg allele size somatic mosaicism and methylation in fmr1 premutation alleles. *J. Med. Genet.* **2014**, *51*, 309–318. [CrossRef] [PubMed]

87. Lee, H.; Kim, J.S. Unexpected crispr on-target effects. *Nat. Biotechnol.* **2018**, *36*, 703–704. [CrossRef] [PubMed]

88. Brokowski, C.; Adli, M. Crispr ethics: Moral considerations for applications of a powerful tool. *J. Mol. Biol.* **2019**, *431*, 88–101. [CrossRef] [PubMed]

Perspective

Repeat Instability in the Fragile X-Related Disorders: Lessons from a Mouse Model

Xiaonan Zhao [1,†], Inbal Gazy [1,†], Bruce Hayward [1], Elizabeth Pintado [2], Ye Hyun Hwang [3], Flora Tassone [3] and Karen Usdin [1,*]

[1] Section on Gene Structure and Disease, Laboratory of Cell and Molecular Biology, National Institute of Diabetes, Digestive and Kidney Diseases, National Institutes of Health, Bethesda, MD 20892, USA; xiaonan.zhao@nih.gov (X.Z.); inbal.gazy@nih.gov (I.G.); bruce.hayward@nih.gov (B.H.)

[2] Department of Medical Biochemistry and Molecular Biology, School of Medicine, University Hospital Virgen Macarena, University of Seville, 41009 Seville, Spain; elizabet@us.es

[3] Department of Biochemistry and Molecular Medicine and MIND Institute, UC Davis Medical Center, Sacramento, CA 95817, USA; yehhwang@ucdavis.edu (Y.H.H.); ftassone@ucdavis.edu (F.T.)

* Correspondence: ku@helix.nih.gov; Tel.: +1-301-496-2189

† These authors contributed equally to this work and thus should be considered co-first authors.

Received: 12 January 2019; Accepted: 27 February 2019; Published: 1 March 2019

Abstract: The fragile X-related disorders (FXDs) are a group of clinical conditions that result primarily from an unusual mutation, the expansion of a CGG-repeat tract in exon 1 of the *FMR1* gene. Mouse models are proving useful for understanding many aspects of disease pathology in these disorders. There is also reason to think that such models may be useful for understanding the molecular basis of the unusual mutation responsible for these disorders. This review will discuss what has been learnt to date about mechanisms of repeat instability from a knock-in FXD mouse model and what the implications of these findings may be for humans carrying expansion-prone *FMR1* alleles.

Keywords: CGG Repeat Expansion Disease; DNA instability; expansion; contraction; mismatch repair (MMR); base excision repair (BER); transcription coupled repair (TCR); double-strand break repair (DSBR); Non-homologous end-joining (NHEJ); mosaicism

---

1. Introduction

The fragile X-related disorders (FXDs) are X-linked disorders that include a form of ovarian dysfunction known as fragile X-associated primary ovarian insufficiency (FXPOI; MIM# 311360), a neurodegenerative condition, fragile X-associated tremor/ataxia syndrome (FXTAS; MIM# 300623) and fragile X syndrome (FXS; MIM# 300624), a major cause of intellectual disability and autism [1]. FXPOI and FXTAS are seen in carriers of *FMR1* premutation (PM) alleles, alleles that have a tandem array of 55–200 CGG repeats in exon 1. Most cases of FXS are seen in carriers of *FMR1* full mutation (FM) alleles, alleles that have >200 repeats, with a minority of individuals having deletions or point mutations that affect the levels or functionality of FMRP, the *FMR1* gene product, an important regulator of translation in the brain. The difference in the clinical consequences of the inheritance of a PM versus a FM allele results from the paradoxical effects of the repeat on *FMR1* expression. Most FM alleles are epigenetically silenced, resulting in the absence of FMRP. In contrast PM alleles are transcriptionally active and can have transcript levels anywhere between 2 and 8 times the levels of normal alleles [2]. Both FM and PM carriers show wide variability in their clinical presentation and both FXTAS and FXPOI show incomplete penetrance suggesting the contribution of other genetic factors to disease severity.

The CGG-repeat tract is unstable and is prone to expand and contract in a manner dependent on repeat number and the number of AGG interruptions present at the 5′ end of the tract [3,4]. Instability occurring in somatic cells can lead to repeat size mosaicism. In fact, it has been estimated that >40% of individuals with alleles >55 repeats are mosaic [5]. Mosaicism for both PM and FM alleles results in some individuals showing symptoms characteristic of both PM and FM alleles [6,7]. The mechanism(s) responsible for the repeat instability is largely unknown. While some instability has been reported in various human cells in culture [8–10], most studies of the instability mechanisms have used mouse models [11–14]. Mouse models offer a number of clear advantages over some of these cell-based systems, including the high frequency of both expansions and contractions and the ability to examine instability in different biologically relevant organs at various stages of development. In addition, since the size of the original allele can be readily established based on the allele size at birth, alleles arising from expansion can be clearly distinguished from those arising by contraction. Of the various mouse models for the FXDs that have been generated, most work on repeat instability has made use of a model in which the short CGG-repeat tract present in the endogenous mouse *FMR1* gene was replaced with ~130 CGG-repeats [14]. In this review we will address what we have learnt to date about repeat instability from this mouse model, as well as from other model systems and other related Repeat Expansion Disorders. We will also discuss some of the implications of this information for diagnosis and disease risk assessment in humans.

2. Instability in Humans and Mice May Share a Common Molecular Basis

There are a number of reasons to think that instability in mice and humans share common mechanisms. Firstly, while both expansions and contractions are seen, in both species expansions predominate over contractions, at least in the PM range. Secondly, in both humans and mice expansion events require transcription or the presence of the PM allele in a region of open chromatin [15,16]. In addition, both mice and humans show a maternal age effect for expansion risk [3,17], suggesting that expansion occurs in the oocyte in both species. While some postnatal oogenesis has been observed in mice whose existing oocytes have been ablated [18], the contribution of a dividing oocyte stem cell population to postnatal oogenesis and the pool of viable oocytes that can be fertilized in normal mammals is still controversial. Thus, since oocytes do not divide, a maternal age effect for expansions is generally considered to reflect events that occur in the absence of cell division. This suggests that the underlying mechanism in both species involves aberrant DNA repair and/or recombination, rather than a problem with chromosomal replication. Finally, many of the same genetic factors that affect expansion risk in the FXD mouse are known to modulate expansion risk in other human Repeat Expansion Diseases [19–21]. Since current evidence supports a common mutational mechanism for all of these diseases, this suggests that the FXD mouse may accurately recapitulate at least some aspects of repeat expansion in the FXDs.

Whilst most intergenerational expansions in mice are relatively small, large expansions characterize the PM to FM transition on maternal transmission in humans. However, there is no clear evidence that the PM to FM transition occurs in a single step in women and, in principle, it is possible to generate an equivalent increase in repeat number by a series of small expansions occurring over the decades that the human oocyte spends in dictyate arrest prior to fertilization [17]. Furthermore, while small expansions predominate in mice, larger intergenerational expansions are seen albeit at a lower frequency [17]. These larger expansions are sensitive to the same genetic factors that affect small expansions and thus likely arise from the same basic mechanism [22,23]. While much less is known about contractions, work in both mice and humans suggest that the underlying mechanism is likely different from the mechanism that gives rise to expansions [4,24,25].

Although available evidence suggests that mice may be useful for understanding the instability of human PM alleles, mouse *Fmr1* alleles with repeat numbers in the FM range do not undergo repeat-mediated epigenetic silencing as do FM alleles in humans. However, there is reason to think that X chromosome inactivation in female mice can provide a window into the factors associated with the instability of silenced alleles [16].

3. Different Cell Types Show Different Propensities to Expand in Mice

As can be seen in Figure 1, in the FXD mice some organs, like heart, show very little post-natal expansion, as evidenced by the fact that the repeat PCR profile of this organ does not change with age and remains indistinguishable from the profile seen in the tail DNA taken at 3 weeks of age [26]. In contrast, most other organs show some expansion as evidenced by the presence of alleles larger than the heart allele (Figure 1) [17,26]. In organs like testes and liver, expansions are apparent as a shift from a unimodal repeat PCR profile as seen in the tail DNA, to one that is more bimodal (Figures 1 and 2a). In young animals that have not yet accumulated many expansions, the second peak can appear more like a "shoulder" rather than a distinct peak, as in the example of the liver of a 3-month old mouse shown in Figure 2a. However, over time, as expansions continue to accumulate, expanding alleles diverge further from the original allele resulting in 2 clearly distinct allele peaks (Figures 1 and 2a). This bimodal peak distribution reflects the fact that, in some organs some cells are expansion-prone whilst others are not.

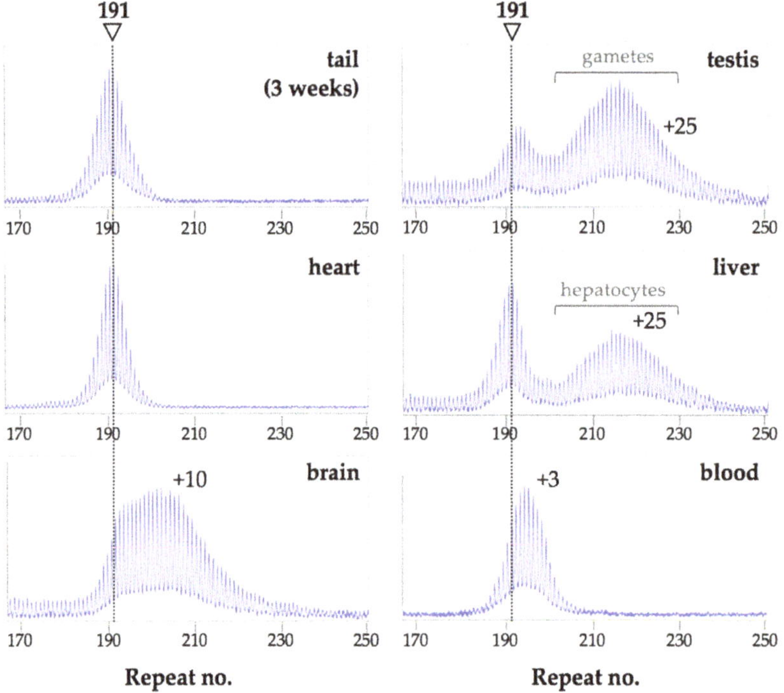

Figure 1. Different mouse organs show different propensities to expand. Repeat PCR was carried out on DNA extracted from different organs of a male mouse with 191 inherited repeats and analyzed as previously described [26]. The tail DNA sample was taken at 3 weeks, the remaining samples at 8 months of age. The arrowhead and dotted lines indicate the repeat number in the original inherited allele as assessed from tail DNA taken at 3 weeks of age. The numbers within each panel indicate the number of repeats added.

For example, in testes expansion is confined to the spermatogonia or primary spermatocytes [17], with the shorter alleles in the testes profile in Figure 1 corresponding to alleles in the somatic cells and the longer alleles corresponding to alleles in the gametes. Similarly, in the liver, expansion is confined to hepatocytes [27], while in the brain expansion is more extensive in the striatum and basolateral amygdala than in the medial prefrontal cortex [26]. Unlike PM alleles in testes and liver, PM alleles in blood show relatively little expansion and what little expansion is seen has a unimodal distribution in males (Figure 1). This may reflect the fact that all white blood cells are equally prone to expansion. Why some cells are expansion-prone and others are not is not fully understood. It does not seem to be simply related to the amount of *Fmr1* transcription since tissues with similar levels of *Fmr1* mRNA can show very different propensities to expand [26]. Rather it may be related to the balance between the levels of expression of proteins involved in the generation of expansions and those involved in the pathway(s) that promotes contraction or error-free repair [26,28]. Computer simulations suggest that the expanded allele profile, even that seen in very expansion-prone tissue, is consistent with the addition of 1–2 repeats with each expansion event [29]. However, as will be discussed below, larger expansions and contractions also occur (Section 7).

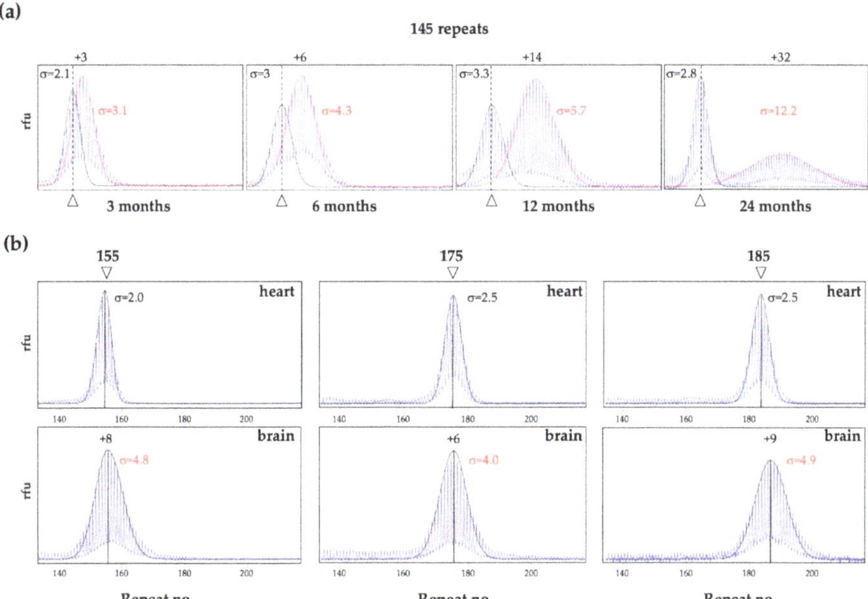

Figure 2. Change in the PM repeat PCR profiles and σ with age and extent of expansion. The PCR profiles and σs were generated from male mice as previously described [26,27]. (**a**) Liver PCR profiles of mice of different ages, all with an original allele of ~145 repeats. The arrowhead on the bottom of each panel and the dotted lines indicate the original inherited allele as assessed from tail DNA taken at 3 weeks of age. The numbers above the panels indicate the number of repeats added to the expanded allele. The σ of the stable allele and the expanded allele are shown in black and red, respectively. The examples shown in this panel are derived from previously published work [27]; (**b**) PCR profiles and corresponding σ for alleles in hearts and brains of 1-year old (155 repeats) and 6-month old (175 and 185 repeats) mice. The number associated with each arrowhead represents the number of repeats in the indicated allele. For the hearts, this number corresponds to the original inherited allele. For the brains, the repeat size reflects a gain of 6–9 repeats from the original allele.

As alleles expand, their PCR profile widens as differences in the timing and size of expansions in different cells transforms the original discrete allele into a more heterogenous mixture of allele sizes (Figure 2a). Thus, the broadness of the allele peak, as reflected in the standard deviation (σ), can be a sensitive metric of expansion [27,29]. In our experience, stable alleles, like those in heart, show a σ of ≤2.5, for a wide range of repeat sizes and mouse ages as illustrated in Figure 2b. In contrast, the σ of alleles in expansion-prone cells is >2.5 and increases as expansion increases as the animals age (Figure 2a,b). Overlapping allele peaks can result in an overestimation of the σ of the smaller allele and an underestimation of the larger one (Figure 2a). However, in unimodal PCR profiles or profiles with distinct peaks, an allele that has expanded has a larger σ than a stable allele of the same size (compare brain to heart in Figure 2b).

It should be noted that while the repeat PCR profiles for alleles in the heart of WT mice shows no evidence of post-natal expansion, they differ from the heart profiles in mice with mutations that abolish expansions completely [30]. Specifically, the profile in WT animals has a normal distribution while the profile in mutant mice is not only sharper, but is also left-skewed [30]. This left skew likely reflects PCR stutter products, while the normal distribution of the WT heart profiles likely reflects some expansion that occurs in these animals during early embryonic development.

Implications for humans: Analysis of the CGG repeat in the human *FMR1* gene is routinely performed using blood where, as in mice, a unimodal PCR profile is commonly seen in males. However, by analogy with what is seen in mice, a unimodal PCR profile may not mean that the allele is stable. As with mice, the σ of an allele profile likely reflects the extent of somatic expansion. This would be predicted to vary with total repeat number, the number of AGG interruptions and the effect of different genetic and/or environmental modifiers of expansion risk. As with mice (Figure 2a), age may also be a factor, at least for very unstable alleles. Figure 3a shows examples of unimodal repeat PCR profiles characteristic of stable (top panel) and unstable (bottom panel) alleles.

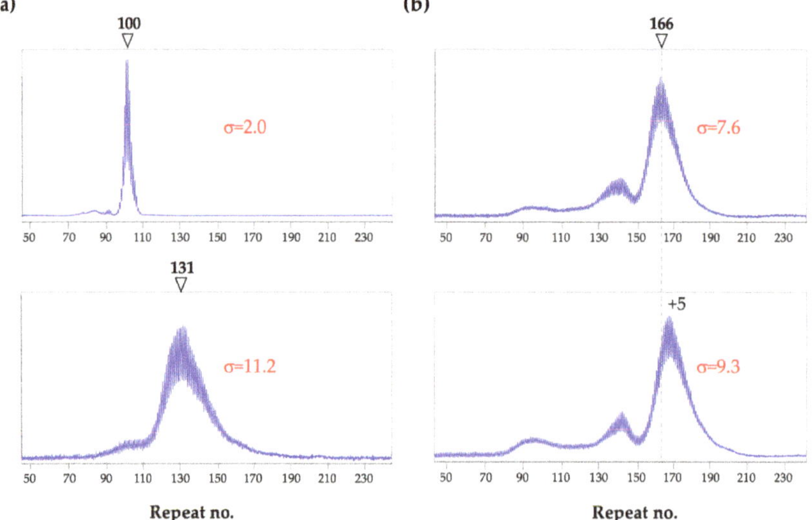

Figure 3. Blood repeat PCR profiles for 3 human male PM carriers. Repeat PCR analysis was carried out as described previously [31]. The number associated with each arrowhead represents the number of repeats in the indicated allele. (**a**) Profiles of two individuals showing a unimodal profile with a sharp peak (top) and a broad peak (bottom); (**b**) Profiles generated from the same individual using samples taken 4 years apart. The dotted line in these panels indicates the size of the major allele at the earlier timepoint. A shift corresponding to the gain of 5 CGG repeats is seen in the later sample. In each case the σ values are for the major allele peak.

As illustrated in Figure 3b, multiple peaks are also sometimes seen in males. These alleles may arise by contraction of larger alleles or from large expansions. As can be seen in the 2 different blood samples from the same individual taken 4 years apart (Figure 3b), these peaks can be broad, indicative of subsequent expansions. Notably, in this individual, both the repeat number and the σ were larger in the second sample taken 4 years later, consistent with an age effect.

PM alleles are relatively stable in mouse blood compared to alleles from other cells. Thus, it is possible that any evidence of expansion in human blood, reflects the presence of even more extensive expansion in organs like brain or gonads. Different allele profiles in different tissues have been reported in some PM carriers [32–39] and 2 different studies support the idea that some men have larger expansions with broader σs in sperm relative to blood [40,41]. Differences between allele sizes in blood and in brain or gonads in expansion-prone individuals could lead to an underestimation of the intergenerational expansion risk. It may be that it also contributes to the apparent variability in the penetrance of FXTAS and FXPOI.

4. Expansions in Females Only Occurs on The Active X Chromosome

While in male mice a bimodal repeat PCR profile is seen in organs that have cell types with different expansion rates, most female mice show a bimodal PCR profile for the PM allele in all expansion-prone tissues [16]. This is a consequence of the fact that expansions in females are confined to alleles on the active X chromosome [16]. Thus, even in expansion-prone cell types, expansion only occurs in ~50% of cells in females with normal X chromosome inactivation (XCI).

Implications for humans: A bimodal PCR profile for the PM allele is also sometimes seen in human females (Figure 4). The larger alleles are lost from the repeat PCR profile when the DNA is digested prior to PCR with a methylation-sensitive enzyme that has one or more cleavage sites within the PCR amplicon. Such an enzyme cuts the *FMR1* allele on the active X chromosome, leaving the allele on the inactive X as the only template for PCR. Thus, in the example shown in Figure 4a, alleles on the active X have gained ~7 repeats relative to alleles on inactive X chromosomes (bottom panel). Notably, unlike the roughly normal distribution seen in the repeat PCR profiles of expanded alleles, the shape of the repeat PCR profile for alleles on the inactive X is asymmetric and closely resembles the shape of the PCR profile seen in mice with mutations that completely block expansions [30,42]. Even among women with a bimodal allele distribution, differences in the extent of instability can result in dramatic differences in their PCR profiles. Figure 4b illustrates the 2 extremes of the possible bimodal PCR profiles, with the woman in the top panel showing a very low level of somatic instability and the woman in the bottom panel showing unusually high levels.

However, not all women show a bimodal profile for the PM allele. For example, as shown in the upper panel of Figure 4c, women with a high activation ratio (AR), that is, a high proportion of cells in which the active X chromosome carries the normal allele, would show a single sharp peak for the PM allele, with little, if any, evidence of a second peak, since no expansion would take place on the inactive X. In contrast, a woman with a low AR, would be more likely to have a profile with more expanded alleles than stable ones as shown in the bottom panel of Figure 4c. An even lower AR might result in unimodal PM profile with a large σ, as reported for a woman with an AR of 0.06 [43]. A single, sharp PM allele profile can also be seen even in the absence of skewed XCI (Figure 4d). Such alleles may result from the presence of AGG interruptions that reduce the expansion frequency, a genetic background that is not prone to expansion or, potentially, to an effect of age, with very young females being more likely to show such a profile. In the cases shown in Figure 4d, both women were of similar age, showed no skewing of XCI and had no AGG interruptions. Thus, the sharp and asymmetric profile seen for the 133 repeat allele may reflect genetic or environmental factors that reduce expansion risk as in mice with mutations that block expansions [24,30,42].

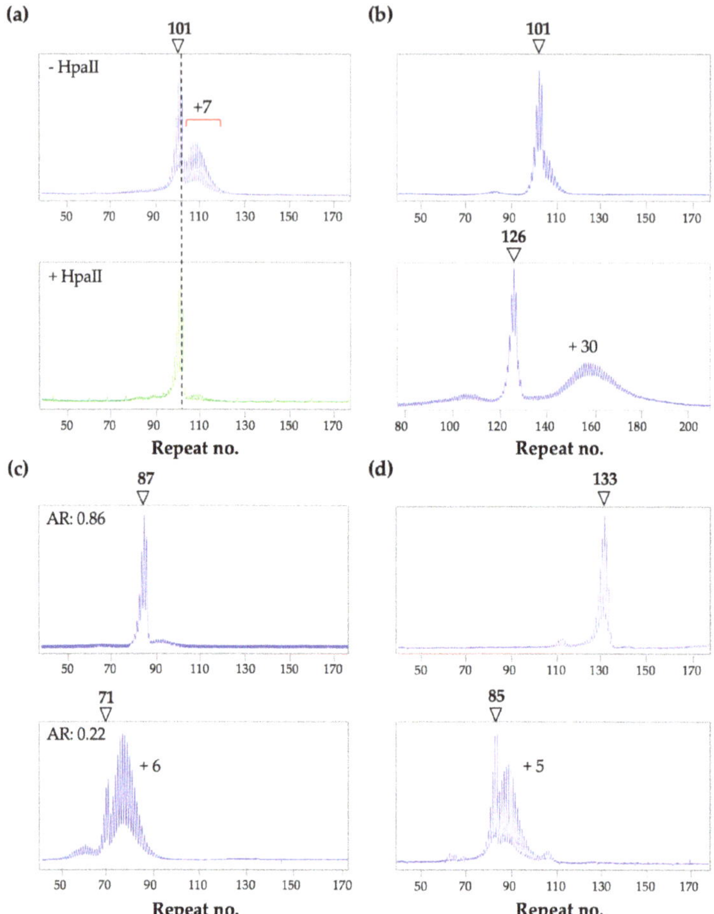

Figure 4. PM Repeat PCR profiles from the blood of human female PM carriers. The arrow in each instance indicates the stable allele with the indicated number of repeats. (**a**) Profiles for a female PM carrier without (top panel) and with (bottom panel) HpaII pre-digestion were generated as previously described [16]. The alleles on both the active and inactive X are shown in blue in the top panel and the alleles on the inactive X in green in the bottom panel; (**b**) Examples of very different bimodal PCR profiles; (**c**) Profiles for 2 women with different activation ratios (ARs); (**d**) Profiles of two females of similar ages and ARs, both with no AGG interruptions showing very different levels of somatic expansion.

5. Expansion in the Male and Female Germline

More expansions are seen in the oocytes of older female mice than younger ones [17]. This is consistent with some expansions occurring postnatally in non-dividing oocytes. In contrast, germ line expansions in male mice occurs in the spermatogonial stem cells (SSCs), cells that undergo multiple rounds of cell division [17]. Furthermore, male mice show a higher frequency of germ line expansions than females [17]. Studies of mice with mutations in different genes shows that the same genetic factors affect expansions in males and females and the same genetic factors also account for large and small expansions [24,25,28,30,42,44]. This would be consistent with single mechanism being responsible for all expansions in both males and females.

Implications for humans: Almost all expansions from a PM to a FM allele in humans occurs on maternal transmission. However, men with PM alleles transmit more expansions at least for smaller PM alleles [4,45] consistent with what is observed in mice. The fact that expansions in male and female mice share a dependence on a common set of genes would argue against a female-specific mechanism for the generation of FM alleles in humans. The FX repeat forms unusual intrastrand secondary structures [46–50], that are thought to make the repeat tract difficult to replicate [49,51,52]. This may result in pressure for larger alleles in dividing gametes to contract over time, as is seen in FX embryonic stem cells [53]. Germ cells in a 30-year old man will have undergone ~400 divisions, compared to 31 in a woman the same age [54,55]. However, by analogy with mice, most expansions in the female germline likely occur during post-natal life, well after cell division is complete. Thus, expanded alleles in female gametes face little pressure to contract, whilst male gametes are under continuous pressure to do so. This might explain why FMRP was detected in primordial germ cells of a 17-week old male FM fetus but not in those of a 13-week old fetus [56] and why older FM males only have PM alleles in their sperm [57,58]. It may also provide an explanation for why male PM carriers do not generally transmit FM alleles to their children.

6. Genetic and Environmental Factors Affecting Instability

A number of genetic and environmental factors have been shown to impact expansion risk in mice. For example, an exogenous source of oxidative stress increases expansions [59]. Mutations in different DNA repair genes also affects the extent of expansion, with some mutations reducing expansions [24,25,28,30,42,44] and others increasing them [27,30,44,60]. For example, mutations in mismatch repair (MMR) proteins, including MSH2, MSH3, MSH6 and MLH3, either eliminate expansions altogether [24,30,42] or severely reduce their incidence [25]. Similarly, a single hypomorphic allele of Polβ, a DNA polymerase that plays an essential role in base excision repair (BER), is sufficient to significantly reduce the expansion frequency [28]. Thus, proteins from multiple DNA repair pathways that normally work to prevent mutations, interact in such a way so as to actually cause the repeat expansion mutation. Work in vitro has shown that the FX repeats form unusual DNA structures including hairpins that have a mixture of Watson-Crick and non-Watson-Crick base pairs [46–50,61,62]. Current thinking is that these structures are the substrates upon which this process acts but the sequence of events and all the factors involved are still not fully understood (see [23] for recent review).

Mutations in other proteins, including two 5'-3' exonucleases, EXO1 and FAN1, lead to an increase in expansions, suggesting that these proteins are protective [30,60]. Loss of EXO1 affects expansions in the germ line and in the small intestine but not in the brain [30]. In contrast, loss of FAN1 affects expansion in multiple organs including brain but does not affect the germ line expansion frequency [60]. ERCC6/CSB plays a paradoxical role in repeat expansion playing a minor role in promoting expansions in some instances [44] and protecting against them in others [63]. This paradoxical effect may reflect this protein's ability to participate in multiple DNA repair pathways. Recently, DNA ligase IV (LIG4) has also been shown to protect against expansion [27]. Since LIG4 is essential for non-homologous end-joining (NHEJ), a form of double strand break (DSB) repair, it suggests that the expansion process competes with the NHEJ pathway for a common substrate. This supports the idea that expansion proceeds through a DSB intermediate, perhaps one generated by MutLγ [27]. A simple gap-filling model for the generation of expansions from a staggered DNA DSB arising during transcription or DNA repair has been suggested [27].

Very little is known about the factors that promote contractions. In the mouse model, factors that abolish expansions do not necessarily reduce contractions [24,25,28,30,42]. In fact, the frequency of contractions usually increases when the expansion frequency drops. This suggests that some, if not all, contractions occur via a mechanism that differs from the expansion mechanism and that when expansions are blocked, a process or processes that favors contractions predominates. This would be consistent with the observation that AGG interruptions, which are an important modifier of expansion risk, do not affect the contraction frequency [4]. Moreover, while transcription or open chromatin is required for expansion, contraction of methylated alleles can be seen in both mouse and humans, particularly in rapidly dividing cells such as those in the early embryo or in cells grown at low cell densities in vitro [10,17,53]. Contractions under these circumstances may reflect the difficulty the cell has in replicating the FX repeats [49,51,52], thus favoring cells in which repeats have been lost. Tandemly repeated sequences often contract via strand-slippage during replication in a variety of organisms [64–66]. Such a mechanism could explain the observed loss of AGG interruptions sometimes associated with contraction [4], if slippage occurred upstream of the interruption with re-priming of DNA synthesis occurring downstream of the interruption. Strand-slippage by the FX repeat also has the potential to generate point mutations within the repeat by frameshifting or limited intra-strand template switching with priming within the hairpin [67–69].

Implications for humans: Genetic factors that impact expansion may contribute to the increased expansion risk seen for PM carriers with a family history of FXS relative to carriers of similar PM alleles in the general population [70]. It may also account for why some individuals show more somatic expansion than others (Figures 3 and 4). Interestingly, single nucleotide polymorphisms (SNPs) in genes including *ERCC6/CSB* [71], *MSH3* [21,72] and *FAN1* [19,20] are thought to modify disease risk in other Repeat Expansion Diseases via their effect on somatic expansion. A SNP in *MLH1*, the binding partner of *MLH3* in the complex MutLγ, has also been shown to have a similar effect [19]. Since most factors that affect expansion in somatic cells also affect expansion in germ cells in mice, it is likely that similar genetic factors would also impact the risk of intergenerational expansion. However, as illustrated by the differences between the effects of FAN1 and EXO1 mutations on expansion in different tissues in mice, some genetic factors may be more important modulators of expansion in some cells than in others and thus, may affect expansion differently in different organs. For example, a polymorphism in *FAN1* may result in increased expansion in brain but not necessarily in oocytes, while *EXO1* polymorphisms may result in increased expansion in gametes, but not liver or other somatic tissue. Thus, a thorough understanding of expansion predisposition may require testing of multiple tissues.

7. The Frequency of Large Contractions and Expansions can be Underestimated

Analysis of repeat length and somatic instability is routinely performed on bulk genomic DNA. Such analysis on mice tissue indicates that somatic instability mostly involves the gain of relatively small number of repeats (Figure 2). However, large expansions and contractions can be seen in intergenerational transmission in mice [17]. They can also be seen if they occur during early embryonic development when they represent a significant fraction of the alleles in the population. These observations indicate that large expansions and contractions can occur not only in humans but also in mice. However, similar events occurring postnatally are difficult to detect using PCR on bulk genomic DNA since the resultant alleles vary considerably in the number of repeats gained or lost. Thus, each of these alleles represents a very small proportion of alleles in the population and likely will not be detected in standard PCR analysis. For example, as seen in Figure 5, when bulk DNA from the brain of a 1-year old mouse is analyzed, the PCR profile is consistent with most changes in repeat number involving the gain of a small number of repeats. However, PCR on single genome equivalents from the same brain sample shows that almost 30% of alleles have lost or gained more than 25 repeats. Thus, larger expansions and contractions actually occur relatively frequently and may ultimately reflect a relatively large fraction of the total alleles in the population.

Implications for humans: This combination of expansion and contraction can result in individuals being highly mosaic for a variety of different alleles. The fact that larger expansions and contractions that occur later in development are difficult to detect in mice, raises the possibility that some humans may be even more mosaic than analysis of their bulk DNA suggests. Thus, careful analysis of the distribution of allele sizes in carriers might be needed to properly assess disease risk.

Small pool PCR of brain PM alleles

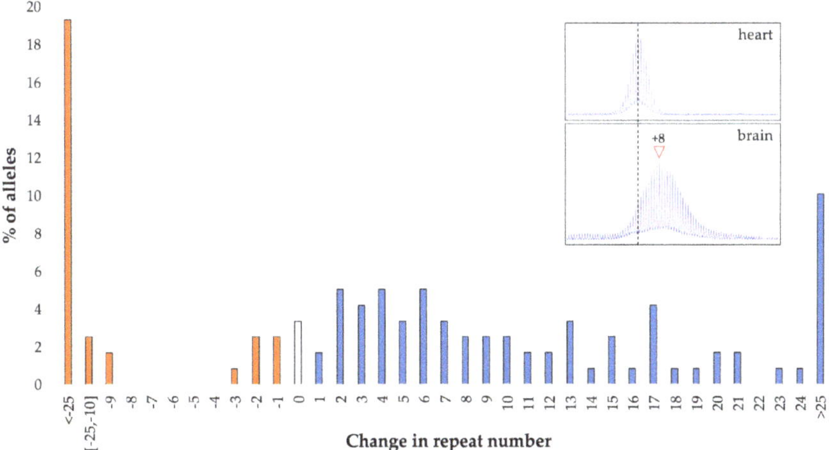

Figure 5. Distribution of the changes in the repeat number seen in the brain of a 1-year old mouse with an inherited allele of 162 repeats. The repeat number of individual alleles was determined by small pool PCR of single genome equivalents as described previously [30]. The data shown represent 119 individual PCR reactions. The inset panels show the bulk PCR profiles for the heart and brain of the same animal.

8. Concluding Remarks

The use of a mouse model allows the dynamics of repeat instability in the *FMR1* gene to be explored over time in multiple tissues. This has resulted in a number of observations that may be relevant to repeat instability in humans. For example, we have learnt that some cell types show more expansion than others [16,17,24,26]. In particular, expansions are more extensive in brain and gonads than in blood. This may mean that, in some people, the size of the repeat or the extent of somatic expansion seen in blood may not reflect what is present in disease-relevant cells. Examination of the repeat PCR profiles from mice over time has also shown that the σ of the allele peak provides a sensitive measure of the extent of somatic instability. Specifically, stable alleles have a low σ (<2.5), while expanded alleles have a larger σ. A wide range of σ values can be seen for human PM alleles in blood that, by analogy with FXD mice, likely reflects a wide variation in the extent of somatic expansion in different people.

A number of genetic factors that promote or protect against expansions in the FXD mouse model have been identified [24,25,27,28,30,42,44,60,63,73,74]. Some of these factors have been implicated in expansion in humans with other Repeat Expansion Diseases [20,21,75–77], suggesting that they may be relevant for human carriers of unstable *FMR1* alleles as well. If so, the prediction would be that polymorphisms in these factors would be modifiers of both germ line and somatic expansion risk in FX families. As such, people who have elevated activities of protective factors or reduced activities of factors that promote expansion may show little, if any, somatic expansion. In contrast, those with elevated activities of expansion-promoting factors or reduced activities of protective factors may show more somatic expansion. Evidence from other Repeat Expansion Diseases suggests that somatic expansion contributes to differences in the age at onset and disease severity [20,21,75–77]. In a recent study, women who showed PM allele mosaicism reported more severe symptoms than women who were not mosaic [39], suggesting that somatic instability may exacerbate PM symptoms. However, additional studies are needed to fully understand the contribution of somatic expansion to disease pathology in PM carriers. In any event, an increased propensity for somatic expansion likely indicates an increased propensity for germ-line expansion and thus an increased risk of intergenerational transmission of larger alleles. Additional work is also needed to assess whether the same genetic factors that affect expansion risk in mice also modulate the risk of somatic and intergenerational expansion of human FXD alleles.

Author Contributions: X.Z, I.G., B.H., E.P., F.T. and K.U. compiled the data and wrote the paper.

Acknowledgments: The work described in this manuscript was funded by a grant from the Intramural Program of the NIDDK to KU (DK057808) which includes funds for covering the costs to publish in open access.

Conflicts of Interest: The authors declare no conflict of interest. The funding sponsors had no role in the design of the study; in the collection, analyses or interpretation of data; in the writing of the manuscript and in the decision to publish the results.

References

1. Lozano, R.; Rosero, C.A.; Hagerman, R.J. Fragile X spectrum disorders. *Intractable Rare Dis. Res.* **2014**, *3*, 134–146. [CrossRef] [PubMed]

2. Tassone, F.; Hagerman, R.J.; Taylor, A.K.; Gane, L.W.; Godfrey, T.E.; Hagerman, P.J. Elevated levels of *FMR1* mRNA in carrier males: A new mechanism of involvement in the fragile-X syndrome. *Am. J. Hum. Genet.* **2000**, *66*, 6–15. [CrossRef] [PubMed]

3. Yrigollen, C.M.; Martorell, L.; Durbin-Johnson, B.; Naudo, M.; Genoves, J.; Murgia, A.; Polli, R.; Zhou, L.; Barbouth, D.; Rupchock, A.; et al. AGG interruptions and maternal age affect *FMR1* CGG repeat allele stability during transmission. *J. Neurodev. Disord.* **2014**, *6*, 24. [CrossRef] [PubMed]

4. Nolin, S.L.; Glicksman, A.; Ersalesi, N.; Dobkin, C.; Brown, W.T.; Cao, R.; Blatt, E.; Sah, S.; Latham, G.J.; Hadd, A.G. Fragile X full mutation expansions are inhibited by one or more AGG interruptions in premutation carriers. *Genet. Med.* **2015**, *17*, 358–364. [CrossRef] [PubMed]

5. Nolin, S.L.; Glicksman, A.; Houck, G.E., Jr.; Brown, W.T.; Dobkin, C.S. Mosaicism in fragile X affected males. *Am. J. Med. Genet.* **1994**, *51*, 509–512. [CrossRef] [PubMed]

6. Basuta, K.; Schneider, A.; Gane, L.; Polussa, J.; Woodruff, B.; Pretto, D.; Hagerman, R.; Tassone, F. High functioning male with fragile X syndrome and fragile X-associated tremor/ataxia syndrome. *Am. J. Med. Genet. A* **2015**, *167*, 2154–2161. [CrossRef] [PubMed]

7. Hwang, Y.T.; Aliaga, S.M.; Arpone, M.; Francis, D.; Li, X.; Chong, B.; Slater, H.R.; Rogers, C.; Bretherton, L.; Hunter, M.; et al. Partially methylated alleles, microdeletion, and tissue mosaicism in a fragile X male with tremor and ataxia at 30 years of age: A case report. *Am. J. Med. Genet. A* **2016**, *170*, 3327–3332. [CrossRef] [PubMed]

8. Wohrle, D.; Salat, U.; Hameister, H.; Vogel, W.; Steinbach, P. Demethylation, reactivation, and destabilization of human fragile X full-mutation alleles in mouse embryocarcinoma cells. *Am. J. Hum. Genet.* **2001**, *69*, 504–515. [CrossRef] [PubMed]

9. Gerhardt, J.; Zaninovic, N.; Zhan, Q.; Madireddy, A.; Nolin, S.L.; Ersalesi, N.; Yan, Z.; Rosenwaks, Z.; Schildkraut, C.L. Cis-acting DNA sequence at a replication origin promotes repeat expansion to fragile X full mutation. *J. Cell Biol.* **2014**, *206*, 599–607. [CrossRef] [PubMed]

10. Brykczynska, U.; Pecho-Vrieseling, E.; Thiemeyer, A.; Klein, J.; Fruh, I.; Doll, T.; Manneville, C.; Fuchs, S.; Iazeolla, M.; Beibel, M.; et al. CGG repeat-Induced *FMR1* silencing depends on the expansion size in human iPSCs and neurons carrying unmethylated full mutations. *Stem Cell Rep.* **2016**, *7*, 1059–1071. [CrossRef] [PubMed]

11. Bontekoe, C.J.; de Graaff, E.; Nieuwenhuizen, I.M.; Willemsen, R.; Oostra, B.A. *FMR1* premutation allele (CGG)81 is stable in mice. *Eur. J. Hum. Genet.* **1997**, *5*, 293–298. [PubMed]

12. Lavedan, C.; Grabczyk, E.; Usdin, K.; Nussbaum, R.L. Long uninterrupted CGG repeats within the first exon of the human *FMR1* gene are not intrinsically unstable in transgenic mice. *Genomics* **1998**, *50*, 229–240. [CrossRef] [PubMed]

13. Willemsen, R.; Hoogeveen-Westerveld, M.; Reis, S.; Holstege, J.; Severijnen, L.A.; Nieuwenhuizen, I.M.; Schrier, M.; van Unen, L.; Tassone, F.; Hoogeveen, A.T.; et al. The *FMR1* CGG repeat mouse displays ubiquitin-positive intranuclear neuronal inclusions; implications for the cerebellar tremor/ataxia syndrome. *Hum. Mol. Genet.* **2003**, *12*, 949–959. [CrossRef] [PubMed]

14. Entezam, A.; Biacsi, R.; Orrison, B.; Saha, T.; Hoffman, G.E.; Grabczyk, E.; Nussbaum, R.L.; Usdin, K. Regional FMRP deficits and large repeat expansions into the full mutation range in a new fragile X premutation mouse model. *Gene* **2007**, *395*, 125–134. [CrossRef] [PubMed]

15. Glaser, D.; Wohrle, D.; Salat, U.; Vogel, W.; Steinbach, P. Mitotic behavior of expanded CGG repeats studied on cultured cells: further evidence for methylation-mediated triplet repeat stability in fragile X syndrome. *Am. J. Med. Genet.* **1999**, *84*, 226–228. [CrossRef]

16. Lokanga, R.; Zhao, X.N.; Entezam, A.; Usdin, K. X inactivation plays a major role in the gender bias in somatic expansion in a mouse model of the fragile X-related disorders: implications for the mechanism of repeat expansion. *Hum. Mol. Genet.* **2014**, *23*, 4985–4994. [CrossRef] [PubMed]

17. Zhao, X.N.; Usdin, K. Timing of expansion of fragile X premutation alleles during intergenerational transmission in a mouse model of the fragile X-related disorders. *Front. Genet.* **2018**, *9*, 314. [CrossRef] [PubMed]

18. Wang, N.; Satirapod, C.; Ohguchi, Y.; Park, E.S.; Woods, D.C.; Tilly, J.L. Genetic studies in mice directly link oocytes produced during adulthood to ovarian function and natural fertility. *Sci. Rep.* **2017**, *7*, 10011. [CrossRef] [PubMed]

19. Genetic Modifiers of Huntington's Disease (GeM-HD) Consortium. Identification of genetic factors that modify clinical onset of Huntington's disease. *Cell* **2015**, *162*, 516–526. [CrossRef]

20. Bettencourt, C.; Hensman-Moss, D.; Flower, M.; Wiethoff, S.; Brice, A.; Goizet, C.; Stevanin, G.; Koutsis, G.; Karadima, G.; Panas, M.; et al. DNA repair pathways underlie a common genetic mechanism modulating onset in polyglutamine diseases. *Ann. Neurol.* **2016**, *79*, 983–990. [CrossRef] [PubMed]

21. Hensman Moss, D.J.; Pardinas, A.F.; Langbehn, D.; Lo, K.; Leavitt, B.R.; Roos, R.; Durr, A.; Mead, S.; TRACK-HD Investigators; REGISTRY Investigators; et al. Identification of genetic variants associated with Huntington's disease progression: A genome-wide association study. *Lancet Neurol.* **2017**, *16*, 701–711. [CrossRef]

22. Zhao, X.N.; Usdin, K. The Repeat Expansion Diseases: The dark side of DNA repair. *DNA Repair (Amst.)* **2015**, *32*, 96–105. [CrossRef] [PubMed]

23. Zhao, X.N.; Usdin, K. Ups and downs: mechanisms of repeat Instability in the fragile X-related disorders. *Genes* **2016**, *7*, 70. [CrossRef] [PubMed]

24. Zhao, X.N.; Kumari, D.; Gupta, S.; Wu, D.; Evanitsky, M.; Yang, W.; Usdin, K. Mutsbeta generates both expansions and contractions in a mouse model of the Fragile X-associated disorders. *Hum. Mol. Genet.* **2015**, *24*, 7087–7096. [CrossRef] [PubMed]

25. Zhao, X.N.; Lokanga, R.; Allette, K.; Gazy, I.; Wu, D.; Usdin, K. A MutSbeta-dependent contribution of MutSalpha to repeat expansions in fragile X premutation mice? *PLoS Genet.* **2016**, *12*, e1006190. [CrossRef] [PubMed]

26. Lokanga, R.A.; Entezam, A.; Kumari, D.; Yudkin, D.; Qin, M.; Smith, C.B.; Usdin, K. Somatic expansion in mouse and human carriers of fragile X premutation alleles. *Hum. Mutat.* **2013**, *34*, 157–166. [CrossRef] [PubMed]

27. Gazy, I.; Hayward, B.; Potapova, S.; Zhao, X.; Usdin, K. Double-strand break repair plays a role in repeat instability in a fragile X mouse model. *DNA Repair (Amst.)* **2019**, *74*, 63–69. [CrossRef] [PubMed]

28. Lokanga, R.A.; Senejani, A.G.; Sweasy, J.B.; Usdin, K. Heterozygosity for a hypomorphic Polbeta mutation reduces the expansion frequency in a mouse model of the Fragile X-related disorders. *PLoS Genet.* **2015**, *11*, e1005181. [CrossRef] [PubMed]

29. Mollersen, L.; Rowe, A.D.; Larsen, E.; Rognes, T.; Klungland, A. Continuous and periodic expansion of CAG repeats in Huntington's disease R6/1 mice. *PLoS Genet.* **2010**, *6*, e1001242. [CrossRef] [PubMed]

30. Zhao, X.; Zhang, Y.; Wilkins, K.; Edelmann, W.; Usdin, K. MutLgamma promotes repeat expansion in a Fragile X mouse model while EXO1 is protective. *PLoS Genet.* **2018**, *14*, e1007719. [CrossRef] [PubMed]

31. Hayward, B.E.; Zhou, Y.; Kumari, D.; Usdin, K. A set of assays for the comprehensive analysis of *FMR1* alleles in the fragile X-related disorders. *J. Mol. Diagn.* **2016**, *18*, 762–774. [CrossRef] [PubMed]

32. Dobkin, C.S.; Nolin, S.L.; Cohen, I.; Sudhalter, V.; Bialer, M.G.; Ding, X.H.; Jenkins, E.C.; Zhong, N.; Brown, W.T. Tissue differences in fragile X mosaics: Mosaicism in blood cells may differ greatly from skin. *Am. J. Med. Genet.* **1996**, *64*, 296–301. [CrossRef]

33. Maddalena, A.; Yadvish, K.N.; Spence, W.C.; Howard-Peebles, P.N. A fragile X mosaic male with a cryptic full mutation detected in epithelium but not in blood. *Am. J. Med. Genet.* **1996**, *64*, 309–312. [CrossRef]

34. Taylor, A.K.; Tassone, F.; Dyer, P.N.; Hersch, S.M.; Harris, J.B.; Greenough, W.T.; Hagerman, R.J. Tissue heterogeneity of the *FMR1* mutation in a high-functioning male with fragile X syndrome. *Am. J. Med. Genet.* **1999**, *84*, 233–239. [CrossRef]

35. MacKenzie, J.J.; Sumargo, I.; Taylor, S.A. A cryptic full mutation in a male with a classical fragile X phenotype. *Clin. Genet.* **2006**, *70*, 39–42. [CrossRef] [PubMed]

36. Pretto, D.I.; Mendoza-Morales, G.; Lo, J.; Cao, R.; Hadd, A.; Latham, G.J.; Durbin-Johnson, B.; Hagerman, R.; Tassone, F. CGG allele size somatic mosaicism and methylation in *FMR1* premutation alleles. *J. Med. Genet.* **2014**, *51*, 309–318. [CrossRef] [PubMed]

37. Jiraanont, P.; Kumar, M.; Tang, H.T.; Espinal, G.; Hagerman, P.J.; Hagerman, R.J.; Chutabhakdikul, N.; Tassone, F. Size and methylation mosaicism in males with fragile X syndrome. *Expert Rev. Mol. Diagn.* **2017**, *17*, 1023–1032. [CrossRef] [PubMed]

38. Fernandez, E.; Gennaro, E.; Pirozzi, F.; Baldo, C.; Forzano, F.; Turolla, L.; Faravelli, F.; Gastaldo, D.; Coviello, D.; Grasso, M.; et al. FXS-like phenotype in two unrelated patients carrying a methylated premutation of the *FMR1* gene. *Front. Genet.* **2018**, *9*, 442. [CrossRef] [PubMed]

39. Mailick, M.R.; Movaghar, A.; Hong, J.; Greenberg, J.S.; DaWalt, L.S.; Zhou, L.; Jackson, J.; Rathouz, P.J.; Baker, M.W.; Brilliant, M.; et al. Health profiles of mosaic versus non-mosaic *FMR1* premutation carrier mothers of children with fragile X syndrome. *Front. Genet.* **2018**, *9*, 173. [CrossRef] [PubMed]

40. Nolin, S.L.; Houck, G.E., Jr.; Gargano, A.D.; Blumstein, H.; Dobkin, C.S.; Brown, W.T. *FMR1* CGG-repeat instability in single sperm and lymphocytes of fragile-X premutation males. *Am. J. Hum. Genet.* **1999**, *65*, 680–688. [CrossRef] [PubMed]

41. Alvarez-Mora, M.I.; Guitart, M.; Rodriguez-Revenga, L.; Madrigal, I.; Gabau, E.; Mila, M. Paternal transmission of a *FMR1* full mutation allele. *Am. J. Med. Genet. A* **2017**, *173*, 2795–2797. [CrossRef] [PubMed]

42. Lokanga, R.A.; Zhao, X.N.; Usdin, K. The mismatch repair protein MSH2 is rate limiting for repeat expansion in a fragile X premutation mouse model. *Hum. Mutat.* **2014**, *35*, 129–136. [CrossRef] [PubMed]

43. Reyes-Quizoz, M.E.; Jesus, S.; Ramos, I.; Garcia, A.E.; Martinez, R.; Mir, P.; Pintado, E. Tissue-specific size and methylation analysis in two fragile X families: Contribution to the clinical phenotype. *J. Mol. Genet. Med.* **2016**, *10*. [CrossRef]

44. Zhao, X.N.; Usdin, K. Gender and cell-type-specific effects of the transcription-coupled repair protein, ERCC6/CSB, on repeat expansion in a mouse model of the fragile X-related disorders. *Hum. Mutat.* **2014**, *35*, 341–349. [CrossRef] [PubMed]

45. Sullivan, A.K.; Crawford, D.C.; Scott, E.H.; Leslie, M.L.; Sherman, S.L. Paternally transmitted *FMR1* alleles are less stable than maternally transmitted alleles in the common and intermediate size range. *Am. J. Hum. Genet.* **2002**, *70*, 1532–1544. [CrossRef] [PubMed]

46. Fry, M.; Loeb, L.A. The fragile X syndrome d(CGG)n nucleotide repeats form a stable tetrahelical structure. *Proc. Natl. Acad. Sci. USA* **1994**, *91*, 4950–4954. [CrossRef] [PubMed]

47. Mitas, M.; Yu, A.; Dill, J.; Haworth, I.S. The trinucleotide repeat sequence d(CGG)15 forms a heat-stable hairpin containing Gsyn. Ganti base pairs. *Biochemistry* **1995**, *34*, 12803–12811. [CrossRef] [PubMed]

48. Nadel, Y.; Weisman-Shomer, P.; Fry, M. The fragile X syndrome single strand d(CGG)n nucleotide repeats readily fold back to form unimolecular hairpin structures. *J. Biol. Chem.* **1995**, *270*, 28970–28977. [CrossRef] [PubMed]

49. Usdin, K.; Woodford, K.J. CGG repeats associated with DNA instability and chromosome fragility form structures that block DNA synthesis in vitro. *Nucleic Acids Res.* **1995**, *23*, 4202–4209. [CrossRef] [PubMed]

50. Yu, A.; Barron, M.D.; Romero, R.M.; Christy, M.; Gold, B.; Dai, J.; Gray, D.M.; Haworth, I.S.; Mitas, M. At physiological pH, d(CCG)15 forms a hairpin containing protonated cytosines and a distorted helix. *Biochemistry* **1997**, *36*, 3687–3699. [CrossRef] [PubMed]

51. Voineagu, I.; Surka, C.F.; Shishkin, A.A.; Krasilnikova, M.M.; Mirkin, S.M. Replisome stalling and stabilization at CGG repeats, which are responsible for chromosomal fragility. *Nat. Struct. Mol. Biol.* **2009**, *16*, 226–228. [CrossRef] [PubMed]

52. Yudkin, D.; Hayward, B.E.; Aladjem, M.I.; Kumari, D.; Usdin, K. Chromosome fragility and the abnormal replication of the *FMR1* locus in fragile X syndrome. *Hum. Mol. Genet.* **2014**, *23*, 2940–2952. [CrossRef] [PubMed]

53. Zhou, Y.; Kumari, D.; Sciascia, N.; Usdin, K. CGG-repeat dynamics and *FMR1* gene silencing in fragile X syndrome stem cells and stem cell-derived neurons. *Mol. Autism* **2016**, *7*, 42. [CrossRef] [PubMed]

54. Drost, J.B.; Lee, W.R. Biological basis of germline mutation: comparisons of spontaneous germline mutation rates among drosophila, mouse, and human. *Environ. Mol. Mutagen.* **1995**, *25* (Suppl. 26), 48–64. [CrossRef]

55. Crow, J.F. The origins, patterns and implications of human spontaneous mutation. *Nat Rev Genet* **2000**, *1*, 40–47. [CrossRef] [PubMed]

56. Malter, H.E.; Iber, J.C.; Willemsen, R.; de Graaff, E.; Tarleton, J.C.; Leisti, J.; Warren, S.T.; Oostra, B.A. Characterization of the full fragile X syndrome mutation in fetal gametes. *Nat. Genet.* **1997**, *15*, 165–169. [CrossRef] [PubMed]

57. Reyniers, E.; Vits, L.; De Boulle, K.; Van Roy, B.; Van Velzen, D.; de Graaff, E.; Verkerk, A.J.; Jorens, H.Z.; Darby, J.K.; Oostra, B.; et al. The full mutation in the *FMR-1* gene of male fragile X patients is absent in their sperm. *Nat. Genet.* **1993**, *4*, 143–146. [CrossRef] [PubMed]

58. Rousseau, F.; Robb, L.J.; Rouillard, P.; Der Kaloustian, V.M. No mental retardation in a man with 40% abnormal methylation at the FMR-1 locus and transmission of sperm cell mutations as premutations. *Hum. Mol. Genet.* **1994**, *3*, 927–930. [CrossRef] [PubMed]

59. Entezam, A.; Lokanga, A.R.; Le, W.; Hoffman, G.; Usdin, K. Potassium bromate, a potent DNA oxidizing agent, exacerbates germline repeat expansion in a fragile X premutation mouse model. *Hum. Mutat.* **2010**, *31*, 611–616. [CrossRef] [PubMed]

60. Zhao, X.N.; Usdin, K. FAN1 protects against repeat expansions in a Fragile X mouse model. *DNA Repair. (Amst.)* **2018**, *69*, 1–5. [CrossRef] [PubMed]

61. Chen, X.; Mariappan, S.V.; Catasti, P.; Ratliff, R.; Moyzis, R.K.; Laayoun, A.; Smith, S.S.; Bradbury, E.M.; Gupta, G. Hairpins are formed by the single DNA strands of the fragile X triplet repeats: structure and biological implications. *Proc. Natl. Acad. Sci. USA* **1995**, *92*, 5199–5203. [CrossRef] [PubMed]

62. Mariappan, S.V.; Catasti, P.; Chen, X.; Ratliff, R.; Moyzis, R.K.; Bradbury, E.M.; Gupta, G. Solution structures of the individual single strands of the fragile X DNA triplets (GCC)n.(GGC)n. *Nucleic Acids Res.* **1996**, *24*, 784–792. [CrossRef] [PubMed]

63. Zhao, X.N.; Usdin, K. The transcription-coupled repair protein ERCC6/CSB also protects against repeat expansion in a mouse model of the fragile X premutation. *Hum. Mutat.* **2015**, *36*, 482–487. [CrossRef] [PubMed]

64. Tran, H.T.; Degtyareva, N.P.; Koloteva, N.N.; Sugino, A.; Masumoto, H.; Gordenin, D.A.; Resnick, M.A. Replication slippage between distant short repeats in Saccharomyces cerevisiae depends on the direction of replication and the RAD50 and RAD52 genes. *Mol. Cell. Biol.* **1995**, *15*, 5607–5617. [CrossRef] [PubMed]

65. Hirst, M.C.; White, P.J. Cloned human *FMR1* trinucleotide repeats exhibit a length- and orientation-dependent instability suggestive of in vivo lagging strand secondary structure. *Nucleic. Acids Res.* **1998**, *26*, 2353–2358. [CrossRef] [PubMed]

66. Bichara, M.; Wagner, J.; Lambert, I.B. Mechanisms of tandem repeat instability in bacteria. *Mutat. Res.* **2006**, *598*, 144–163. [CrossRef] [PubMed]

67. Bissler, J.J. DNA inverted repeats and human disease. *Front. Biosci.* **1998**, *3*, 408–418. [CrossRef]

68. Lovett, S.T. Encoded errors: Mutations and rearrangements mediated by misalignment at repetitive DNA sequences. *Mol. Microbiol.* **2004**, *52*, 1243–1253. [CrossRef] [PubMed]

69. Kim, N.; Cho, J.E.; Li, Y.C.; Jinks-Robertson, S. RNA/DNA hybrids initiate quasi-palindrome-associated mutations in highly transcribed yeast DNA. *PLoS Genet* **2013**, *9*, e1003924. [CrossRef] [PubMed]

70. Nolin, S.L.; Sah, S.; Glicksman, A.; Sherman, S.L.; Allen, E.; Berry-Kravis, E.; Tassone, F.; Yrigollen, C.; Cronister, A.; Jodah, M.; et al. Fragile X AGG analysis provides new risk predictions for 45–69 repeat alleles. *Am. J. Med. Genet. A* **2013**, *161A*, 771–778. [CrossRef] [PubMed]

71. Martins, S.; Pearson, C.E.; Coutinho, P.; Provost, S.; Amorim, A.; Dube, M.P.; Sequeiros, J.; Rouleau, G.A. Modifiers of (CAG)(n) instability in Machado-Joseph disease (MJD/SCA3) transmissions: an association study with DNA replication, repair and recombination genes. *Hum. Genet.* **2014**, *133*, 1311–1318. [CrossRef] [PubMed]

72. Morales, F.; Vasquez, M.; Santamaria, C.; Cuenca, P.; Corrales, E.; Monckton, D.G. A polymorphism in the *MSH3* mismatch repair gene is associated with the levels of somatic instability of the expanded CTG repeat in the blood DNA of myotonic dystrophy type 1 patients. *DNA Repair (Amst.)* **2016**, *40*, 57–66. [CrossRef] [PubMed]

73. Entezam, A.; Usdin, K. ATR protects the genome against CGG.CCG-repeat expansion in fragile X premutation mice. *Nucleic Acids Res.* **2008**, *36*, 1050–1056. [CrossRef] [PubMed]

74. Entezam, A.; Usdin, K. ATM and ATR protect the genome against two different types of tandem repeat instability in fragile X premutation mice. *Nucleic Acids Res.* **2009**, *37*, 6371–6377. [CrossRef] [PubMed]

75. Wheeler, V.C.; Lebel, L.A.; Vrbanac, V.; Teed, A.; te Riele, H.; MacDonald, M.E. Mismatch repair gene *Msh2* modifies the timing of early disease in Hdh(Q111) striatum. *Hum. Mol. Genet.* **2003**, *12*, 273–281. [CrossRef] [PubMed]

76. Morales, F.; Couto, J.M.; Higham, C.F.; Hogg, G.; Cuenca, P.; Braida, C.; Wilson, R.H.; Adam, B.; del Valle, G.; Brian, R.; et al. Somatic instability of the expanded CTG triplet repeat in myotonic dystrophy type 1 is a heritable quantitative trait and modifier of disease severity. *Hum. Mol. Genet.* **2012**, *21*, 3558–3567. [CrossRef] [PubMed]

77. Budworth, H.; Harris, F.R.; Williams, P.; Lee, D.Y.; Holt, A.; Pahnke, J.; Szczesny, B.; Acevedo-Torres, K.; Ayala-Pena, S.; McMurray, C.T. Suppression of somatic expansion delays the onset of pathophysiology in a mouse model of Huntington's disease. *PLoS Genet.* **2015**, *11*, e1005267. [CrossRef] [PubMed]

Article

Comparative Behavioral Phenotypes of *Fmr1* KO, *Fxr2* Het, and *Fmr1* KO/*Fxr2* Het Mice

Rachel Michelle Saré, Christopher Figueroa, Abigail Lemons, Inna Loutaev and Carolyn Beebe Smith *

Section on Neuroadaptation and Protein Metabolism, National Institute of Mental Health, National Institutes of Health, Department of Health and Human Services, Bethesda, MD 20814, USA; Rachel.Sare@nih.gov (R.M.S.); Christopher.Figueroa@nih.gov (C.F.); Abigail.Lemons@nih.gov (A.L.); Loutaev.inna@nih.gov (I.L.)
* Correspondence: beebe@mail.nih.gov; Tel.: +1-301-402-3120

Received: 6 November 2018; Accepted: 10 January 2019; Published: 16 January 2019

Abstract: Fragile X syndrome (FXS) is caused by silencing of the *FMR1* gene leading to loss of the protein product fragile X mental retardation protein (FMRP). FXS is the most common monogenic cause of intellectual disability. There are two known mammalian paralogs of FMRP, FXR1P, and FXR2P. The functions of FXR1P and FXR2P and their possible roles in producing or modulating the phenotype observed in FXS are yet to be identified. Previous studies have revealed that mice lacking *Fxr2* display similar behavioral abnormalities as *Fmr1* knockout (KO) mice. In this study, we expand upon the behavioral phenotypes of *Fmr1* KO and $Fxr2^{+/-}$ (Het) mice and compare them with *Fmr1* KO/*Fxr2* Het mice. We find that *Fmr1* KO and *Fmr1* KO/*Fxr2* Het mice are similarly hyperactive compared to WT and *Fxr2* Het mice. *Fmr1* KO/*Fxr2* Het mice have more severe learning and memory impairments than *Fmr1* KO mice. *Fmr1* KO mice display significantly impaired social behaviors compared to WT mice, which are paradoxically reversed in *Fmr1* KO/*Fxr2* Het mice. These results highlight the important functional consequences of loss or reduction of FMRP and FXR2P.

Keywords: Fragile X; FMRP; *Fxr2*; *Fmr1*

---

1. Introduction

Fragile X syndrome (FXS) is the leading heritable cause of intellectual disability in humans, affecting about 1 in 4000 males [1]. In addition to intellectual disability, FXS patients often display multiple behavioral phenotypes including hyperactivity, attention deficits, susceptibility to seizures, hypersensitivity, sleep abnormalities, and social anxiety/autism-like behaviors [2]. FXS is primarily caused by a CGG repeat expansion in the 5′UTR of *FMR1* which leads to gene silencing and the consequent loss of its protein product, FMRP [3]. FMRP is an RNA-binding protein with over 800 mRNA targets [4]. FMRP is highly expressed in the brain, and is thought to act as a translational suppressor [5]. The loss of translational regulation by FMRP has been shown to lead to excessive brain protein synthesis in *Fmr1* knockout (KO) mice, a mouse model of FXS [6]. In addition to its presumed role in mRNA regulation, FMRP is also thought to be involved in nuclear export and cytoplasmic transport [7], ion channel activity [8], and participates in the DNA damage response [9].

Fmr1 KO mice exhibit many of the behavioral symptoms seen in humans with FXS, including hyperactivity, deficits in learning and memory, reduced preference for social novelty, repetitive behaviors, and reduced sleep [10–12]. In mammals, there are two autosomal paralogs of FMRP, FXR1P and FXR2P, which together comprise the fragile-X related (FXR) family of proteins [13]. FXR1P and FXR2P share a conserved structure and amino acid sequence (86% and 70% respective conservation in the N-terminus and central regions) with FMRP; this conservation includes the presumed RNA-binding sites [13]. Due to sequence homology and similar expression patterns, these proteins are hypothesized to have similar functions [14]. Therefore, it is thought that these proteins can, at least partially, compensate for loss of FMRP, thereby playing an as yet undiscovered role in FXS [15]. The FXR family of genes has been investigated in a clinical setting. The *FXR2* gene is located on chromosome 17. Microdeletions in the 17p13.1 comprise a syndrome that is often associated with dysmorphic features and developmental delay [16,17]. Given that this microdeletion syndrome is accompanied by the loss of multiple genes, it is hard to identify the specific role of *FXR2*. Additionally, it has been suggested that accumulation of single nucleotide polymorphisms in the fragile x gene family (including *FMR1*, *FXR1*, and *FXR2*) are associated with autistic phenotypes [18].

To further study the roles of the FXR mutations, *Fxr1* and *Fxr2* KO mice have been generated [15,19]. *Fxr1* KO is lethal shortly after birth [19]. In addition to brain expression, FXR1P is highly expressed in cardiac and skeletal muscle (more so than FMRP or FXR2P) and is likely involved in RNA translation/transport regulation in muscle; it is this role that may lead to early lethality in knockout models [19]. Like FMRP, FXR2P is expressed in brain with a similar regional distribution and mice deficient in FXR2P (*Fxr2* KO) exhibit some of the same behavioral deficits seen in *Fmr1* KO mice such as hyperactivity and learning and memory impairments [15].

Fmr1/Fxr2 double KOs animals have been shown to have a greater enhancement in metabotropic glutamate receptor activated long-term depression than *Fmr1* or *Fxr2* KO mice [20]. Behaviorally, *Fmr1/Fxr2* double KOs exhibited increased hyperactivity and greater impairments in contextual fear conditioning compared to single *Fmr1* or *Fxr2* KO mice [21]. Importantly, it was noted that *Fmr1/Fxr2* double KO mice had reduced survival rates and were often runted in size [21], so the adult animals studied were a select population of survivors with likely higher functioning than the total population. To get a better idea of the overlapping and novel functions of FMRP and FXR2P, we chose to perform studies in *Fmr1* KO/ *Fxr2* Het mice to ensure a non-biased sample. Both *Fmr1* KO/*Fxr2* KO and *Fmr1* KO/*Fxr2* Het have been shown to have similar impairments in circadian rhythm compared to *Fmr1* KO mice [22], indicating that, in this modality, *Fxr2* haploinsufficiency is sufficient to induce an impairment in the context of *Fmr1* deletion. Furthermore, *Fmr1* KO/*Fxr2* Het mice had a further decrease in sleep than *Fmr1* KO animals [11].

In this manuscript, we present behavioral studies of *Fmr1* KO, *Fxr2* Het, and *Fmr1* KO/*Fxr2* Het mice. We hypothesized that *Fmr1* KO/*Fxr2* Het mice would have an exaggerated behavioral phenotype compared to either single mutation. We found that both *Fmr1* KO and *Fmr1* KO/*Fxr2* Het mice are similarly hyperactive compared to wild-type (WT) and *Fxr2* Het mice. Additionally, *Fmr1* KO mice have statistically significantly impaired social behavior compared with WT mice, which is paradoxically reversed in *Fmr1* KO/*Fxr2* Het mice.

2. Materials and Methods

2.1. Animals

These studies were conducted on male WT, *Fmr1* KO, *Fxr2* Het, and *Fmr1* KO/*Fxr2* Het mice on a C57BL/6J background maintained in house. The original breeders were obtained from David Nelson at Baylor [22] and genotyped as previously described [11]. Breeders were either male *Fxr2* Het and female *Fmr1* Het mice or male WT mice and female *Fmr1* Het/*Fxr2* Het mice. Mice were group housed in a central climate-controlled facility with a standard 12:12 light:dark (lights on at 6:00 a.m.) facility. Food and water were available ad libitum. All procedures were carried out in accordance with the National Institutes of Health Guidelines on the Care and Use of Animals and approved by the National Institute of Mental Health Animal Care and Use Committee.

2.2. Behavior Testing

Mice were subjected to a battery of behavior tests starting between 63–77 days of age. Behavior tests were conducted from the least stressful to the most stressful in the following order: open field, novel object recognition, zero maze, marble burying, social behavior, and passive avoidance. No more than two tests were conducted in the same week with at least two days between tests. Our initial cohort of animals consisted of WT ($n = 33$), *Fmr1* KO ($n = 26$), *Fxr2* Het ($n = 22$), and *Fmr1* KO/ *Fxr2* Het ($n = 24$).

2.3. Open Field

Open field testing was done to assay activity and anxiety-like behaviors in response to a novel environment. Mice were placed in the center of a plexiglass open field arena (Coulbourn Instruments, Holliston, MA, USA) and allowed to explore for 30 min. Total distance traveled as well as distance traveled in the center of the chamber were recorded by TruScan software (Coulbourn Instruments) in five-minute epochs. Anxiety-like behavior was assessed by the ratio of distance traveled in the center to the total distance traveled; the center to total distance ratio is inversely proportional to anxiety levels. For open field testing, nine WT mice were not included (one due to missing data, three due to equipment malfunction, and five were statistical outliers); eight *Fmr1* KO mice were not included (one due to missing data, three due to equipment malfunction, and four were statistical outliers); eight *Fxr2* Het mice were not included (three due to equipment malfunction and six were statistical outliers); and ten *Fmr1* KO/ *Fxr2* Het mice were not included (two due to missing data, three due to equipment malfunction, and five were statistical outliers.

2.4. Novel Object Recognition (NOR)

NOR was performed to assay learning and memory capability. Testing was performed in the open field arena (Coulbourn Instruments). We used open field testing as the habituation phase for the NOR on Day 1. On Days 2 and 3, two identical objects were placed in the arena and the mouse was allowed to explore the objects for 5 min. Any animal that showed a preference for one of the two identical objects on these training days was eliminated from the study. Any animal that did not spend more than 10 seconds sniffing objects was also eliminated. On Day 4, one of the objects was replaced by a novel object and the mouse was allowed to explore for 5 min. The behavior of the mice was recorded by a video camera for later analysis of sniffing behavior. A discrimination index was calculated as the (sniffing time of novel—sniffing time of familiar)/total sniffing time. For NOR, 14 WT mice were not included (one was not run due to scheduling conflicts and 13 did not sniff for enough time); seven *Fmr1* KO mice were not included (one was not run due to scheduling conflicts, four did not sniff for enough time, and two knocked over the objects); seven *Fxr2* Het mice were not included (six did not sniff for enough time and one knocked over the objects); and seven *Fmr1* KO/*Fxr2* Het mice were not included (five were not run because of scheduling conflicts, one demonstrated a side preference, and one did not sniff for enough time).

2.5. Zero Maze

Anxiety-like behavior was assayed by means of the zero maze. Mice were placed inside the closed portion of the maze facing toward the open portion. They were then allowed to explore the maze for 5 min and the total time spent in the open portion was recorded. A mouse was counted as being in the open/closed portion when both front paws had crossed the threshold. If the mouse fell off the maze during the test period, its results were disqualified from the analysis. For zero maze testing, five WT mice were not included (one due to missing data, two fell off, and two were statistical outliers); two *Fmr1* KO mice were not included (one due to missing data and one was a statistical outlier); 0 *Fxr2* Het mice were not included; and two *Fmr1* KO/*Fxr2* Het mice were not included due to missing data.

2.6. Marble Burying

Repetitive behaviors were assayed by means of the marble burying test. A standard cage with bedding at a depth of 4.5 cm was prepared and 20 marbles were evenly arranged in a 4×5 grid on top of the bedding. The test mouse was placed in the cage and allowed to explore for 30 min. The number of marbles buried (>50% covered) was recorded. For marble burying, four WT mice were not included (one due to missing data, one not properly set up, and two were outliers); three *Fmr1* KO mice were not included (one due to missing data, one was not tested due to scheduling issues, and one was a statistical outlier); one *Fxr2* Het was not included as it was a statistical outlier; and four *Fmr1* KO/*Fxr2* Het mice were not included (two due to missing data, one was not tested due to scheduling issues, and one was a statistical outlier).

2.7. Social Behavior

Sociability and preference for social novelty were tested by means of the three chamber social behavior apparatus [23]. The test mouse was placed in the center chamber of a three-chamber apparatus and allowed to explore all three chambers for a 5 min habituation period. If an animal spent more than 3 min in one chamber or did not explore a chamber during the habituation period, it was excluded. Following habituation, an empty holding enclosure (10 cm diameter) was introduced to one of the side chambers and another identical enclosure with a stranger mouse inside was introduced to the other side. The test mouse was allowed to explore for another 5 min. In the third part of the test, a novel stranger mouse was introduced to the previously empty enclosure. The test mouse was then allowed to explore for another 5 min. Time in each chamber was recorded automatically by photobeam breaks. Video recording was performed for later assessment of sniffing behavior which was assessed by means of an automated software (TopScan) (Clever Systems, Reston, VA, USA). Sniffing was defined as the animal being within 2.0 cm of the enclosure with his nose directed toward the enclosure. For social behavior, 12 WT mice were not included (one due to missing data, three because of a side preference during habituation, one was not run due to scheduling issues, and six were statistical outliers); seven *Fmr1* KO mice were not included (two due to missing data, one due to equipment malfunction, and three were statistical outliers); eight *Fxr2* Het mice were not included (one because of missing data, three because of a side preference during habituation and four were statistical outliers); eight *Fmr1* KO/*Fxr2* Het mice were not included (two due to missing data, three because of a side preference during habituation, and three were statistical outliers).

2.8. Passive Avoidance

The passive avoidance test was done to assess memory. The apparatus is a light/dark shuttle box with a shocker in the floor and a guillotine door (Coulbourn Instruments). The test was conducted over three consecutive days. On Day 1, habituation to the shuttle box, the test mouse was placed on the lighted side; after 30 s, the guillotine door opened to the dark chamber and the mouse was free to enter. Once the mouse entered the dark chamber, it was removed from the apparatus. Day 2 was the training day. The mouse was placed in the lighted chamber, and after 30 s the door opened to the dark chamber. Once the mouse entered the dark chamber, it received a mild foot shock (0.3 mA, 1 s). The mouse was then placed in a recovery cage for 2 min before being placed back in the lighted chamber with the guillotine door closed and the training repeated. On Day 3, the mouse was placed in the lighted chamber, and after 30 s the door opened to the dark chamber. The latency (maximum of 10 min) for the mouse to enter the dark chamber was recorded. If the animal did not enter the dark, the maximum value was assigned. For passive avoidance, 10 WT mice were not included (three were not run due to scheduling issues and seven due to equipment issues); six *Fmr1* KO were not included (three were not run due to scheduling issues and three due to equipment issues); six *Fxr2* Het were not included (three were not run due to scheduling issues and three due to equipment issues); seven *Fmr1* KO/*Fxr2* Het were not included (two were not run due to scheduling issues and five due to equipment issues).

2.9. Protein Expression

Adult male *Fmr1* KO and *Fmr1* KO/*Fxr2* Het mice were decapitated and brains were rapidly removed and the hippocampus dissected. Tissue was placed into Precellys lysis kits (Bertin Corportation, Rockville, MD, USA) and stored at −80 °C until further processing. Protein was extracted using the Precellys Homogenizer as previously described [24] and 15 µg of protein was loaded per lane. We used the Bio-Rad mini-protein stain-free gel technology for Western blotting as previously described [24].

We chose the stain free technology in order to normalize to total protein loaded rather than a standard housekeeping gene. In disorders like fragile X syndrome, where protein translation is thought to be affected, normalizing to a housekeeping protein might alter the results. The blot was incubated in primary antibody solution (1:1000 FXR2 (A303-894A) (Bethyl Laboratories, Montgomery, TX, USA)) overnight at 4 °C.

2.10. Statistical Analysis

Data reported are the means ± standard error of the mean (SEM). Data for a given behavioral test were excluded if one data point from that mouse was more than two standard deviations away from the mean. Zero maze, marble burying, and passive avoidance were analyzed by a one-way ANOVA with genotype as the variable. Open field (epoch) and social behavior (chamber) were analyzed by repeated measures ANOVA with the corresponding repeated measure as a within subjects' variable. When appropriate, post-hoc *t*-tests were performed with Bonferroni correction. The results of all ANOVA tests are presented in Table 1. We considered tests with $p \leq 0.05$ statistically significant. In the figures, these effects are denoted with a "*". We also indicate effects that are approaching statistical significance $0.05 \leq p \leq 0.10$ with a "~".

Table 1. Results of repeated measures ANOVA for behavior testing with corresponding F-values and *p*-values. *, $p < 0.05$. ~, $0.05 \leq p \leq 0.10$.

ANOVA Results				
Behavior	**Interaction**	**Main Effect**	$F_{(df,\text{error})}$ **Value**	*p*-**Value**
Open Field				
Distance	Genotype × epoch		$F_{(14,300)} = 0.796$	0.671
		Genotype	$F_{(3,66)} = 7.477$	<0.001 *
		Epoch	$F_{(5,300)} = 105.254$	<0.001 *
Center/Total Distance	Genotype × epoch		$F_{(13,279)} = 2.390$	0.005 *
		Genotype	$F_{(3,66)} = 13.296$	<0.001 *
		Epoch	$F_{(4,279)} = 2.034$	0.074 ~
Zero Maze		Genotype	$F_{(3,92)} = 2.283$	0.084 ~
Marble Burying		Genotype	$F_{(3,89)} = 2.675$	0.052 ~
Sociability				
Time in Chamber	Genotype × chamber		$F_{(3,69)} = 0.252$	0.860
		Genotype	$F_{(3,69)} = 0.236$	0.871
		Chamber	$F_{(1,69)} = 47.857$	<0.001 *
Sniffing Time	Genotype × chamber		$F_{(3,67)} = 0.026$	0.994
		Genotype	$F_{(3,67)} = 1.186$	0.322
		Chamber	$F_{(1,67)} = 53.519$	<0.001 *
Social Novelty				
Time in Chamber	Genotype × chamber		$F_{(3,69)} = 0.427$	0.734
		Genotype	$F_{(3,69)} = 1.981$	0.125
		Chamber	$F_{(1,69)} = 0.206$	0.651
Sniffing Time	Genotype × chamber		$F_{(3,67)} = 2.871$	0.043 *
		Genotype	$F_{(3,67)} = 1.516$	0.218
		Chamber	$F_{(1,67)} = 6.344$	0.014 *
NOR		Genotype	$F_{(3,70)} = 0.519$	0.671
Passive Avoidance		Genotype	$F_{(3,72)} = 3.421$	0.022 *

3. Results

3.1. Expression of FXR2

We measured FXR2 protein expression in *Fmr1* KO and *Fmr1* KO/*Fxr2* Het mice. FXR2 protein expression was reduced by 60% in Fxr2 heterozygous animals compared to controls ($p = 0.0001$, student's *t*-test) (Figure 1). Protein expression was normalized to total protein loaded.

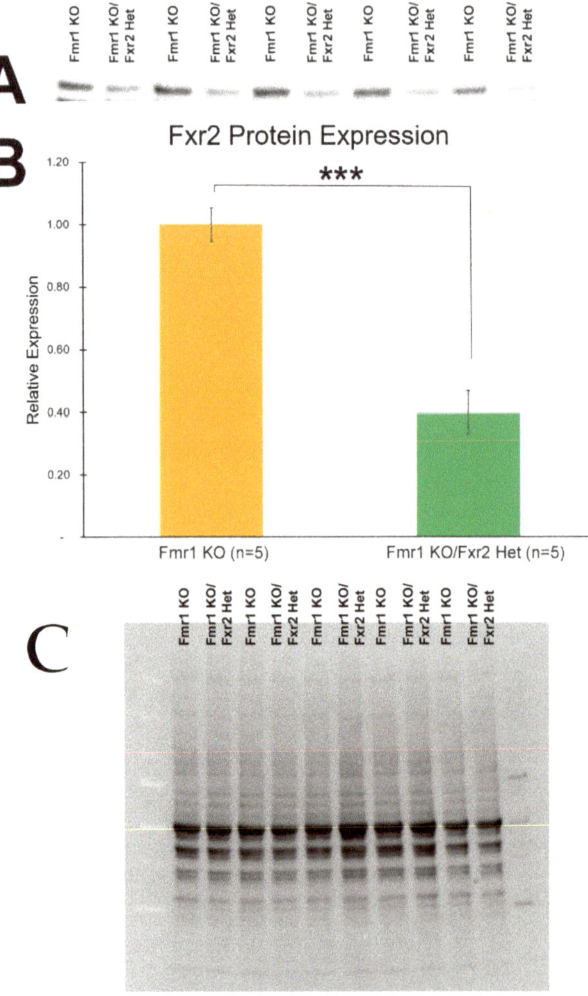

Figure 1. FXR2 protein expression is reduced in *Fxr2* heterozygous mice. (**A**) Western blots showing FXR2 protein expression in *Fmr1* KO and *Fmr1* KO/*Fxr2* Het hippocampus. (**B**) FXR2 protein expression is reduced by 60% in *Fmr1* KO/*Fxr2* Het animals (*** $p = 0.0001$, student's *t*-test). Each bar represents the mean ± SEM for the number of mice indicated on the figure. (**C**) Stain-free image of total protein loaded for FXR2 protein expression. FXR2 protein expression (seen in Figure 1) was normalized to the total protein loaded shown here. The image was acquired by UV Trans illumination, exposed for 2 s.

3.2. Activity in the Open Field

We measured activity in response to a novel environment by analyzing distance traveled in the open-field test. There was no genotype × epoch interaction indicating that, regardless of genotype, all mice showed a typical burst of activity in the beginning of exposure to the open field followed by adaptation to the environment (measured by decreased activity with time) (Figure 2). We found a statistically significant main effect of genotype ($p < 0.001$) (Table 1). Both *Fmr1* KO ($p = 0.017$) and *Fmr1* KO/*Fxr2* Het mice ($p < 0.001$) were statistically significantly hyperactive compared to WT mice (Figure 2). In addition, *Fmr1* KO/*Fxr2* Het mice were more active compared to *Fxr2* Hets ($p = 0.025$). WT and *Fxr2* Het mice showed similar levels of activity (Figure 2).

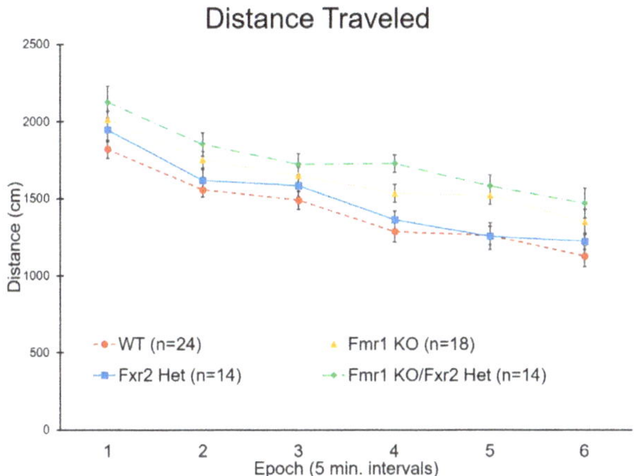

Figure 2. Distance traveled in the open field across the 30 min testing period. We found a statistically significant main effect of genotype ($p < 0.001$). Both *Fmr1* KO and *Fmr1* KO/*Fxr2* Het mice were statistically significantly hyperactive compared to WT ($p = 0.017$ and $p < 0.001$, respectively). Additionally, *Fmr1* KO/*Fxr2* Het mice were more active than *Fxr2* Het mice ($p = 0.025$). Each point represents the mean ± SEM for the number of mice indicated on the figure.

3.3. Anxiety-Like Behavior

The ratio of distance traveled in the center to total distance traveled was analyzed as an inverse measure of anxiety-like behavior. We found a statistically significant ($p = 0.005$) genotype × epoch interaction (Table 1). Post-hoc tests are shown in Table 2. In general, both *Fmr1* KO and *Fmr1* KO/*Fxr2* Het mice traveled more relative distance in the center compared to WT indicating lower anxiety. *Fxr2* Het mice had similar distance traveled in the center compared to WT mice (Figure 3A). These results suggest that both *Fmr1* KO and *Fmr1* KO/*Fxr2* Het mice are statistically significantly less anxious than either WT or *Fxr2* Hets.

Table 2. Results of post-hoc *t*-tests, Bonferroni corrected, following statistically significant genotype × epoch interaction in the ratio of center distance to total distance in the open field. *, $p < 0.05$. ~, $0.05 \leq p \leq 0.10$.

	p-Values for Post Hoc Pairwise Comparisons—Center: Total Distance Ratio Open Field					
Epoch	**WT × *Fmr1***	**WT × *Fxr2***	**WT × *Fmr1*/*Fxr2***	***Fmr1* × *Fxr2***	***Fmr1* × *Fmr1*/*Fxr2***	***Fxr2* × *Fmr1*/*Fxr2***
1	0.973	1.000	0.085 ~	1.000	1.000	1.000
2	0.036 *	1.000	<0.001 *	0.312	0.567	0.006 *
3	0.261	1.000	<0.001 *	1.000	0.031 *	0.003 *
4	0.014 *	1.000	0.001 *	0.710	1.000	0.128
5	0.001 *	1.000	0.003 *	0.006 *	1.000	0.010 *
6	<0.001 *	1.000	<0.001 *	0.005 *	1.000	0.005 *

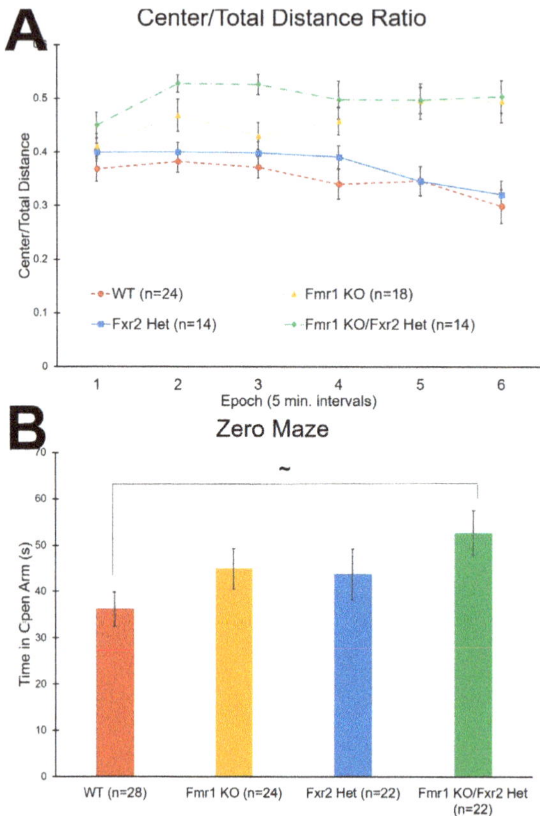

Figure 3. Tests for anxiety-like behavior. (**A**) In the open field test, the ratio of center distance to total distance traveled is an inverse measure of anxiety-like behavior. The genotype × epoch interaction was statistically significant ($p = 0.005$). Overall, *Fmr1* KO/*Fxr2* Het mice traveled more relative distance in the center than WT mice throughout the test. Similarly, *Fmr1* KO also traveled more relative distance in the center than WT mice in almost every epoch of the test (Table 2). *Fmr1* KO (Epochs 5 and 6) and *Fmr1* KO/*Fxr2* Het mice (Epochs 2, 3, 5, and 6) also traveled more relative distance in the center than *Fxr2* Het mice. In epoch 3, *Fmr1* KO/*Fxr2* Het mice traveled more relative distance in the center than *Fmr1* KO mice. Each point represents the mean ± SEM for the number of mice indicated on the figure. (**B**) In the zero maze, the time spent in the open portions of the maze is an inverse measure of anxiety-like behavior. The effect of genotype approached statistical significance ($p = 0.084$) so we proceeded with post-hoc t-tests. *Fmr1* KO/*Fxr2* Het mice tended to spend more time in the open portions than WT mice ($p = 0.065$). Each bar represents the mean ± SEM for the number of mice indicated on the figure. ~, $0.05 \leq p \leq 0.10$.

We also used the time spent in the open portion of the zero maze as another inverse measure of anxiety-like behavior. We found that the main effect of genotype approached statistical significance ($p = 0.084$). Post-hoc t-tests showed a trend for the *Fmr1* KO/*Fxr2* Het mice to spend more time in the open portion than WT mice ($p = 0.065$) (Figure 3B). These results are consistent with a reduced anxiety phenotype also demonstrated in the open field in *Fmr1* KO/*Fxr2* Het mice.

3.4. Repetitive Behavior

The number of marbles buried during this 30 min test is considered a measure of repetitive behavior. We found a near statistically significant effect of genotype ($p = 0.052$) (Table 1). Post-hoc comparisons revealed trends with *Fmr1* KO mice burying more marbles than both WT ($p = 0.095$) and *Fmr1 KO/Fxr2* Hets ($p = 0.098$) (Figure 4). These trends suggest that, of these four genotypes, only *Fmr1* KO mice show elevated repetitive behavior.

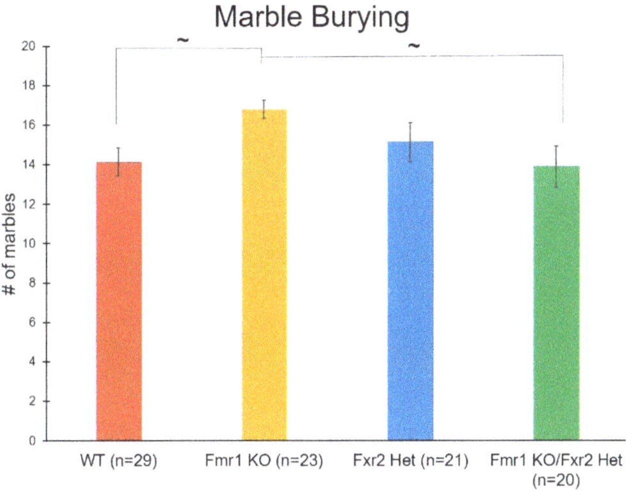

Figure 4. The number of marbles buried is a measure of repetitive behavior. We found a near statistically significant effect of genotype ($p = 0.052$). Post-hoc tests revealed that *Fmr1* knockout (KO) mice tended to bury more marbles than wild-type (WT) ($p = 0.095$) and *Fmr1 KO/Fxr2* Het ($p = 0.098$) mice. Each bar represents the mean ± SEM for the number of mice indicated on the figure. ~, $0.05 \leq p \leq 0.10$.

3.5. Social Behavior

We tested social behavior by means of the three-chambered apparatus. In this test, we measured both time in chamber and time sniffing either the enclosure or the mouse. In the first phase of the task, the sociability phase, all genotypes showed a preference for the stranger mouse compared to the object with respect to both the time in chamber (Figure 5A) and sniffing time (Figure 5B) (Table 1).

For the second phase of the task, the preference for social novelty, all four genotypes spent about the same amount of time in the two chambers showing no preference for either the novel or familiar mouse (Figure 5C) (Table 1). The time sniffing either the familiar or novel mouse differentiated the genotypes (Figure 5D). The genotype × chamber interaction was statistically significant ($p = 0.043$) (Table 1), and post-hoc t-tests showed that *Fmr1 KO/Fxr2* Het ($p = 0.006$) mice showed a clear preference for the novel mouse compared with the familiar mouse. This phenotype was also seen in WT, but to a lesser degree ($p = 0.062$). Sniffing times for the novel and familiar mice in *Fmr1* KO and *Fxr2* Het mice were not significantly different, suggesting no preference for social novelty in these genotypes (Figure 5D).

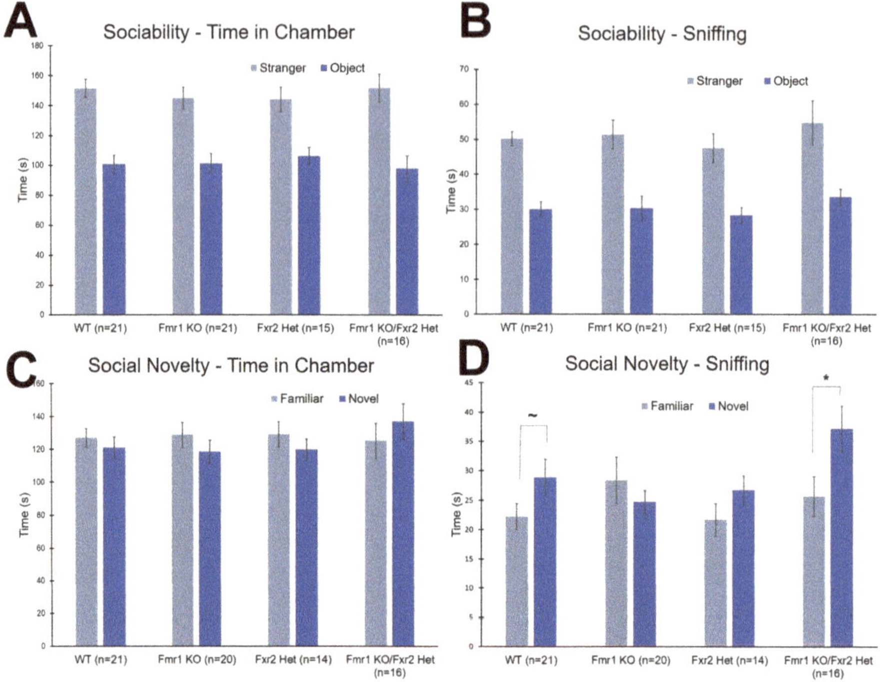

Figure 5. The three-chambered apparatus was used to assess social behavior. (**A**) In the sociability phase, there were no effects of genotype on time in chamber. All mice spent more time in the chamber with the stranger mouse compared to the chamber with the object. (**B**) In the sociability, phase, there were also no effects of genotype on time spent sniffing either the object or the stranger mouse. All mice sniffed the stranger mouse statistically significantly more than the object. (**C**) In the preference for social novelty phase, there were no effects of genotype on time in chamber. None of the four genotypes showed a preference for either the chamber with the novel mouse or the chamber with the familiar mouse. (**D**) In the preference for social novelty phase, there was a statistically significant genotype x chamber interaction for sniffing time ($p = 0.043$). WT and *Fmr1* KO/*Fxr2* Het mice spent more time sniffing the novel mouse compared with the familiar mouse ($p = 0.062$, and $p = 0.006$, respectively), whereas *Fmr1* KO and *Fxr2* Het mice did not. Each bar represents the mean ± SEM for the number of mice indicated on the figure. *, $p < 0.05$. ~, $0.05 \leq p \leq 0.10$.

3.6. Learning and Memory

To assay learning and memory, we performed NOR. Our data had high variability and we did not see any statistically significant effects (Table 1, Figure 6).

For learning and memory, we also performed passive avoidance testing. With this test, we found a statistically significant main effect of genotype ($p = 0.022$) (Table 1). Post-hoc *t*-tests indicate that *Fmr1* KO/*Fxr2* Het mice had statistically significantly ($p = 0.014$) shorter latencies to enter the dark than WT mice. This result suggests that learning and memory is more compromised in mice with the double mutation (Figure 7).

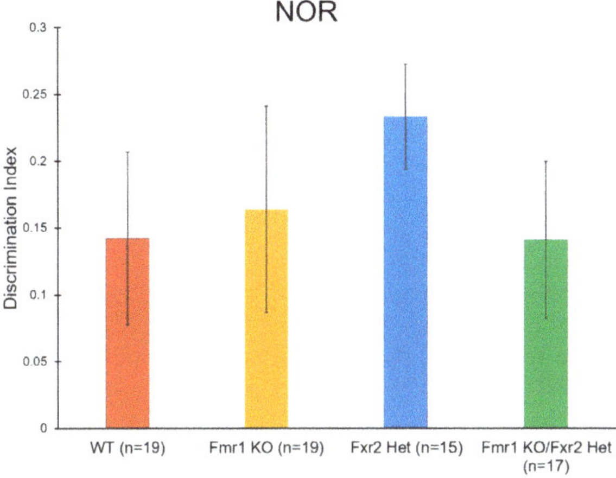

Figure 6. To assay learning and memory, we performed novel object recognition (NOR). We did not find any statistically significant effects. Bars represent the means ± SEMs for the number of mice indicated in parentheses.

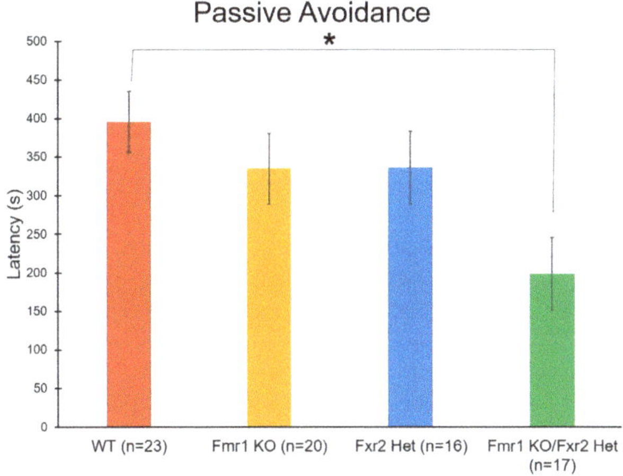

Figure 7. To assay learning and memory, we performed the passive avoidance test. We found a statistically significant effect of genotype on latency to enter the dark chamber ($p = 0.022$). Post-hoc tests revealed that *Fmr1* KO/*Fxr2* Het mice had statistically significantly shorter latencies to enter the dark than WT mice ($p = 0.014$), indicating a deficit in memory. Each bar represents the mean ± SEM for the number of mice indicated on the figure. *, $p < 0.05$.

4. Discussion

Here, we present behavioral similarities and differences between WT, *Fmr1* KO, *Fxr2* Het, and *Fmr1* KO/*Fxr2* Het mice. Our data suggest that, in the context of *Fmr1* deletion, deleting one copy of *Fxr2* does not have a detrimental effect on activity and therefore one copy may be sufficient for maintaining this behavior. With respect to learning and memory, deleting both *Fmr1* and *Fxr2* has a more severe effect than either single mutation suggesting that one protein can compensate for loss of the other. Finally, paradoxically, these proteins seem to have opposite roles in social behavior. Deleting both *Fmr1* and one copy of *Fxr2* results in improvements in behavior compared to single *Fmr1* deletion.

These data highlight possible differing roles of these proteins depending on the behavior examined. Our results add to our understanding of the functions of fragile X related proteins.

Fmr1 KO/*Fxr2* KO mice have been previously described to have worse learning and memory (measured by fear conditioning) than either of the single mutations alone [21]. In passive avoidance, we find that *Fmr1* KO/*Fxr2* Het mice have more profound learning and memory impairments than do the other genotypes studied. This behavior may be a reflection of either learning and memory or of impulsivity [25] and impulsivity is also known to be affected in *Fmr1* KO mice [26]. We did not find deficiencies in the passive avoidance test in single *Fmr1* KO mice, which contrasts with several published studies [12]. Importantly, whereas we do not see statistically significant genotype differences between WT and *Fmr1* KO mice, the mean values do reflect the expected trend that *Fmr1* KO mice have shorter latencies to enter the dark suggesting impaired memory. The differences between this and previous studies might have more to do with increased variability in the current results. During the course of this study, we had to replace our passive avoidance equipment. Although our new equipment was from the same company (Coulbourn Instruments), it is possible that changing the system affected the results. Another possible caveat is that we capped the latency to enter the dark compartment at 10 min. A large number of animals timed out of the study and were assigned the maximum value of 10 min (43.5% WT, 12.5% *Fxr2* Het, 35% *Fmr1* KO, 5.9% *Fmr1* KO/*Fxr2* Het). If the threshold were higher, it might be possible to detect more genotype differences.

One important note to the behavior testing conducted in this study is the fact that two different experimenters performed the testing, one male and one female. Prior studies have shown that the sex of the experimenter can have effects on anxiety behavior in WT mice. The mice respond with increased anxiety in the presence of a male experimenter [27], and effects on anxiety may have affected other behavioral measures. It is not known how *Fmr1* KO and *Fmr1* KO/*Fxr2* Het mice may respond to this difference, but we note that in our study, the male experimenter tested more *Fmr1* KO/*Fxr2* Het mice than the female experimenter. Whereas this may have biased our results, the fact that *Fmr1* KO/*Fxr2* Het mice demonstrated the least anxiety-related behavior suggests that the direction of the bias would be to underestimate the genotype difference.

Previously published results suggested that anxiety-like behavior was the same between the *Fmr1* KO and *Fmr1* KO/*Fxr2* KO mice based on the open field test and the light:dark box [21]. Similarly, our open field results suggest that *Fmr1* and *Fmr1* KO/*Fxr2* Het mice have similar reductions in anxiety compared to WT mice. However, another measure of anxiety-like behavior, the zero maze, showed a statistical trend indicating that *Fmr1* KO/*Fxr2* Het mice had even lower anxiety than *Fmr1* KO mice. It should be noted that, although studies have reported a phenotype in *Fmr1* KO mice on the zero maze [12], we did not detect any statistically significant differences between WT and *Fmr1* KO mice in the current study.

We also assessed repetitive behaviors and social behaviors, two behaviors that are relevant to mouse models of autism. This is important given that autism is reported to be present in about 50% of patients with fragile X syndrome [28]. In agreement with previous studies [12], we found that *Fmr1* KO mice had a trend toward increased repetitive behaviors and impaired preference for social novelty. Paradoxically, these phenotypes were reversed in *Fmr1* KO/*Fxr2* Het mice, suggesting contrasting functions between these two proteins in repetitive and social behaviors.

Although FMRP and FXR2P have similar regional distributions in brain, the results of our studies indicate that the functional roles of these proteins may differ. Other studies have also indicated functional differences between FMRP and FXR2P. For example, at the synapse, the fragile X related proteins form granules with ribonucleoprotein particles [29]. FXR2P is always expressed in these granules, but FMRP is only co-expressed in the granules in certain brain regions (the granules in much of the brain stem and cerebellum do not contain FMRP) [30]. For the most part, FMRP and FXR2P seem to have overlapping, cooperative roles in regulating metabolism. It had been shown previously that the phenotype of *Fmr1* KO/*Fxr2* Het mice was not as severe as that of *Fmr1* KO/*Fxr2* KO mice [31]. To fully assess the potentially compensatory functions of these proteins for each other, it would be best to study full *Fxr2* KOs with and without the deletion of *Fmr1*. However, we chose to study only the *Fxr2* Het because we know from prior studies that these animals have a high mortality and we were concerned that we would be studying a selected population of survivors.

5. Conclusions

Overall, our data point to important functions of *Fmr1* and *Fxr2* in behavior. Based on these results, there are some domains (like learning and memory) where *Fmr1* and *Fxr2* may have overlapping functions and can partially compensate for loss of the other. However, there are other domains (like social behavior) in which they appear to have different, even opposite roles in behavior.

Author Contributions: R.M.S. conceived and designed the experiments. C.F., A.L., and I.L. performed the experiments, R.M.S., C.F., A.L., and C.B.S. analyzed the data, R.M.S., C.F., and C.B.S. wrote the paper.

Acknowledgments: R.M.S. was supported by a postdoctoral fellowship from FRAXA. This work was also supported by the Intramural research program of the National Institute of Mental Health (ZIA MH000889). The authors would like to thank Zengyan Xia for help in genotyping the mice. The authors would also like to acknowledge Merlin Levine, Alex Song, Isabella Maita, Anna Cook, and Lee Harkless for their role in preliminary experiments. The authors would like to acknowledge the editorial assistance of the Fellows Editorial Board.

Conflicts of Interest: The authors declare no conflict of interest.

References

1. Turner, G.; Webb, T.; Wake, S.; Robinson, H. Prevalence of fragile X syndrome. *Am. J. Med. Genet.* **1996**, *64*, 196–197. [CrossRef]
2. Garber, K.B.; Visootsak, J.; Warren, S.T. Fragile X syndrome. *Eur. J. Hum. Genet. EJHG* **2008**, *16*, 666–672. [CrossRef]
3. Verheij, C.; Bakker, C.E.; de Graaff, E.; Keulemans, J.; Willemsen, R.; Verkerk, A.J.; Galjaard, H.; Reuser, A.J.; Hoogeveen, A.T.; Oostra, B.A. Characterization and localization of the *FMR-1* gene product associated with fragile X syndrome. *Nature* **1993**, *363*, 722–724. [CrossRef]
4. Darnell, J.C.; Van Driesche, S.J.; Zhang, C.; Hung, K.Y.; Mele, A.; Fraser, C.E.; Stone, E.F.; Chen, C.; Fak, J.J.; Chi, S.W.; et al. Fmrp stalls ribosomal translocation on mrnas linked to synaptic function and autism. *Cell* **2011**, *146*, 247–261. [CrossRef] [PubMed]
5. Darnell, J.C.; Klann, E. The translation of translational control by *Fmrp*: Therapeutic targets for fxs. *Nat. Neurosci.* **2013**, *16*, 1530–1536. [CrossRef]
6. Qin, M.; Kang, J.; Burlin, T.V.; Jiang, C.; Smith, C.B. Postadolescent changes in regional cerebral protein synthesis: An in vivo study in the *Fmr1* null mouse. *J. Neurosci. Off. J. Soc. Neurosci.* **2005**, *25*, 5087–5095. [CrossRef] [PubMed]
7. Bardoni, B.; Schenck, A.; Mandel, J.L. The fragile X mental retardation protein. *Brain Res. Bull.* **2001**, *56*, 375–382. [CrossRef]
8. Ferron, L. Fragile x mental retardation protein controls ion channel expression and activity. *J. Physiol.* **2016**, *594*, 5861–5867. [CrossRef]
9. Liu, W.; Jiang, F.; Bi, X.; Zhang, Y.Q. Drosophila fmrp participates in the DNA damage response by regulating g2/m cell cycle checkpoint and apoptosis. *Hum. Mol. Genet.* **2012**, *21*, 4655–4668. [CrossRef] [PubMed]
10. Bakker, C.E.; Verheij, C.; Willemsen, R.; van der Helm, R.; Oerlemans, F.; Vermey, M.; Bygrave, A.; Hoogeveen, A.; Oostra, B.A.; Reyniers, E.; et al. *Fmr1* knockout mice: A model to study fragile X mental retardation. The dutch-belgian fragile X consortium. *Cell* **1994**, *78*, 23–33.
11. Sare, R.M.; Harkless, L.; Levine, M.; Torossian, A.; Sheeler, C.A.; Smith, C.B. Deficient sleep in mouse models of fragile X syndrome. *Front. Mol. Neurosci.* **2017**, *10*, 280. [CrossRef]
12. Kazdoba, T.M.; Leach, P.T.; Silverman, J.L.; Crawley, J.N. Modeling fragile X syndrome in the *Fmr1* knockout mouse. *Intractable Rare Dis. Res.* **2014**, *3*, 118–133. [CrossRef] [PubMed]
13. Zhang, Y.; O'Connor, J.P.; Siomi, M.C.; Srinivasan, S.; Dutra, A.; Nussbaum, R.L.; Dreyfuss, G. The fragile X mental retardation syndrome protein interacts with novel homologs *FXR1* and *FXR2*. *EMBO J.* **1995**, *14*, 5358–5366. [CrossRef] [PubMed]

14. Bakker, C.E.; de Diego Otero, Y.; Bontekoe, C.; Raghoe, P.; Luteijn, T.; Hoogeveen, A.T.; Oostra, B.A.; Willemsen, R. Immunocytochemical and biochemical characterization of FMRP, FXR1P, and FXR2P in the mouse. *Exp. Cell Res.* **2000**, *258*, 162–170. [CrossRef] [PubMed]

15. Bontekoe, C.J.; McIlwain, K.L.; Nieuwenhuizen, I.M.; Yuva-Paylor, L.A.; Nellis, A.; Willemsen, R.; Fang, Z.; Kirkpatrick, L.; Bakker, C.E.; McAninch, R.; et al. Knockout mouse model for *Fxr2*: A model for mental retardation. *Hum. Mol. Genet.* **2002**, *11*, 487–498. [CrossRef] [PubMed]

16. Giordano, L.; Palestra, F.; Giuffrida, M.G.; Molinaro, A.; Iodice, A.; Bernardini, L.; La Boria, P.; Accorsi, P.; Novelli, A. 17p13.1 microdeletion: Genetic and clinical findings in a new patient with epilepsy and comparison with literature. *Am. J. Med. Genet. Part A* **2014**, *164A*, 225–230. [CrossRef] [PubMed]

17. Schluth-Bolard, C.; Sanlaville, D.; Labalme, A.; Till, M.; Morin, I.; Touraine, R.; Edery, P. 17p13.1 microdeletion involving the *TP53* gene in a boy presenting with mental retardation but no tumor. *Am. J. Med. Genet. Part A* **2010**, *152A*, 1278–1282. [CrossRef]

18. Stepniak, B.; Kastner, A.; Poggi, G.; Mitjans, M.; Begemann, M.; Hartmann, A.; van der Auwera, S.; Sananbenesi, F.; Krueger-Burg, D.; Matuszko, G.; et al. Accumulated common variants in the broader fragile X gene family modulate autistic phenotypes. *EMBO Mol. Med.* **2015**, *7*, 1565–1579. [CrossRef]

19. Mientjes, E.J.; Willemsen, R.; Kirkpatrick, L.L.; Nieuwenhuizen, I.M.; Hoogeveen-Westerveld, M.; Verweij, M.; Reis, S.; Bardoni, B.; Hoogeveen, A.T.; Oostra, B.A.; et al. *Fxr1* knockout mice show a striated muscle phenotype: Implications for *Fxr1p* function in vivo. *Hum. Mol. Genet.* **2004**, *13*, 1291–1302. [CrossRef]

20. Zhang, J.; Hou, L.; Klann, E.; Nelson, D.L. Altered hippocampal synaptic plasticity in the *Fmr1* gene family knockout mouse models. *J. Neurophysiol.* **2009**, *101*, 2572–2580. [CrossRef]

21. Spencer, C.M.; Serysheva, E.; Yuva-Paylor, L.A.; Oostra, B.A.; Nelson, D.L.; Paylor, R. Exaggerated behavioral phenotypes in *Fmr1/Fxr2* double knockout mice reveal a functional genetic interaction between fragile X-related proteins. *Hum. Mol. Genet.* **2006**, *15*, 1984–1994. [CrossRef]

22. Zhang, J.; Fang, Z.; Jud, C.; Vansteensel, M.J.; Kaasik, K.; Lee, C.C.; Albrecht, U.; Tamanini, F.; Meijer, J.H.; Oostra, B.A.; et al. Fragile X-related proteins regulate mammalian circadian behavioral rhythms. *Am. J. Hum. Genet.* **2008**, *83*, 43–52. [CrossRef] [PubMed]

23. Nadler, J.J.; Moy, S.S.; Dold, G.; Trang, D.; Simmons, N.; Perez, A.; Young, N.B.; Barbaro, R.P.; Piven, J.; Magnuson, T.R.; et al. Automated apparatus for quantitation of social approach behaviors in mice. *Genes Brain Behav.* **2004**, *3*, 303–314. [CrossRef]

24. Sare, R.M.; Song, A.; Loutaev, I.; Cook, A.; Maita, I.; Lemons, A.; Sheeler, C.; Smith, C.B. Negative effects of chronic rapamycin treatment on behavior in a mouse model of fragile X syndrome. *Front. Mol. Neurosci.* **2017**, *10*, 452. [CrossRef]

25. Corr, P.J. J.A. Gray's reinforcement sensitivity theory: Tests of the joint subsystems hypothesis of anxiety and impulsivity. *Personal. Individ. Differ.* **2002**, *33*, 511–532. [CrossRef]

26. Moon, J.; Beaudin, A.E.; Verosky, S.; Driscoll, L.L.; Weiskopf, M.; Levitsky, D.A.; Crnic, L.S.; Strupp, B.J. Attentional dysfunction, impulsivity, and resistance to change in a mouse model of fragile X syndrome. *Behav. Neurosci.* **2006**, *120*, 1367–1379. [CrossRef] [PubMed]

27. Sorge, R.E.; Martin, L.J.; Isbester, K.A.; Sotocinal, S.G.; Rosen, S.; Tuttle, A.H.; Wieskopf, J.S.; Acland, E.L.; Dokova, A.; Kadoura, B.; et al. Olfactory exposure to males, including men, causes stress and related analgesia in rodents. *Nature Methods* **2014**, *11*, 629–632. [CrossRef]

28. Hagerman, R.J.; Rivera, S.M.; Hagerman, P.J. The fragile X family of disorders: A model for autism and targeted treatments. *Curr. Pediatr. Rev.* **2008**, *4*, 40–52. [CrossRef]

29. Christie, S.B.; Akins, M.R.; Schwob, J.E.; Fallon, J.R. The fxg: A presynaptic fragile X granule expressed in a subset of developing brain circuits. *J. Neurosci. Off. J. Soc. Neurosci.* **2009**, *29*, 1514–1524. [CrossRef]

30. Chyung, E.; LeBlanc, H.F.; Fallon, J.R.; Akins, M.R. Fragile X granules are a family of axonal ribonucleoprotein particles with circuit-dependent protein composition and mrna cargos. *J. Comp. Neurol.* **2018**, *526*, 96–108. [CrossRef]

31. Lumaban, J.G.; Nelson, D.L. The fragile X proteins Fmrp and Fxr2p cooperate to regulate glucose metabolism in mice. *Hum. Mol. Genet.* **2015**, *24*, 2175–2184. [CrossRef] [PubMed]

Review

Molecular Biomarkers in Fragile X Syndrome

Marwa Zafarullah [1] and Flora Tassone [1,2,*

[1]   Department of Biochemistry and Molecular Medicine, University of California Davis, School of Medicine, Sacramento, CA 95817, USA; mzafarullah@ucdavis.edu

[2]   MIND Institute, University of California Davis Medical Center, Sacramento, CA 95817, USA

*   Correspondence: ftassone@ucdavis.edu; Tel.: +1-(916)-703-0463

Received: 7 March 2019; Accepted: 24 April 2019; Published: 27 April 2019

Abstract: Fragile X syndrome (FXS) is the most common inherited form of intellectual disability (ID) and a known monogenic cause of autism spectrum disorder (ASD). It is a trinucleotide repeat disorder, in which more than 200 CGG repeats in the 5′ untranslated region (UTR) of the fragile X mental retardation 1 (*FMR1*) gene causes methylation of the promoter with consequent silencing of the gene, ultimately leading to the loss of the encoded fragile X mental retardation 1 protein, FMRP. FMRP is an RNA binding protein that plays a primary role as a repressor of translation of various mRNAs, many of which are involved in the maintenance and development of neuronal synaptic function and plasticity. In addition to intellectual disability, patients with FXS face several behavioral challenges, including anxiety, hyperactivity, seizures, repetitive behavior, and problems with executive and language performance. Currently, there is no cure or approved medication for the treatment of the underlying causes of FXS, but in the past few years, our knowledge about the proteins and pathways that are dysregulated by the loss of FMRP has increased, leading to clinical trials and to the path of developing molecular biomarkers for identifying potential targets for therapies. In this paper, we review candidate molecular biomarkers that have been identified in preclinical studies in the FXS mouse animal model and are now under validation for human applications or have already made their way to clinical trials.

Keywords: fragile X syndrome; molecular biomarkers; *FMR1*; FMRP; intellectual disability; *Fmr1* KO mouse; ASD

---

1. Introduction

A biomarker is "a characteristic that is objectively measured and evaluated as an indicator of normal biological processes, pathogenic processes, or pharmacologic responses to a therapeutic intervention" [1]. Biomarkers can be found in blood, plasma, or other tissues and are generally viewed as a molecular signature able to identify individuals who are at high risk for a specific condition. They can also be detected before disease symptoms and therefore used to predict the occurrence of a condition or the nature and severity of disease outcomes in an individual. Importantly, they can be used to evaluate the efficacy of response to pharmacological intervention.

Fragile X syndrome (FXS) is the most prevalent inherited cause of intellectual disability and the single leading monogenic known cause of autism, as 60% of those with a full mutation present with autism spectrum disorder (ASD) [2]. The clinical symptoms include anxiety, impairment in cognitive, executive and language performance, hyperactivity, impulsivity, insomnia, seizures and physical features such as hypotonia, flat feet, hyperextensible joints, and macroorchidism [3]. FXS is caused by the abnormal expansion, greater than 200 units of a naturally occurring CGG repeat in the 5′ untranslated region (UTR) of the fragile X mental retardation 1 (*FMR1*) gene, located on the X chromosome. This expansion, named full mutation, results in hypermethylation and transcriptional silencing of the gene, leading to the loss or reduction of fragile X mental retardation 1 protein (FMRP) expression and to the diagnosis of fragile X syndrome [4–6]. Individuals carrying expansion of 55–200 CGG repeat are premutation carriers and at risk of developing the late-onset neurodegenerative syndrome, fragile X-associated tremor/ataxia syndrome (FXTAS), the fragile X-associated primary ovarian insufficiency (FXPOI) [7] and the fragile X-associated neuropsychiatric disorders (FXAND) [8].

FMRP is an RNA-binding protein and a translational regulator, whose function affects synaptic plasticity, spine morphology, and several cellular signaling pathways. Reduced expression of FMRP leads to the abnormalities in neurodevelopmental processes and the disturbed neuronal communications observed in FXS [9]. Young adults and adolescents with FXS show neuroanatomical abnormalities [10], and the regions of the brain that are significantly impacted by the loss of FMRP are the hippocampus (a structure that plays a critical role in the learning and memory and the regulation of mood and cognition [11]), the cerebellum, and the basal forebrain (nucleus basalis) [12]. Several studies in the *Fmr1* knockout (KO) mouse model suggest that FMRP plays a critical role during specific periods of cortical development with regional brain volume changes occurring in adult mouse brain [13,14]. Brain volume changes have also been observed in children with FXS, specifically in the temporal lobe, cerebellum, caudate nucleus, and amygdala regions of the brain [15,16].

FMRP function appears to be mostly inhibitory as it prevents the activity of various biochemical pathways in a "controlled" manner [17]. In a sense, reduced FMRP leads to exaggerated or reduced biochemical reactions that can adversely affect neural function. The past two decades of research have shown defects in the central excitatory glutamatergic and inhibitory GABAergic pathways and in several other neurotransmitter systems including serotonin and dopamine [18,19]. Thus, the development of molecular measures that reflect the impact of a drug on one or more of the FMRP-regulated pathways (Figure 1), including the activity or the expression level of proteins in the translational activation pathway and particularly of those regulated by FMRP, could potentially act as molecular biomarkers for FXS.

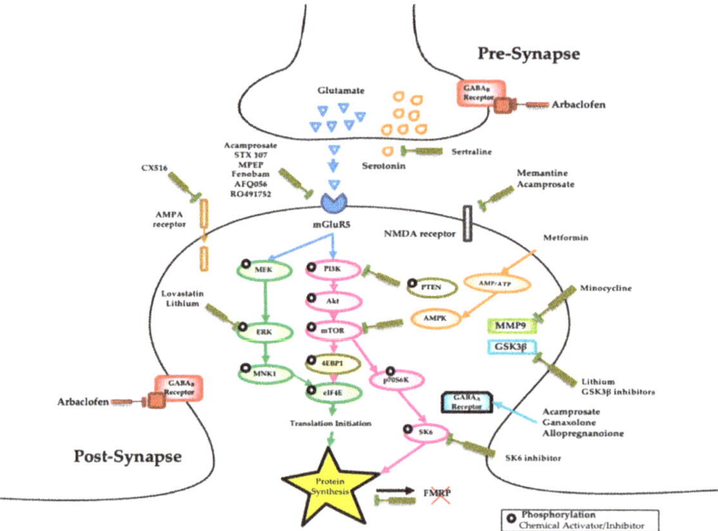

Figure 1. Potential therapeutic targets for fragile X syndrome (FXS). Diagram of the mechanisms implicated in FXS leading to altered synaptic plasticity. The figure also shows the molecular pathways targeted or understudy, for the reversal of cognitive and behavioral impairments in FXS patients. Several types of drugs, modulators, and compounds (inhibitor, agonist, and antagonist) can interfere with different pathways disturbed in FXS and have been used in a number of pharmacological treatments some of which are currently under investigation and are indicated in the figure.

The *Fmr1* knockout (KO) mouse model [20], lacks a functional *FMR1* gene and therefore does not express FMRP. Many studies have shown that the *Fmr1* KO mouse presents with some phenotypes that resemble the human disorder, including biochemistry [21], electrophysiology [22], neuropathology [23], and spine morphology [24]. Although the observed patterns of brain activity, including audiogenic seizures, are similar to those in individuals affected by FXS [25], these mice poorly mimic human behavior. Indeed, the strains of the *Fmr1* KO mouse that are often used to test drugs for FXS do not show the cognitive problems seen in patients with FXS [26]. Nevertheless, a large body of literature on the *Fmr1* KO mouse has paved the way to preclinical studies which have shown to rescue several of the FXS phenotypes [27] and have ultimately led to clinical trials in patients with FXS.

Hope has been tempered by the lack of translating the positive results observed in the *Fmr1* KO mouse model into therapy in a clinical setting. Currently, nonpharmacological and behavioral treatments are symptomatic, and they can be coupled with pharmacological treatments of anxiety, aggression, and attention deficit hyperactivity disorder (ADHD).

To date, there is no cure for FXS, and the recent failures of multiple clinical trials have highlighted the need for the development and validation of new biomarkers to better measure the clinical outcome of these treatments [28,29]. Many studies aimed to a better understanding of the underlying molecular mechanisms and pathways involved in FXS have led to the development of specific biomarkers for defining targeted therapeutic strategies intended to reverse the intellectual and behavioral problems of patients with FXS. In this paper, we will review the proposed candidate molecular biomarkers (Figure 2) that have been identified in *Fmr1* KO mouse as an early sign of drug promise and in some cases, later moved to a clinical trial in patients with FXS.

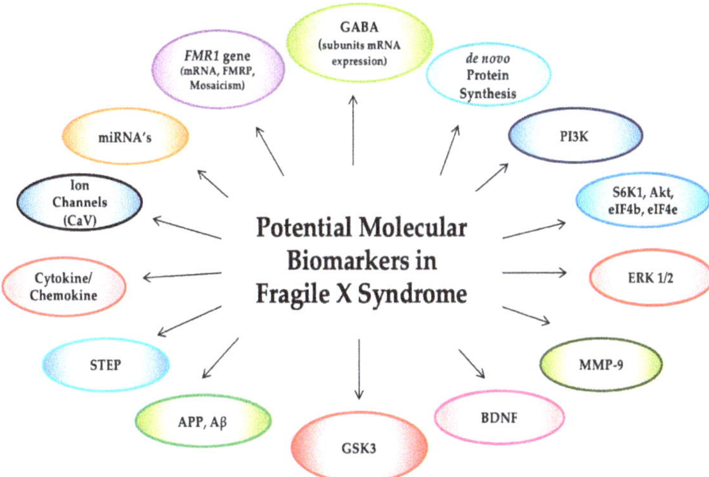

Figure 2. Candidate molecular biomarkers for FXS include a number of targets and substrates of several signaling pathways, in addition to fragile X mental retardation 1 (*FMR1*) molecular measures and metabolites, of which expression levels or activity have been found dysregulated in FXS animal models and in human FXS tissues. *Fmr1* mRNA and fragile X mental retardation 1 protein (FMRP) expression, de novo protein synthesis, γ-aminobutyric acid (GABA) receptors (GABA$_A$ and GABA$_B$), phosphoinositide 3-kinase (PI3K), extracellular-regulated kinase (ERK), matrix metalloproteinase-9 (MMP-9), brain-derived neurotrophic factor (BDNF), mammalian target of rapamycin (mTOR), p70 ribosomal S6 kinase (S6K1), ion channels (KNa, BKCa, CaV, Kv, HCN1), bone morphogenetic protein receptor Type 2 (BMPR2), Diacylglycerol Kinase Kappa (Dgkκ), endocannabinoid system (eCS), amyloid-β protein precursor (APP), microRNA's (miRNA's), striatal-enriched protein tyrosine phosphatase (STEP), glycogen synthase kinase-3 (GSK-3) cytokine and chemokine profiles, metabotropic glutamate receptor (mGluRs).

2. *FMR1* Molecular Measures

FMR1-related measures, including CGG repeat number, percent of methylation, *FMR1* mRNA and FMRP expression levels have been correlated to neurocognitive and social–affective functioning assessments and mental health problems in individuals with FXS [30–38]. The magnitude of the observed correlations generally suggests that these molecular biomarkers are likely accounting only for a proportion of the phenotypic variability of this disorder.

Variation in CGG repeat size and methylation, so-called mosaicism, could be a useful biomarker of various types of risks that could affect subjects with FXS. Mosaicism defines differences in gene expression between those with fully hypermethylated *FMR1* alleles and those carrying unmethylated alleles, and ultimately reflects the levels of expression of *FMR1* mRNA and FMRP. Generally, mosaicism refers to the presence of a full mutation allele(s) and a premutation allele (size mosaicism) or the presence of a full methylated allele(s) and unmethylated alleles (methylation mosaicism), throughout the CGG repeat size range.

Sex differences undoubtedly contribute to the severity of the FXS phenotype; indeed, intellectual and developmental disability is observed in 85% of males and only in 25% of females [27,39,40]. In females, who have two X chromosomes, the process of X inactivation, early during embryonic development, leads to methylation and therefore inactivity of one X chromosome in each cell. However, due to the presence of the chromosome carrying the normal allele, the impact of the *FMR1* mutation in females is reduced relative to males, who have only one X chromosome [41]. The relative proportion of the normal allele on the active and inactive X chromosomes, so-called activation ratio (AR), has shown to contribute to differences in affectedness among females, making the AR a useful biomarker for determining the severity of the phenotype. It should be noted that since X inactivation is a random process, it could be different in different tissues, such as blood and brain [42–44].

3. Metabotropic Glutamate Receptors (mGluRs)

The "mGluR theory of FXS" states that the absence of FMRP leads to excessive metabotropic glutamate receptors (mGluRs, mGluR1 and mGluR5) activated long-term depression (LTD) and reduced responsiveness to signals in the hippocampus and other parts of the brain involved in memory and learning. Together, they are contributing to the neurological and psychiatric symptoms of FXS [45–47]. Reduction of mGluR signaling has demonstrated a reversal of the fragile X phenotypes providing substantial support to the involvement of the mGluR5 pathway in FXS [48]. For more than a decade, our understanding of the molecular pathophysiology of FXS has been substantially advanced by the corroboration of "mGluR theory of FXS" in a wide range of experiments with a number of different mGluR5 inhibitors tested in both the *Fmr1* KO mouse [49–53] and in the Drosophila models of FXS [54–62]. *Fmr1* mutant mouse with a 50% reduction in mGluR5 expression was generated to demonstrate that a range of FXS phenotypes could be corrected by downregulating signaling through group 1 mGluRs [45]. Their findings showed that the decrease in mGlu5 expression levels from early embryonic development effectively prevented the onset of a broad range of FXS phenotypes, including audiogenic seizures, increased basal protein synthesis, spine density, although no effect on macroorchidism was observed.

MPEP (2-methyl-6-phenylethynyl-pyridine) was the first mGluR5-antagonist tested in the *Fmr1* KO mouse, which demonstrated rescue of behavioral defects, including open field performance [63], the rescue of the spine/filopodia ratio in *Fmr1* KO neurons to the levels observed in wild-type neurons [64]. Further, MPEP treated *Fmr1* KO mouse showed improved behavior by significantly fewer errors, less perseveration, and impulsivity when navigating mazes, in addition to reverse postsynaptic density-95 (PSD-95) protein deficits which, if confirmed, could be considered a molecular biomarker [65]. Finally, MPEP prevented an abnormal clustering of DHPG (group I mGluR agonist (S)-3,5-dihydroxyphenylglycine) responsive cells (responsible for activation of ionotropic receptors in mouse FXS neurospheres) and corrected morphological defects of differentiated cells [66].

Similarly, a study on chronic treatment of *Fmr1* KO mouse with the long-acting mGlu5 inhibitor 2-chloro-4-((2,5-dimethyl-1-(4-(trifluoromethoxy)phenyl)-1H-imidazol-4-yl)ethynyl) pyridine (CTEP), fully corrected numerous phenotypes including the increased synaptic spine density, protein synthesis rate, aberrant synaptic plasticity, learning and memory deficits, increased body growth rate, and sensitivity to audiogenic seizures. In addition, this study shows a reduction of both extracellular signal-regulated kinase (ERK) activity and mTOR phosphorylation levels in the *Fmr1* KO but not in wild-type (WT) animals, suggesting that they could represent potential biomarkers in FXS [53]. These studies have shown that the long-term, uninterrupted mGluR5 inhibition is essential for a successful pharmacological intervention as a single dose of the mGluR5 inhibitors was not sufficient to correct the mouse phenotypes. [50,53]. One of the potential molecular mechanisms for mGluR5 dysfunction in FXS is the decreased association of mGluR5 with the Homer family of scaffolding proteins. Indeed, genetic deletion of H1 (an activity-inducible isoform of Homer1) restored regular mGluR5-long Homer association in the *Fmr1* KO and corrected much of the mGluR5 dysfunction as well as behavioral phenotypes, including anxiety and audiogenic seizures [67]. Further, the disruption of mGluR5-Homer resulted in phenotypes of FXS including reduced mGluR5 association with the postsynaptic density, deficits in agonist-induced translational control, protein synthesis-independent LTD, neocortical hyperexcitability, audiogenic seizures, and altered behaviors, such as anxiety and sensorimotor gating [68].

The Drosophila genome encodes only a single mGluR (DmGluRA), compared to the eight separate receptors in mammals [69]. The simplicity of the Drosophila system, coupled with the evolutionary conservation of the activation pathways, has provided an excellent model to test the mGluR hypothesis. Treatment with lithium and MPEP restored normal courtship behavior, mushroom, body defects, and short-term memory, but not β-lobe crossing, suggesting that other morphological abnormalities are responsible for the memory defects [54,70].

Molecular analyses reveal an inverse relationship between dFMRP and DmGluRA, with the latter overexpressed in *dFmr1* null animals and dFMRP overexpressed in DmGluRA nulls [57]. The DmGluRA null also shows more striking defects in activity-dependent synaptic function, including high transmission amplitudes during high-frequency stimulation and abnormally strong hyper potentiation following high-frequency stimulation [57,58]. The successful unbiased screen for small molecules that can rescue the lethality of glutamate-treated larvae and adults *dFmr1* mutants, using the mGluR5 noncompetitive antagonist MPEP or LiCl has been reported to rescue naïve courtship behavior, immediate recall memory, and short-term memory of *dFmr1* mutants [56,59]. The compelling results of these preclinical studies, showing evidence of benefits in rodent and Drosophila disease models, have prompted the application of mGlu5 inhibitors as potential target treatments in human clinical trials for FXS. Thus, clinical trials in FXS patients have been conducted to explore the safety, tolerability, and efficacy of a number of different mGluR5 antagonists.

Fenobam [71], the first mGlur5 antagonist drug evaluated in a single-dose open-label study of 12 male and female adults with FXS (mean age 23.9 years), showed trends of improvement in a prepulse inhibition deficit relative to controls who did not receive the drug [72]. Subsequently, in an exploratory study, the efficacy of mavoglurant (AFQ056) [73] was tested in a randomized double-blind crossover study of 30 FXS males. In this study, seven patients with a hypermethylated full mutation with no detectable *FMR1* mRNA expression, improved stereotypic behavior, hyperactivity, and inappropriate speech, while no improvement found in 18 patients with partial promoter methylation [74]. Thus, it appears that those with full methylation responded best, whereas those who were mosaics with partial methylation had a variable response with a lack of overall efficacy in that group. Although methylation is often regarded as a biomarker, results to date do not explain why some of those with lack of methylation responded and others did not [74]. In addition, the reported behavioral effects of stereotypic behavior, hyperactivity, and inappropriate speech were not replicated with FXS male and female adolescents and adults either full or partial *FMR1* methylation in subsequent 12-week double-blind mavoglurant studies [75].

Similarly, extensive proof of concept study was conducted with basimglurant, a potent and selective mGluR5-negative allosteric modulator (NAM) [76,77] and mavoglurant in male and female adults with FXS. In spite of their promising results in preclinical studies [77–80] these studies ended because no improvement in the clinical phenotype of patients enrolled in the clinical trials using these modulators were observed [29,81,82]. Recently, in a phase 2 12-weeks double-blind clinical trial, basimglurant did not demonstrate improvement over placebo in a parallel-group study of 183 adults and adolescents (aged 14–50, mean 23.4 years) with FXS [83]. Later, the study reported the long-term safety and efficacy of mavoglurant in the two open-label extensions in adolescent ($n = 119$, aged 12–19 years) and adult ($n = 148$, aged 18–45 years). In both studies, mavoglurant was well tolerated, and moderate behavioral improvements were observed in FXS as compared to the placebo control group. Thus, the compelling preclinical evidence for the therapeutic potential of mGlu5 inhibitors in the mouse and the Drosophila disease models has not translated in the anticipated benefits and improvement of the phenotype in FXS patients [84].

4. γ-Aminobutyric Acid (GABA) Receptors

GABA is the most prominent inhibitory neurotransmitter that acts through three receptors in the brain. $GABA_A$ receptors are ligand-regulated chloride channels that upon activation cause hyperpolarization in mature neurons; $GABA_B$ receptors are heterodimeric G protein-coupled receptors (GPCRs) which are mostly expressed presynaptically in the brain; and, $GABA_C$ is CYS-loop ligand-gated ion channels receptors with a similar pentameric structure to $GABA_A$ but are homomeric. FMRP directly binds several $GABA_A$ receptor ($\alpha1$, $\alpha2$, $\alpha3$, δ, and $\gamma2$) mRNAs of which expression is reduced in the cortex and cerebellum of young *Fmr1* KO mouse. Thus, the mRNA expression level of these subunits could be used as biomarker; however, they have not been studied in clinical trials for FXS with any GABA agonists [85]. The introduction of a yeast artificial chromosome (YAC) containing the "healthy" human *FMR1* genomic region into *Fmr1* KO mouse rescued the expression of these specific subunits of $GABA_A$ receptors [86]. A recent electrophysiological study supported the notion that the δ subunit of the aminobutyric acid type A receptors ($GABA_A$Rs) is compromised in the *Fmr1* KO mouse, by reporting a 4-fold decrease in tonic inhibition [87].

The delay in switching from depolarizing to hyperpolarizing GABA has also been observed in the cortex of *Fmr1* KO mouse during development [88]. Moreover, the oxytocin-mediated, GABA excitation–inhibition shift that occurs in newborn rodents during delivery is absent from the hippocampal neurons of *Fmr1* KO mouse. As a result, the hippocampal neurons have elevated intracellular chloride levels, increased excitatory GABA, enhanced the glutamatergic activity, and elevated gamma oscillations [89].

In a study, the response of the FXS neurons (differentiated in vitro from human embryonic stem cells lacking synaptic activity) has been investigated by pulse application of the neurotransmitter GABA. The results confirmed that human FXS neurons do not respond to GABA as FMRP plays a role in the development of the GABAergic synapse during neurogenesis, and that might be one of the potential reasons of the observed default synaptic activity in FXS patients. [90]. Some GABA agonists have been used in the *Fmr1* KO mouse to rescue behavioral abnormalities. The primary neuron excitability deficits in the amygdala of the *Fmr1* KO mouse was restored by gaboxadol (THIP), a $GABA_A$ receptor agonist, which also improved some specific behavioral characteristics, including hyperactivity and auditory seizures [91]. The treatment of the *Fmr1* KO mouse with bumetanide (specific NKCC1 chloride importer antagonist) normalized electrophysiological abnormalities in the mutant offspring as well as hyperactivity and autistic behaviors [89]. Finally, arbaclofen, a $GABA_B$ agonist, improved protein synthesis, the abnormal auditory-evoked gamma oscillations, working memory and anxiety-related behavior in *Fmr1* KO mouse [92–94].

Thus, these findings from different studies in the FXS animal models confirmed that GABA receptors are suitable targets for target treatment in FXS [18,39,95–103]. Indeed, two phase 3 placebo-controlled trials were conducted (with subjects aged 12–50 and in subjects aged 5–11) to determine the safety and efficacy of arbaclofen for improving social behavior in FXS patients. Although, arbaclofen did not meet the primary outcome measures of improved social avoidance in FXS in either study [104], in a double-blind placebo-controlled crossover trial [105], improved social function and behavior were reported in FXS patients. Acamprosate, which activates GABA$_B$ and GABA$_A$ receptors, also improved several phenotypes like cortical upstate duration, behavioral improvement, anxiety, locomotor tests in *Fmr1* KO mouse and reduced ERK1/2 activation in brain tissue [106]. Acamprosate has also been tested in an open-label 10-week trial of 12 young children aged 6–17 years with FXS. It was found safe and well-tolerated and resulted in better social behavior and reduced hyperactivity [107]. Ganaxolone is a neurosteroid and a positive GABA$_A$ modulator that rescued several phenotypes in the *Fmr1* KO mouse, like increased marble-burying assay, sensory and sensorimotor gating in the acoustic startle response, and prepulse inhibition [86]. Tested in a recent randomized double-blind placebo-controlled crossover trial in children with FXS, aged 6–17, years, ganaxolone was found to be safe and have beneficial effects in some patients, particularly for those with higher anxiety or lower cognitive abilities [108]. These preclinical and clinical studies strengthen the hypothesis of GABA receptors involved in the pathology of FXS and as they are the major inhibitory receptors in the brain, they point to the therapeutic potential of the GABA receptor particularly for the behavioral and epileptic phenotypes associated with fragile X syndrome.

5. De Novo Protein Synthesis

Synaptic strength plays a crucial role in learning and memory and it is compromised in many neurodevelopmental disorders. One of the molecular mechanisms that regulate spine morphology, and therefore synaptic strength, is local de novo protein synthesis that enables synapses to alter their function and structure autonomously [109]. FMRP, an RNA binding protein which acts as a translational repressor of many synaptic proteins, is crucial in regulating this process, and the partial or complete lack of FMRP in FXS leads to increased protein translation at the synapses. The metabotropic glutamate receptor subtype 5 (mGluR5) theory of FXS also indicate that the imbalance of mechanisms involved in synaptic shaping and protein translation are responsible for many of the symptoms observed in FXS patients [49]. The lack of FMRP also leads to a loss of translational control and to increased rates of cerebral protein synthesis (rCPS) in some regions of the brain including the hippocampus, thalamus, and hypothalamus of the *Fmr1* KO mouse model of FXS [110].

Fibroblasts from FXS patients also showed significantly elevated rates of basal protein synthesis along with increased levels of the phosphorylated target of rapamycin (p-mTOR), phosphorylated extracellular signal-regulated kinase $\frac{1}{2}$ (ERK1/2), and phosphorylated p70 ribosomal S6 kinase 1 (p-S6K1) [111]. Similarly, a recent study reported that the level of protein synthesis increased in fibroblast cell lines derived from individuals with FXS and from *Fmr1* KO mouse. However, this cellular phenotype displayed a broad distribution with a proportion of individuals with FXS and in the *Fmr1* KO mouse, showing a basal de novo protein synthesis within the normal range. These findings indicate that the molecular mechanisms that control protein synthesis are the primary targets in FXS. However, altered protein synthesis may not be the cause of all symptoms observed in FXS and, therefore, those with normal levels of protein synthesis are not likely going to benefit from target treatments aimed to lower protein synthesis [112]. Thus, de novo protein synthesis could be a very useful biomarker to predict phenotypic subgroups, symptoms severity, and treatment response. Further, as the treatment of fibroblast cells derived from FXS patients, with small molecules that block S6K1 and phosphoinositide 3-kinase (PI3K) catalytic subunit p110β, decreased the rates of protein synthesis in both control and patient fibroblasts; the role of these targets as a potential biomarker should be considered [111]. FXS subjects, under propofol sedation, showed a reduced rCPS in whole brain, cerebellum, and cortex compared to sedated controls. Similar results have been observed in most regions examined in sedative *Fmr1* KO mouse as compared to the WT mouse suggesting that changes in synaptic signaling can correct increased rCPS in FXS [113]. Chronic dietary lithium treatment also demonstrated to be efficacious in reversing the increased rCPS in the *Fmr1* KO mouse [114].

Some studies have shown that the mechanisms regulating the levels of protein synthesis, can be restored by reducing the mGluR5 signaling genetically or with pharmacological treatments [46,53,100,115–118]. Moreover, haploinsufficiency of mGluR5, reduction of MMP9, of striatal-enriched tyrosine phosphatase (STEP) signaling, or of S6K signaling can not only restore the levels of protein synthesis but also restore the synaptic and behavioral phenotypes in the FXS mouse model [50,119–126]. Recently, a study showed that treatment of the *Fmr1* KO mouse with a cell-permeable peptide able to modulate ADAM metallopeptidase domain 10 (ADAM10) activity and amyloid-β protein precursor (APP) processing, restored protein synthesis to the wild-type (WT) level [127].

These preclinical and clinical studies suggest that basal protein synthesis could be considered as a potential biomarker and a molecular hallmark for FXS, but unfortunately, replicating this optimal translational scenario into reality has not been fully successful [27]. The extent to which excessive protein synthesis associated with cognitive and behavioral impairments also remained unknown. More importantly, none of the human studies have shown an effect on the primary outcome measures which were mainly behavioral questionnaires in children, adolescents, or adults with FXS [74,104,105]. Finally, although FMRP modulates protein synthesis, there are other factors (environmental and genetic) that may contribute to the modulation of homeostasis of molecules involved in synaptic plasticity.

6. Phosphoinositide 3-Kinase (PI3K)

Phosphoinositide 3-kinase (PI3K) is the signaling molecule involved in cell motility, survival, growth, and proliferation. PI3K class I catalytic subunits, p110α, p110β, p110γ, and p110δ, have their specific dysregulation in FXS [128]. FMRP regulates the synthesis and synaptic localization of p110β, which is a crucial signaling molecule downstream of group 1 metabotropic glutamate receptor (gp1 mGluRs) and other membrane receptors. Lack of FMRP in the *Fmr1* KO mouse leads to excess mRNA translation and synaptic protein expression of p110β [123]. Treatment with a p110β-selective antagonist was effective in rescuing the excess of protein synthesis in the *Fmr1* KO mouse synaptoneurosomes and in lymphoblastoid cells derived from FXS patients [123,129]. Further, a prefrontal cortex (PFC) selective knockdown of p110β, reversed deficits in higher cognition, normalized excessive PI3K activity, restored stimulus-induced protein synthesis, and corrected increased dendritic spine density in the *Fmr1* KO mouse [130,131]. Thus, PI3K activity in patient cells might be a biomarker and could be used to assess the efficacy of drug response in target treatment in FXS.

7. Mammalian Target of Rapamycin (mTOR) and Substrate p70 Ribosomal S6 Kinase (S6K1)

Mammalian target of rapamycin (mTOR) is a 289 kDa serine/threonine kinase protein that controls various energetic functions at both the cellular and organism level and an essential regulator of cell proliferation, autophagy, translation, and growth. In neuronal cells, protein synthesis plays a fundamental role in the regulation of lasting alterations in synaptic strength or plasticity, and of long-term potentiation (LTP), processes that are important in learning and memory [132,133]. The components of the mTOR signaling cascade, which is involved in protein synthesis-dependent phase of synaptic strengthening, are present in dendrites suggesting a role for mTOR in local translation and synaptic plasticity. mTOR is activated in dendrites by stimulation of group I mGluRs and it is required for mGluR-LTD [134,135]. It has been reported that increased activity in these systems can lead to repetitive and perseverative behavior patterns [132].

The best-characterized function of mTOR is the regulation of translation. mTOR regulates two critical and core components of the translational initiation machinery, p70 ribosomal S6 kinase 1 and 2 (S6K1/2), and the eIF4E-binding proteins (4E-BPs), and it is also known to regulate the activity of phosphatases such as protein phosphatase 2A (PP2A). These phosphatases, in turn, regulate mTOR substrates, thereby generating mTOR-dependent feedback loops that control initiation rates. Increased phosphorylation of (mTOR) substrate, p70 ribosomal subunit 6 kinase 1 (S6K1) along with the high expression of mTOR regulator, and the serine/threonine protein kinase (Akt) was also observed in lymphocytes and brain tissues derived from subjects with FXS [136].

The enhanced mTOR signaling observed in the hippocampus of the *Fmr1* KO mouse associates with the increase eukaryotic initiation factor complex F4 (eIF4F) [137] and with the increased phosphorylation of the cap-binding protein eukaryotic initiation factor 4E (eIF4E) [136] to further support the increased protein synthesis observed in FXS. These findings, in both FXS mice and humans, are consistent with the idea that the loss of FMRP results in the dysregulation of mechanisms of translational initiation control rather than transcriptional regulation and provide the direct evidence that mTOR dysregulation may be useful for designing targeted treatments in FXS [136]. Therefore, targets and substrates in the mTOR signaling pathways can act as potential molecular biomarkers. Since the molecular signaling effects resulting from FMRP loss are likely causal in the wide-range of the severity of the FXS symptoms, including autism, identifying the effects of FMRP loss on molecular signaling pathways, like those governing translation, is key to advancing our ability to treat the disorder.

Finally, metformin, a type 2 diabetes medication that can improve obesity and excessive appetite, has emerged as a candidate drug for targeted treatment of FXS based on preclinical studies. These studies have shown rescue of a number of FXS phenotypes including memory deficits, social novelty, grooming, dendritic spine morphology, and electrophysiology in the CA1 of the hippocampus [138,139]. Metformin suppresses mRNA translation via inhibition of ERK and mTOR pathways, which are overactive in FXS, supporting their potential role as molecular biomarkers, and therefore, may contribute to normalizing signaling pathways in the CNS of FXS patients. In humans, metformin has been used in the clinical treatment of several individuals with FXS and showed benefits not only in lowering weight gain but also in improving language and behavior [138]. Thus, metformin shows promises for targeting several signaling pathways disrupted in FXS and possibly rescuing some of the clinical symptoms observed in individuals with FXS. Interestingly, a double-blind placebo-controlled trial of metformin in individuals with FXS is currently ongoing which will assess safety and benefit of metformin in the treatment of language deficits, behavioral problems, and obesity in individuals with FXS.

8. Extracellular-Regulated Kinase (ERK)

The ERK pathway is a chain of proteins in the cell that acts as a nodal point for cell signaling cascades. The absence of FMRP in *Fmr1* KO mouse results in rapid dephosphorylation of ERK upon mGluR1/5 stimulation suggesting that over-activation of phosphatases in synapses affects the synaptic translation, transcription, and synaptic receptor regulation in FXS [53,119,140,141]. Delayed early-phase phosphorylation of ERK is observed in both neurons and thymocytes of the *Fmr1* KO mouse. Likewise, the early-phase kinetics of ERK activation in lymphocytes from human peripheral blood is also delayed in individuals with FXS, as compared to controls [142]. The correction of the delayed ERK activation time, resulting in a faster activation, was observed after 2 months of treatment with lithium in a pilot open-label trial in FXS or with riluzole treatment [143,144]. These findings, based on a small number of subjects, suggest ERK activity as a potential biomarker for measuring the metabolic status of the disease in FXS.

Recently, the significant FMRP-dependent over-activation of ERK was observed in both FXS mouse and humans. ERK activity was normalized in FXS platelets [145], and correlated with clinical response to lovastatin, pointing this inhibitor of ERK pathway signaling cascade as a promising treatment for FXS [146]. The findings by Pellerin et al. [145] suggest that the use of platelet's ERK activity represents a new potentially interesting biomarker for future clinical trials. It may also pave the way for other promising and very exciting discoveries that will eventually improve FXS patients' assessment in future clinical trials where either lovastatin or other ERK-targeting drugs is applied.

9. Matrix Metalloproteinase-9 (MMP-9)

FMRP deficit is associated with alterations in the expression of a number of proteins, including matrix metalloproteinase 9 (MMP-9). MMP-9 is an extracellular operating Zn^{2+} dependent endopeptidase that is expressed in neurons and locally translated and released at the dendrites in response to enhanced neuronal activity driven by glutamate. MMP-9 plays an essential role in both establishing synaptic connections during development and in restructuring synaptic networks in the adult brain [147]. MMP-9 mRNA is part of the FMRP complex and localizes in dendrites. Translation of MMP-9 is increased at synapses in *Fmr1* KO mice suggesting its contribution to the aberrant dendritic spine morphology observed in the *Fmr1* KO mice and in FXS patients [148,149]. The genetic disruption of MMP-9 in the *Fmr1* KO mouse rescued key aspects of *Fmr1* abnormalities, including abnormal mGluR5-dependent LTD and dendritic spine abnormalities [150], providing evidence that MMP-9 is necessary to the development of FXS-associated defects in the *Fmr1* KO mouse. Interestingly, a high level of MMP-9 has been observed in the auditory cortex of adult *Fmr1* KO mice and the deletion of MMP-9 reversed the habituation defects [151]. A decreased MMP-9 activity in the hippocampus of the *Fmr1* KO mouse, dendritic spine maturation, improvement in anxiety, and strategic exploratory behavior were observed after treatment with the antibiotic minocycline [152]. These findings prompted the use of minocycline as a targeted treatment in humans with FXS through open-label trials which have demonstrated benefits with improvements in language, attention, social communication, and anxiety [153,154]. More recently, a controlled double-blind crossover study of minocycline for FXS treatment provided evidence for the safety of minocycline and showed benefits in global functioning in children with FXS [155]. In addition, as expected, the higher plasma activity of MMP-9 observed in FXS patients was lowered by minocycline in some patients [156], as minocycline is known to be a MMP-9 inhibitor [152]. On the other hand, no changes in plasma MMP 9 activity was found after treatment with sertraline [157], a selective serotonin-reuptake inhibitor which selectively blocks the uptake of serotonin at the presynaptic membrane, resulting in an increased synaptic concentration of serotonin in the central nervous system (CNS), and therefore, to an intensified serotonergic neurotransmission. Interestingly, a reduction of the MMP-9 levels was also reported in the *Fmr1* KO mouse following metformin treatment [139]. The results of the preclinical and clinical studies indicate that minocycline, through its mechanism of action as an MMP inhibitor, may be an additional potential effective avenue as FXS therapeutic treatment and MMP-9 activity, a potential biomarker in FXS.

10. Brain-Derived Neurotrophic Factor (BDNF)

Brain-derived neurotrophic factor is involved in the regulation of various processes of normal neural circuit function and development. Dysregulation in BDNF/TrkB signaling in the *Fmr1* KO mouse leads to altered brain development, including excessive sponginess, dendritic arborization [158], and impaired synaptic plasticity [159]. These neural alterations are promoted by activity-dependent variation in the sensitivity to BDNF-TrkB signaling, compensating postsynaptic activity [158].

The effects of reduced BDNF expression on the learning and behavioral phenotypes, including fear conditioning, pain behaviors, and hyperactivity, was examined in the *Fmr1* KO mouse crossed with a mouse carrying a deletion of one copy of the *Bdnf* gene (*Bdnf* +/−) [160]. The authors reported age-dependent alterations in the expression of BDNF in the hippocampus, reduced locomotor hyperactivity, deficits in sensorimotor learning, and startle responses typical of *Fmr1* KO mice. In addition, altered BDNF signaling in FMRP-deficient neural progenitor cells (NPCs) suggested that perturbations of brain development in FXS occur at very early stages of development [161].

A single-nucleotide polymorphism (SNP) in the human BDNF gene, leading to a methionine (Met) substitution for valine (Val) at amino acid 66, interferes with the intercellular trafficking and the activity-dependent secretion of BDNF in cortical neurons. One study found that the Val66Met BDNF polymorphism associates with epilepsy in a Finnish FXS male [162] but was not confirmed in a group of 77 patients with FXS [157]. However, a significant association between the BDNF polymorphism and improvements of several clinical measures was observed in a double-blind randomized placebo-controlled clinical trial of sertraline in FXS aimed to determine the efficacy of treatment in young children with FXS [157]. In addition, an open-label study showed a significant increase in BDNF level after treatment with the GABA$_A$ agonist acamprosate [107]. Although more studies are warranted, these findings point to BDNF genotype as a potential molecular biomarker in FXS.

11. Amyloid-β Protein Precursor and Amyloid-β (APP, Aβ)

FMRP protein binds to the coding region of the APP mRNA and results in increased translation of the encoded product, the amyloid precursor protein (APP), which plays a vital role in the developing brain during synapse formation, while β-amyloid (Aβ) accumulation, results in synaptic loss and impaired neurotransmission. APP is processed by β- and γ-secretases to produce amyloid-β (Aβ), which is the prominent peptide found in the case of Alzheimer's disease (AD).

A study by Westmark et al. found a 1.7-fold increase APP expression in the *Fmr1* KO mouse versus WT using western blot analysis and showed that the genetic knockdown of one APP allele in the *Fmr1* KO mouse rescued the FXS phenotypes including anxiety, seizures, mGLuR-LTD, and the ratio of mature versus immature dendritic spines [163]. APP and Aβ were evaluated as blood-based biomarkers and in a prospective open-label trial of acamprosate in pediatric subjects with FXS-associated autism spectrum disorder and found that acamprosate treatment significantly reduced sAPP and sAPPα [164].

Although blood levels of APP metabolites may not correlate with brain levels, which is one of the limitations of these studies, altogether these findings support a role for dysregulated APP production and processing in FXS and indicate that the APP metabolites may be viable biomarkers for FXS treatment.

12. Additional Potential Biomarkers

12.1. Ion Channels (CaV)

Voltage-gated ion channels are involved in neural transmission and some recent past studies showed their involvement in the FXS pathology [165]; more specifically, with the voltage-gated calcium channels (VGCC) family, namely Cav2.1 and Cav2.2 [166]. Synaptic transmission depends critically on presynaptic calcium entry via voltage-gated calcium (CaV) channels. FMRP regulates the expression of neuronal N-type CaV channels (CaV2.2) [166] and dysregulated calcium homeostasis, in addition to the decreased expression of the pore-forming subunit of the Cav2.1 channel, the Cacna1a gene, in *Fmr1*-KO cultured neurons [167]. Their findings indicate that FMRP plays a key role in calcium homeostasis during brain development; furthermore, the authors suggest that calcium homeostasis could be used as a cellular biomarker and for the identification of new drugs for target treatment in FXS.

12.2. Glycogen Synthase Kinase-3 (GSK-3)

Glycogen synthase kinase-3 (GSK-3) is a serine/threonine protein kinase that, when phosphorylated, regulates a variety of developmental processes, such as cell migration, cell morphology, neurogenesis, and gliogenesis via interaction with a variety of signaling pathways [168]. The lack of FMRP results in an abnormal increase in GSK-3β mRNA and protein levels in several regions of the brain [169] of the *Fmr1* KO mouse and in decreased hippocampal neurogenesis that likely contributes to the pathogenesis of FXS [170].

Several studies have demonstrated that lithium treatment rescued the FXS-associated impairments sustainable throughout the aging process in the Drosophila model of FXS [54,59]. In addition, GSK-3 inhibitors and lithium treatment provided the direct evidence of GSK-3 involvement in the pathology of FXS by reducing audiogenic seizure activity, improved performance on the open field elevated plus maze and passive avoidance tests [171], improved social defects [172], rescue of the hippocampus-dependent learning deficits [173], and improved cognitive deficits [174] in the *Fmr1* KO mouse. Additionally, the attenuation of reactive astrocytes, which has been observed in many brain regions of the *Fmr1* KO mouse with lithium treatment, provides further evidence of the involvement of GSK-3 in FXS [175]. These findings raise the possibility that GSK-3 activity may represent a biochemical mediator biomarker of impaired cognitive function in FXS and that modulators of its activity may have potential as therapeutic agents [176].

12.3. Striatal-Enriched Protein Tyrosine Phosphatase (STEP)

Striatal-enriched protein tyrosine phosphatase (STEP) is a brain-specific tyrosine phosphatase that plays a significant role in the development of the CNS by regulating dendritic proteins involved in synaptic plasticity [177,178]. STEP dysregulation is involved in the pathophysiology of several neuropsychiatric disorders [179], including FXS, likely by dephosphorylating both NMDARs and AMPARs [177]. While the enhanced activity of mGluRs in the absence of FMRP upregulates the translation of STEP [178,180,181] in the hippocampus of the *Fmr1* KO mouse, genetic reduction of STEP significantly diminishes some FXS-associated behaviors in *Fmr1* KO including seizures and restores select social and nonsocial anxiety-related behaviors [181]. Benzopentathiepin 8-(trifluoromethyl)-1,2,3,4,5-benzopentathiepin-6-amine hydrochloride (known as TC-2153) is a newly discovered STEP inhibitor [182]. A recent study [183] reported that this STEP inhibitor reduces seizure incidence and hyperactivity, anxiety and improves sociability, electrophysiological deficits in acute brain slices and spine morphology in *Fmr1* KO mouse. These observations suggest that STEP's expression and activity could be useful for evaluating the clinical benefits of pharmacological therapeutic approaches in FXS targeting STEP.

12.4. Plasma Cytokines and Chemokines

Cytokines are the most important mediators of cell–cell communication in the human immune system. They perform a variety of functions like modulation of the central nervous system (CNS), brain functioning, and responses to infections or injury. Significant differences in plasma cytokine and chemokines levels were reported in patients with FXS including a high level of IL-1α, RANTES, and IP-10 [184]. It is currently unknown if the changes in the cytokine and chemokines are determinant in the development of FXS and if they occur throughout the lifetime of FXS patients, and therefore, their potential use as biomarkers needs more investigation.

12.5. Diacylglycerol Kinase Kappa (Dgkκ)

Diacylglycerol kinase kappa (Dgkκ) is a master regulator that controls two critical signaling pathways involved in protein synthesis. Lack of FMRP in the *Fmr1* KO mouse neurons results in the loss of Dgkκ expression along with mGluR1 receptor-dependent DGK activity, leading to synaptic plasticity alterations, dendritic spine abnormalities, and behavior disorders. These findings support the involvement of Dgkκ deregulation in FXS pathology and suggest that overexpression of Dgkκ in neurons could rescue the dendritic spine defects of the *Fmr1* KO mouse. Thus, DGKκ expression levels could represent a biomarker and targeting DGKκ signaling might provide new therapeutic approaches for FXS [185].

12.6. MicroRNAs (miRNAs)

MicroRNAs (miRNAs) are known as a class of small noncoding RNA molecules (19–23 nucleotides) that regulate almost 30% of genes at the post-transcriptional level in eukaryotic organisms [186]. Several studies have provided evidence of miRNA involvement in the pathogenesis of FXS by identifying and isolating several r(CGG)-derived miRNAs, including miR-fmr1-27 and miR-fmr1-42 in the zebrafish FXS model [187,188]. Their brain exhibits long dendrites and disconnected synapses, similar to those found in the human FXS hippocampal–neocortical junction [189]. Further, microarray analyses of miRNAs associated with FMRP in the *Fmr1* KO mouse brain identified miR-125a, miR-125b, and miR-132, and disruption of the regulating of these miRNA-mediated protein translation results in early neural development and synaptic physiology [190,191]. Another microarray study [192] in the *Fmr1* KO mouse showed the interaction of miR-34b, miR-340, and miR-148a with the Met 3′ UTR of the *FMR1* gene, suggesting that alterations in the miRNA expression resulted from the absence of FMRP, could contribute to the molecular pathology of FXS. Enhanced expression of miR-510, located on chromosome X in the 27.3Xq region, flanking to a fragile X site, was reported in full mutation female carriers [193]. Thus, although more studies are necessary to confirm their utility, many pieces of evidence indicate that miRNAs could be attractive candidate biomarkers in FXS.

13. Conclusions

FXS is a challenging disorder in terms of drug development and clinical implementation. An extensive preclinical work, carried out in the FXS animal models, has provided ways to improve the behavior, language, and cognitive ability, but several factors (complex clinical phenotype, genetic variability, gender differences, and use of multiple medications, limitations in the outcome measures and of tools) might have contributed to a lack of translation from the preclinical to clinical outcomes. When looking at the design of the preclinical studies to date, some limitations can be identified. Most of the FXS research in mammalian model systems is limited to two disease models, the *Fmr1* KO mouse and *Fmr1* KO drosophila animal model, but the central issue in using these models is variability and small effect size of the phenotype particularly in the area of cognitive defects. Moreover, overlapping phenotypes in these animal models sometimes may lead to over-prediction of the therapeutic potential of novel drug treatments.

Research to date on FXS has provided us with several potential candidate biomarkers that can, in principle, be used to assess efficacy; molecular biomarkers are promising, simple, and minimally invasive diagnostic tools that can objectively measure the biologically relevant effects of targeted treatments on the underlying molecular defects observed in FXS. However, the current research on molecular biomarkers in FXS suffers from a number of limitations. FXS is a neurological disorder, but brain tissue is not easily accessible. Therefore, biomarkers must be developed in a tissue that can be obtained easily, such as PBMCs, platelets, and fibroblasts. No single consistent molecule or modification state (i.e., phosphorylation or acetylation) has been reported to be differentially regulated in FXS patients versus controls consistently across multiple testing sites. Although many molecular biomarkers have been proposed in FXS (Figure 2), no one is accurate enough as changes according to disease modifications in a trial setting. No clinical history for any marker is available, and lengthy expensive processing and time consumption are required to generate test substrates such as primary fibroblasts (and induced pluripotent stem cells).

In summary, there is an urgent need to establish novel and reliable biomarkers in FXS, particularly blood-based biomarkers, essential to the development of new treatments. They can provide measures of disease severity and can be used to develop personalized treatments. Interestingly, when monitored over time, they can be used to evaluate treatment outcomes and help to identify responders, and therefore those individuals that following treatment have shown real benefit with phenotype improvements.

Author Contributions: All of the authors participated in drafting the manuscript and critically editing it for important intellectual content.

Funding: This project was supported by the National Institute of Health, Grant R01GM113929-01.

Conflicts of Interest: F.T. received funds from Asuragen and Zynerba. The remaining authors have no disclosures.

References

1. Biomarkers Definitions Working Group. Biomarkers and surrogate endpoints: Preferred definitions and conceptual framework. *Clin. Pharmacol. Ther.* **2001**, *69*, 89–95. [CrossRef]
2. Harris, S.; Goodlin-Jones, B.; Nowicki, S.; Bacalman, S.; Tassone, F.; Hagerman, R. Autism Profiles of Young Males with Fragile X Syndrome. *J. Dev. Behav. Pediatr.* **2005**, *26*, 464. [CrossRef]
3. Landowska, A.; Rzonca, S.; Bal, J.; Gos, M. [Fragile X syndrome and *FMR1*-dependent diseases-clinical presentation, epidemiology and molecular background]. *Dev. Period Med.* **2018**, *22*, 14–21. [PubMed]
4. Verkerk, A.J.; Pieretti, M.; Sutcliffe, J.S.; Fu, Y.-H.; Kuhl, D.P.; Pizzuti, A.; Reiner, O.; Richards, S.; Victoria, M.F.; Zhang, F.; et al. Identification of a gene (FMR-1) containing a CGG repeat coincident with a breakpoint cluster region exhibiting length variation in fragile X syndrome. *Cell* **1991**, *65*, 905–914. [CrossRef]

5. Oberlé, I.; Rousseau, F.; Heitz, D.; Kretz, C.; Devys, D.; Hanauer, A.; Boué, J.; Bertheas, M.; Mandel, J. Instability of a 550-base pair DNA segment and abnormal methylation in fragile X syndrome. *Science* **1991**, *252*, 1097–1102. [CrossRef]

6. Pieretti, M.; Zhang, F.; Fu, Y.-H.; Warren, S.T.; Oostra, B.A.; Caskey, C.; Nelson, D.L. Absence of expression of the FMR-1 gene in fragile X syndrome. *Cell* **1991**, *66*, 817–822. [CrossRef]

7. Tassone, F.; Hall, D.A. *FXTAS, FXPOI, and Other Premutation Disorders*; Springer Nature: Basel, Switzerland, 2016.

8. Hagerman, R.J.; Protic, D.; Rajaratnam, A.; Salcedo-Arellano, M.J.; Aydin, E.Y.; Schneider, A. Fragile X-Associated Neuropsychiatric Disorders (FXAND). *Front. Psychol.* **2018**, *9*, 564. [CrossRef]

9. Santos, A.R.; Bagni, C.; Kanellopoulos, A.K. Learning and behavioral deficits associated with the absence of the fragile X mental retardation protein: What a fly and mouse model can teach us. *Learn. Mem.* **2014**, *21*, 543–555. [CrossRef] [PubMed]

10. Sandoval, G.M.; Shim, S.; Hong, D.S.; Garrett, A.S.; Quintin, E.-M.; Marzelli, M.J.; Patnaik, S.; Lightbody, A.A.; Reiss, A.L. Neuroanatomical abnormalities in fragile X syndrome during the adolescent and young adult years. *J. Psychiatr.* **2018**, *107*, 138–144. [CrossRef]

11. Bostrom, C.; Yau, S.-Y.; Majaess, N.; Vetrici, M.; Gil-Mohapel, J.; Christie, B.R.; Yau, S.S.Y. Hippocampal dysfunction and cognitive impairment in Fragile-X Syndrome. *Neurosci. Biobehav. Rev.* **2016**, *68*, 563–574. [CrossRef]

12. Hoeft, F.; Carter, J.C.; Lightbody, A.A.; Hazlett, H.C.; Piven, J.; Reiss, A.L. Region-specific alterations in brain development in one- to three-year-old boys with fragile X syndrome. *Proc. Natl. Acad. Sci. USA* **2010**, *107*, 9335–9339. [CrossRef]

13. Harlow, E.G.; Till, S.M.; Russell, T.A.; Wijetunge, L.S.; Kind, P.; Contractor, A. Critical period plasticity is disrupted in the barrel cortex of *Fmr1* knockout mice. *Neuron* **2010**, *65*, 385–398. [CrossRef]

14. Lai, J.; Lerch, J.; Docring, L.; Foster, J.; Ellegood, J., Lai, J. Regional brain volumes changes in adult male *FMR1*-KO mouse on the FVB strain. *Neuroscience* **2016**, *318*, 12–21. [CrossRef]

15. Hazlett, H.C.; Poe, M.D.; Lightbody, A.A.; Styner, M.; MacFall, J.R.; Reiss, A.L.; Piven, J. Trajectories of Early Brain Volume Development in Fragile X Syndrome and Autism. *J. Am. Acad. Child Adolesc. Psychiatry* **2012**, *51*, 921–933. [CrossRef]

16. Measurement of Cerebral and Cerebellar Volumes in Children with Fragile X Sundrome. Available online: https://paperpile.com/app/p/2405c439-1b64-0314-96b7-37ba0f0ea488 (accessed on 12 January 2019).

17. Sunamura, N.; Iwashita, S.; Enomoto, K.; Kadoshima, T.; Isono, F. Loss of the fragile X mental retardation protein causes aberrant differentiation in human neural progenitor cells. *Sci. Rep.* **2018**, *8*, 11585. [CrossRef]

18. Yang, Y.-M.; Arsenault, J.; Bah, A.; Krzeminski, M.; Fekete, A.; Chao, O.Y.; Pacey, L.K.; Wang, A.; Forman-Kay, J.; Hampson, D.R.; et al. Identification of a molecular locus for normalizing dysregulated GABA release from interneurons in the Fragile X brain. *Mol. Psychiatry* **2018**. [CrossRef]

19. Hanson, A.C.; Hagerman, R.J. Serotonin dysregulation in Fragile X Syndrome: Implications for treatment. *Intractable Rare Dis.* **2014**, *3*, 110–117. [CrossRef]

20. The Dutch-Belgian Fragile X Consorthium; Bakker, C.E.; Verheij, C.; Willemsen, R.; van der Helm, R.; Oerlemans, F.; Vermey, M.; Bygrave, A.; Hoogeveen, A.; Oostra, B.A.; et al. *Fmr1* Knockout Mice: A Model to Study Fragile X Mental Retardation. *Cell* **1994**, *78*, 22–23.

21. Dahlhaus, R. Of Men and Mice: Modeling the Fragile X Syndrome. *Front. Neurosci.* **2018**, *11*, 41. [CrossRef]

22. Rais, M.; Binder, D.K.; Razak, K.A.; Ethell, I.M. Sensory Processing Phenotypes in Fragile X Syndrome. *ASN Neuro* **2018**, *10*, 1759091418801092. [CrossRef]

23. Greco, C.M.; Berman, R.F.; Martin, R.M.; Tassone, F.; Schwartz, P.H.; Chang, A.; Trapp, B.D.; Iwahashi, C.; Brunberg, J.; Grigsby, J.; et al. Neuropathology of Fragile X-Associated Tremor/ataxia Syndrome (FXTAS). *Brain* **2006**, *129*, 243–255. [CrossRef] [PubMed]

24. Irwin, S.A.; Galvez, R.; Greenough, W.T. Dendritic Spine Structural Anomalies in Fragile-X Mental Retardation Syndrome. *Cereb. Cortex* **2000**, *10*, 1038–1044. [CrossRef] [PubMed]

25. Bernardet, M.; Crusio, W.E. *Fmr1* KO Mice as a Possible Model of Autistic Features. *Sci. World J.* **2006**, *6*, 1164–1176. [CrossRef]

26. Wright, J. Questions for Elizabeth Berry-Kravis: Dodging mouse traps | Spectrum | Autism Research News. Available online: https://www.spectrumnews.org/opinion/q-and-a/questions-for-elizabeth-berry-kravis-dodging-mouse-traps/ (accessed on 9 January 2019).

27. Berry-Kravis, E.M.; Lindemann, L.; Jønch, A.E.; Apostol, G.; Bear, M.F.; Carpenter, R.L.; Crawley, J.N.; Curie, A.; Des Portes, V.; Hossain, F.; et al. Drug Development for Neurodevelopmental Disorders: Lessons Learned from Fragile X Syndrome. *Nat. Rev. Drug Discov.* **2018**, *17*, 280–299. [CrossRef]

28. Scharf, S.H.; Jaeschke, G.; Wettstein, J.G.; Lindemann, L. Metabotropic glutamate receptor 5 as drug target for Fragile X syndrome. *Curr. Opin. Pharmacol.* **2015**, *20*, 124–134. [CrossRef] [PubMed]

29. Mullard, A. Fragile X drug development flounders. *Nat. Rev. Drug Discov.* **2016**, *15*, 77. [CrossRef]

30. De Caro, J.J.; Dominguez, C.; Sherman, S.L. Reproductive Health of Adolescent Girls Who Carry the *FMR1* Premutation: Expected Phenotype Based on Current Knowledge of Fragile X-Associated Primary Ovarian Insufficiency. *Ann. N. Y. Acad. Sci.* **2008**, *1135*, 99–111. [CrossRef] [PubMed]

31. Roberts, J.E.; Bailey, D.B.; Mankowski, J.; Ford, A.; Sideris, J.; Weisenfeld, L.A.; Heath, T.M.; Golden, R.N. Mood and anxiety disorders in females with the *FMR1* premutation. *Am. J. Med. Genet. Part B Neuropsychiatr. Genet.* **2009**, *150*, 130–139. [CrossRef]

32. Hamlin, A.; Liu, Y.; Nguyen, D.V.; Tassone, F.; Zhang, L.; Hagerman, R.J. Sleep apnea in fragile X premutation carriers with and without FXTAS. *Am. J. Med. Genet. Part B Neuropsychiatr. Genet.* **2011**, *156*, 923–928. [CrossRef]

33. Hamlin, A.A.; Sukharev, D.; Campos, L.; Mu, Y.; Tassone, F.; Hessl, D.; Nguyen, D.V.; Loesch, D.; Hagerman, R.J. Hypertension in *FMR1* Premutation Males with and without Fragile X-Associated Tremor/Ataxia Syndrome (FXTAS). *Am. J. Med. Genet. Part A* **2012**, *158*, 1304–1309. [CrossRef]

34. Bailey, D.B.; Raspa, M.; Bishop, E.; Mitra, D.; Martin, S.; Wheeler, A.; Sacco, P. Health and Economic Consequences of Fragile X Syndrome for Caregivers. *J. Dev. Behav. Pediatr.* **2012**, *33*, 705–712. [CrossRef] [PubMed]

35. Ji, N.Y.; Findling, R.L. Pharmacotherapy for mental health problems in people with intellectual disability. *Curr. Opin.* **2016**, *29*, 103–125. [CrossRef] [PubMed]

36. Lieb-Lundell, C.C.E. Three Faces of Fragile X. *Phys. Ther.* **2016**, *96*, 1782–1790. [CrossRef]

37. Hoyos, L.R.; Thakur, M. Fragile X Premutation in Women: Recognizing the Health Challenges beyond Primary Ovarian Insufficiency. *J. Assist. Reprod. Genet.* **2017**, *34*, 315–323. [CrossRef] [PubMed]

38. Napoli, E.; Schneider, A.; Hagerman, R.; Song, G.; Wong, S.; Tassone, F.; Giulivi, C. Impact of *FMR1* Premutation on Neurobehavior and Bioenergetics in Young Monozygotic Twins. *Front. Genet.* **2018**, *9*, 9. [CrossRef]

39. Ligsay, A.; Hagerman, R.J. Review of targeted treatments in fragile X syndrome. *Intractable Rare Dis.* **2016**, *5*, 158–167. [CrossRef]

40. Hagerman, P.J. The Fragile X Prevalence Paradox. *J. Med. Genet.* **2008**, *45*, 498–499. [CrossRef]

41. Stembalska, A.; Łaczmańska, I.; Gil, J.; Pesz, K.A. Fragile X Syndrome in Females—A Familial Case Report and Review of the Literature. *Dev. Period. Med.* **2016**, *20*, 99–104.

42. Berry-Kravis, E.; Potanos, K.; Weinberg, D.; Zhou, L.; Goetz, C.G. Fragile X-Associated Tremor/ataxia Syndrome in Sisters Related to X-Inactivation. *Ann. Neurol.* **2005**, *57*, 144–147. [CrossRef]

43. Van Esch, H. The Fragile X premutation: New insights and clinical consequences. *Eur. J. Med Genet.* **2006**, *49*, 1–8. [CrossRef]

44. Seltzer, M.M.; Abbeduto, L.; Greenberg, J.S.; Almeida, D.; Hong, J.; Witt, W. Chapter 7 Biomarkers in the Study of Families of Children with Developmental Disabilities. *Families* **2009**, *37*, 213–249.

45. Bear, M.F.; Huber, K.M.; Warren, S.T. The mGluR theory of fragile X mental retardation. *Trends Neurosci.* **2004**, *27*, 370–377. [CrossRef]

46. Dölen, G.; Bear, M.F. Role for Metabotropic Glutamate Receptor 5 (mGluR5) in the Pathogenesis of Fragile X Syndrome. *J. Physiol.* **2008**, *586*, 1503–1508. [CrossRef]

47. Kim, M.; Ceman, S. Fragile X Mental Retardation Protein: Past, Present and Future. *Curr. Pept. Sci.* **2012**, *13*, 358–371. [CrossRef]

48. Darnell, J.C.; Klann, E. The Translation of Translational Control by FMRP: Therapeutic Targets for FXS. *Nat. Neurosci.* **2013**, *16*, 1530–1536. [CrossRef]

49. Huber, K.M.; Gallagher, S.M.; Warren, S.T.; Bear, M.F. Altered Synaptic Plasticity in a Mouse Model of Fragile X Mental Retardation. *Proc. Natl. Acad. Sci. USA* **2002**, *99*, 7746–7750. [CrossRef]

50. Dölen, G.; Osterweil, E.; Rao, B.S.S.; Smith, G.B.; Auerbach, B.D.; Chattarji, S.; Bear, M.F. Correction of fragile X syndrome in mice. *Neuron* **2007**, *56*, 955–962. [CrossRef] [PubMed]

51. Krueger, D.D.; Bear, M.F. Toward Fulfilling the Promise of Molecular Medicine in Fragile X Syndrome. *Annu. Med.* **2011**, *62*, 411–429. [CrossRef]

52. Bhakar, A.L.; Dölen, G.; Bear, M.F. The Pathophysiology of Fragile X (and What It Teaches Us about Synapses). *Annu. Neurosci.* **2012**, *35*, 417–443. [CrossRef] [PubMed]

53. Michalon, A.; Sidorov, M.; Ballard, T.M.; Ozmen, L.; Spooren, W.; Wettstein, J.G.; Jaeschke, G.; Bear, M.F.; Lindemann, L. Chronic Pharmacological mGlu5 Inhibition Corrects Fragile X in Adult Mice. *Neuron* **2012**, *74*, 49–56. [CrossRef]

54. McBride, S.M.; Choi, C.H.; Wang, Y.; Liebelt, D.; Braunstein, E.; Ferreiro, D.; Sehgal, A.; Siwicki, K.K.; Dockendorff, T.C.; Nguyen, H.T.; et al. Pharmacological Rescue of Synaptic Plasticity, Courtship Behavior, and Mushroom Body Defects in a Drosophila Model of Fragile X Syndrome. *Neuron* **2005**, *45*, 753–764. [CrossRef] [PubMed]

55. Pan, L.; Broadie, K.S. Drosophila Fragile X Mental Retardation Protein and Metabotropic Glutamate Receptor A Convergently Regulate the Synaptic Ratio of Ionotropic Glutamate Receptor Subclasses. *J. Neurosci.* **2007**, *27*, 12378–12389. [CrossRef]

56. Chang, S.; Bray, S.M.; Li, Z.; Zarnescu, D.C.; He, C.; Jin, P.; Warren, S.T. Identification of small molecules rescuing fragile X syndrome phenotypes in Drosophila. *Nat. Methods* **2008**, *4*, 256–263. [CrossRef]

57. Pan, L.; Woodruff, E., 3rd; Liang, P.; Broadie, K. Mechanistic Relationships between Drosophila Fragile X Mental Retardation Protein and Metabotropic Glutamate Receptor A Signaling. *Mol. Cell. Neurosci.* **2008**, *37*, 747–760. [CrossRef] [PubMed]

58. Repicky, S.; Broadie, K. Metabotropic Glutamate Receptor-Mediated Use-Dependent down-Regulation of Synaptic Excitability Involves the Fragile X Mental Retardation Protein. *J. Neurophysiol.* **2009**, *101*, 672–687. [CrossRef] [PubMed]

59. Choi, C.H.; McBride, S.M.J.; Schoenfeld, B.P.; Liebelt, D.A.; Ferreiro, D.; Ferrick, N.J.; Hinchey, P.; Kollaros, M.; Rudominer, R.L.; Terlizzi, A.M.; et al. Age dependent cognitive impairment in a Drosophila Fragile X model and its pharmacological rescue. *Biogerontology* **2009**, *11*, 347–362. [CrossRef] [PubMed]

60. Kanellopoulos, A.K.; Semelidou, O.; Kotini, A.G.; Anezaki, M.; Skoulakis, E.M.C. Learning and Memory Deficits Consequent to Reduction of the Fragile X Mental Retardation Protein Result from Metabotropic Glutamate Receptor-Mediated Inhibition of cAMP Signaling in Drosophila. *J. Neurosci.* **2012**, *32*, 13111–13124. [CrossRef]

61. Tessier, C.R.; Broadie, K. Molecular and Genetic Analysis of the Drosophila Model of Fragile X Syndrome. *Results Probl. Cell Differ.* **2012**, *54*, 119–156.

62. Drozd, M.; Bardoni, B.; Capovilla, M. Modeling Fragile X Syndrome in Drosophila. *Front. Neurosci.* **2018**, *11*, 124. [CrossRef]

63. Yan, Q.; Rammal, M.; Tranfaglia, M.; Bauchwitz, R.; Bauchwitz, R. Suppression of two major Fragile X Syndrome mouse model phenotypes by the mGluR5 antagonist MPEP. *Neuropharmacology* **2005**, *49*, 1053–1066. [CrossRef]

64. De Vrij, F.M.; Levenga, J.; Van Der Linde, H.C.; Koekkoek, S.K.; De Zeeuw, C.I.; Nelson, D.L.; Oostra, B.A.; Willemsen, R. Rescue of behavioral phenotype and neuronal protrusion morphology in *Fmr1* KO mice. *Neurobiol. Dis.* **2008**, *31*, 127–132. [CrossRef] [PubMed]

65. Gandhi, R.M.; Kogan, C.S.; Messier, C.; Gandhi, R.M. 2-Methyl-6-(phenylethynyl) pyridine (MPEP) reverses maze learning and PSD-95 deficits in *Fmr1* knock-out mice. *Front. Cell. Neurosci.* **2014**, *8*, 70. [CrossRef] [PubMed]

66. Achuta, V.S.; Grym, H.; Putkonen, N.; Louhivuori, V.; Kärkkäinen, V.; Koistinaho, J.; Roybon, L.; Castrén, M.L. Metabotropic Glutamate Receptor 5 Responses Dictate Differentiation of Neural Progenitors to NMDA-Responsive Cells in Fragile X Syndrome. *Dev. Neurobiol.* **2017**, *77*, 438–453. [CrossRef] [PubMed]

67. Ronesi, J.A.; Collins, K.A.; Hays, S.A.; Tsai, N.-P.; Guo, W.; Birnbaum, S.G.; Hu, J.-H.; Worley, P.F.; Gibson, J.R.; Huber, K.M. Disrupted Homer scaffolds mediate abnormal mGluR5 function in a mouse model of fragile X syndrome. *Nat. Neurosci.* **2012**, *15*, 431–440. [CrossRef]

68. Guo, W.; Molinaro, G.; Collins, K.A.; Hays, S.A.; Paylor, R.; Worley, P.F.; Szumlinski, K.K.; Huber, K.M. Selective Disruption of Metabotropic Glutamate Receptor 5-Homer Interactions Mimics Phenotypes of Fragile X Syndrome in Mice. *J. Neurosci.* **2016**, *36*, 2131–2147. [CrossRef]

69. Bogdanik, L.; Mohrmann, R.; Ramaekers, A.; Bockaert, J.; Grau, Y.; Broadie, K.; Parmentier, M.-L. The Drosophila Metabotropic Glutamate Receptor DmGluRA Regulates Activity-Dependent Synaptic Facilitation and Fine Synaptic Morphology. *J. Neurosci.* **2004**, *24*, 9105–9116. [CrossRef]

70. Michel, C.I.; Kraft, R.; Restifo, L.L. Defective Neuronal Development in the Mushroom Bodies of Drosophila Fragile X Mental Retardation 1 Mutants. *J. Neurosci.* **2004**, *24*, 5798–5809. [CrossRef]

71. Porter, R.H.P.; Jaeschke, G.; Spooren, W.; Ballard, T.M.; Büttelmann, B.; Kolczewski, S.; Peters, J.-U.; Prinssen, E.; Wichmann, J.; Vieira, E.; et al. Fenobam: A Clinically Validated Nonbenzodiazepine Anxiolytic Is a Potent, Selective, and Noncompetitive mGlu5 Receptor Antagonist with Inverse Agonist Activity. *J. Pharmacol. Exp. Ther.* **2005**, *315*, 711–721. [CrossRef]

72. Berry-Kravis, E.; Hessl, D.; Coffey, S.; Hervey, C.; Schneider, A.; Yuhas, J.; Hutchison, J.; Snape, M.; Tranfaglia, M.; Nguyen, D.V.; et al. A pilot open label, single dose trial of fenobam in adults with fragile X syndrome. *J. Med Genet.* **2009**, *46*, 266–271. [CrossRef]

73. Vranesic, I.; Ofner, S.; Flor, P.J.; Bilbe, G.; Bouhelal, R.; Enz, A.; Desrayaud, S.; McAllister, K.; Kuhn, R.; Gasparini, F. AFQ056/mavoglurant, a Novel Clinically Effective mGluR5 Antagonist: Identification, SAR and Pharmacological Characterization. *Bioorg. Med. Chem.* **2014**, *22*, 5790–5803. [CrossRef]

74. Jacquemont, S.; Curie, A.; Portes, V.D.; Torrioli, M.G.; Berry-Kravis, E.; Hagerman, R.J.; Ramos, F.J.; Cornish, K.; He, Y.; Paulding, C.; et al. Epigenetic Modification of the *FMR1* Gene in Fragile X Syndrome Is Associated with Differential Response to the mGluR5 Antagonist AFQ056. *Sci. Transl. Med.* **2011**, *3*, 64ra1. [CrossRef]

75. Berry-Kravis, E.; Hagerman, R.; Jacquemont, S.; Charles, P.; Visootsak, J.; Brinkman, M.; Rerat, K.; Koumaras, B.; Zhu, L.; Barth, G.M.; et al. Mavoglurant in fragile X syndrome: Results of two randomized, double-blind, placebo-controlled trials. *Sci. Transl. Med.* **2016**, *8*, 321. [CrossRef] [PubMed]

76. Jaeschke, G.; Kolczewski, S.; Spooren, W.; Vieira, E.; Bitter-Stoll, N.; Boissin, P.; Borroni, E.; Büttelmann, B.; Ceccarelli, S.; Clemann, N.; et al. Metabotropic Glutamate Receptor 5 Negative Allosteric Modulators: Discovery of 2-Chloro-4-[1-(4-fluorophenyl)-2,5-dimethyl-1H-imidazol-4-ylethynyl]pyridine (Basimglurant, RO4917523), a Promising Novel Medicine for Psychiatric Diseases. *J. Med. Chem.* **2015**, *58*, 1358–1371. [CrossRef]

77. Lindemann, L.; Porter, R.H.; Scharf, S.H.; Kuennecke, B.; Bruns, A.; Von Kienlin, M.; Harrison, A.C.; Paehler, A.; Funk, C.; Gloge, A.; et al. Pharmacology of Basimglurant (RO4917523, RG7090), a Unique Metabotropic Glutamate Receptor 5 Negative Allosteric Modulator in Clinical Development for Depression. *J. Pharmacol. Exp. Ther.* **2015**, *353*, 213–233. [CrossRef] [PubMed]

78. Gantois, I.; Pop, A.S.; De Esch, C.E.; Buijsen, R.A.; Pooters, T.; Gomez-Mancilla, B.; Gasparini, F.; Oostra, B.A.; D'Hooge, R.; Willemsen, R. Chronic administration of AFQ056/Mavoglurant restores social behaviour in *Fmr1* knockout mice. *Behav. Brain* **2013**, *239*, 72–79. [CrossRef]

79. Pop, A.S.; Levenga, J.; de Esch, C.E.F.; Buijsen, R.A.M.; Nieuwenhuizen, I.M.; Li, T.; Isaacs, A.; Gasparini, F.; Oostra, B.A.; Willemsen, R. Rescue of Dendritic Spine Phenotype in *Fmr1* KO Mice with the mGluR5 Antagonist AFQ056/Mavoglurant. *Psychopharmacology* **2014**, *231*, 1227–1235. [CrossRef]

80. Levenga, J.; Hayashi, S.; De Vrij, F.M.; Koekkoek, S.K.; Van Der Linde, H.C.; Nieuwenhuizen, I.; Song, C.; Buijsen, R.A.; Pop, A.S.; GomezMancilla, B.; et al. AFQ056, a new mGluR5 antagonist for treatment of fragile X syndrome. *Neurobiol. Dis.* **2011**, *42*, 311–317. [CrossRef]

81. Tranfaglia, M. Roche Reports Fragile X Clinical Trial Negative Results. Available online: https://www.fraxa.org/roche-reports-clinical-trial-negative-results/ (accessed on 10 January 2019).

82. FRAXA Research Foundation. Novartis Discontinues Development of mavoglurant (AFQ056) for Fragile X Syndrome. Fragile X Research—FRAXA Research Foundation. Available online: https://www.fraxa.org/novartis-discontinues-development-mavoglurant-afq056-fragile-x-syndrome/ (accessed on 10 January 2019).

83. A Youssef, E.; Berry-Kravis, E.; Czech, C.; Hagerman, R.J.; Hessl, D.; Wong, C.Y.; Rabbia, M.; Deptula, D.; John, A.; Kinch, R.; et al. Effect of the mGluR5-NAM Basimglurant on Behavior in Adolescents and Adults with Fragile X Syndrome in a Randomized, Double-Blind, Placebo-Controlled Trial: FragXis Phase 2 Results. *Neuropsychopharmacol* **2017**, *43*, 503–512. [CrossRef] [PubMed]

84. Hagerman, R.; Jacquemont, S.; Berry-Kravis, E.; Des Portes, V.; Stanfield, A.; Koumaras, B.; Rosenkranz, G.; Murgia, A.; Wolf, C.; Apostol, G.; et al. Mavoglurant in Fragile X Syndrome: Results of Two Open-Label, Extension Trials in Adults and Adolescents. *Sci. Rep.* **2018**, *8*, 16970. [CrossRef] [PubMed]

85. D'Hulst, C.; Heulens, I.; Brouwer, J.R.; Willemsen, R.; De Geest, N.; Reeve, S.P.; De Deyn, P.P.; Hassan, B.A.; Kooy, R.F. Expression of the GABAergic system in animal models for fragile X syndrome and fragile X associated tremor/ataxia syndrome (FXTAS). *Brain Res.* **2009**, *1253*, 176–183. [CrossRef]

86. Braat, S.; D'Hulst, C.; Heulens, I.; De Rubeis, S.; Mientjes, E.; Nelson, D.L.; Willemsen, R.; Bagni, C.; Van Dam, D.; De Deyn, P.P.; et al. The GABAA receptor is an FMRP target with therapeutic potential in fragile X syndrome. *Cell Cycle* **2015**, *14*, 2985–2995. [CrossRef]

87. Zhang, N.; Peng, Z.; Tong, X.; Lindemeyer, A.K.; Cetina, Y.; Huang, C.S.; Olsen, R.W.; Otis, T.S.; Houser, C.R. Decreased Surface Expression of the δ Subunit of the GABAA Receptor Contributes to Reduced Tonic Inhibition in Dentate Granule Cells in a Mouse Model of Fragile X Syndrome. *Exp. Neurol.* **2017**, *297*, 168–178. [CrossRef] [PubMed]

88. He, Q.; Nomura, T.; Xu, J.; Contractor, A. The Developmental Switch in GABA Polarity Is Delayed in Fragile X Mice. *J. Neurosci.* **2014**, *34*, 446–450. [CrossRef] [PubMed]

89. Tyzio, R.; Nardou, R.; Ferrari, D.C.; Tsintsadze, T.; Shahrokhi, A.; Eftekhari, S.; Khalilov, I.; Tsintsadze, V.; Brouchoud, C.; Chazal, G.; et al. Oxytocin-Mediated GABA Inhibition During Delivery Attenuates Autism Pathogenesis in Rodent Offspring. *Science* **2014**, *343*, 675–679. [CrossRef] [PubMed]

90. Telias, M.; Segal, M.; Ben-Yosef, D. Immature Responses to GABA in Fragile X Neurons Derived from Human Embryonic Stem Cells. *Front. Cell. Neurosci.* **2016**, *10*, 167. [CrossRef]

91. Olmos-Serrano, J.L.; Corbin, J.G.; Burns, M.P. The GABAA Receptor Agonist THIP Ameliorates Specific Behavioral Deficits in the Mouse Model of Fragile X Syndrome. *Dev. Neurosci.* **2011**, *33*, 395–403. [CrossRef] [PubMed]

92. Sinclair, D.; Featherstone, R.; Naschek, M.; Nam, J.; Du, A.; Wright, S.; Pance, K.; Melnychenko, O.; Weger, R.; Akuzawa, S.; et al. GABA-B Agonist Baclofen Normalizes Auditory-Evoked Neural Oscillations and Behavioral Deficits in the *Fmr1* Knockout Mouse Model of Fragile X Syndrome. *Eneuro* **2017**, *4*, ENEURO.0380–16.2017. [CrossRef]

93. Qin, M.; Huang, T.; Kader, M.; Krych, L.; Xia, Z.; Burlin, T.; Zeidler, Z.; Zhao, T.; Smith, C.B. R-Baclofen Reverses a Social Behavior Deficit and Elevated Protein Synthesis in a Mouse Model of Fragile X Syndrome. *Int. J. Neuropsychopharmacol.* **2015**, *18*, 18. [CrossRef]

94. Henderson, C.; Wijetunge, L.; Kinoshita, M.N.; Shumway, M.; Hammond, R.S.; Postma, F.R.; Brynczka, C.; Rush, R.; Thomas, A.; Paylor, R.; et al. Reversal of Disease-Related Pathologies in the Fragile X Mouse Model by Selective Activation of GABAB Receptors with Arbaclofen. *Sci. Transl. Med.* **2012**, *4*, 152. [CrossRef]

95. Kang, J.-Y.; Chadchankar, J.; Vien, T.N.; Mighdoll, M.I.; Hyde, T.M.; Mather, R.J.; Deeb, T.Z.; Pangalos, M.N.; Brandon, N.J.; Dunlop, J.; et al. Deficits in the activity of presynaptic γ-aminobutyric acid type B receptors contribute to altered neuronal excitability in fragile X syndrome. *J. Boil. Chem.* **2017**, *292*, 6621–6632. [CrossRef] [PubMed]

96. Wang, L.; Wang, Y.; Zhou, S.; Yang, L.; Shi, Q.; Li, Y.; Zhang, K.; Yang, L.; Zhao, M.; Yang, Q.; et al. Imbalance between Glutamate and GABA in *Fmr1* Knockout Astrocytes Influences Neuronal Development. *Genes* **2016**, *7*, 45. [CrossRef] [PubMed]

97. Zhao, W.; Wang, J.; Song, S.; Li, F.; Yuan, F. Reduction of α1GABAA receptor mediated by tyrosine kinase C (PKC) phosphorylation in a mouse model of fragile X syndrome. *Int. J. Clin. Exp. Med.* **2015**, *8*, 13219–13226.

98. Fatemi, S.H.; Folsom, T.D. GABA receptor subunit distribution and FMRP–mGluR5 signaling abnormalities in the cerebellum of subjects with schizophrenia, mood disorders, and autism. *Schizophr. Res.* **2015**, *167*, 42–56. [CrossRef]

99. Braat, S.; Kooy, R.F. Fragile X syndrome neurobiology translates into rational therapy. *Drug Discov. Today* **2014**, *19*, 510–519. [CrossRef] [PubMed]

100. Lozano, R.; Hare, E.B.; Hagerman, R.J. Modulation of the GABAergic pathway for the treatment of fragile X syndrome. *Neuropsychiatr. Dis. Treat.* **2014**, *10*, 1769–1779. [PubMed]

101. Olmos-Serrano, J.L.; Paluszkiewicz, S.M.; Martin, B.S.; Kaufmann, W.E.; Corbin, J.G.; Huntsman, M.M. Defective GABAergic Neurotransmission and Pharmacological Rescue of Neuronal Hyperexcitability in the Amygdala in a Mouse Model of Fragile X Syndrome. *J. Neurosci.* **2010**, *30*, 9929–9938. [CrossRef] [PubMed]

102. Curia, G.; Papouin, T.; Séguéla, P.; Avoli, M. Downregulation of Tonic GABAergic Inhibition in a Mouse Model of Fragile X Syndrome. *Cereb. Cortex* **2008**, *19*, 1515–1520. [CrossRef]

103. Centonze, D.; Rossi, S.; Mercaldo, V.; Napoli, I.; Ciotti, M.T.; De Chiara, V.; Musella, A.; Prosperetti, C.; Calabresi, P.; Bernardi, G.; et al. Abnormal Striatal GABA Transmission in the Mouse Model for the Fragile X Syndrome. *Boil. Psychiatry* **2008**, *63*, 963–973. [CrossRef] [PubMed]

104. Berry-Kravis, E.; Hagerman, R.; Visootsak, J.; Budimirovic, D.; Kaufmann, W.E.; Cherubini, M.; Zarevics, P.; Walton-Bowen, K.; Wang, P.; Bear, M.F.; et al. Arbaclofen in fragile X syndrome: Results of phase 3 trials. *J. Neurodev. Disord.* **2017**, *9*, 3. [CrossRef]

105. Berry-Kravis, E.M.; Hessl, D.; Rathmell, B.; Zarevics, P.; Cherubini, M.; Walton-Bowen, K.; Mu, Y.; Nguyen, D.V.; Gonzalez-Heydrich, J.; Wang, P.P.; et al. Effects of STX209 (Arbaclofen) on Neurobehavioral Function in Children and Adults with Fragile X Syndrome: A Randomized, Controlled, Phase 2 Trial. *Sci. Transl. Med.* **2012**, *4*, 152ra127. [CrossRef]
106. Schaefer, T.L.; Davenport, M.H.; Grainger, L.M.; Robinson, C.K.; Earnheart, A.T.; Stegman, M.S.; Lang, A.L.; Ashworth, A.A.; Molinaro, G.; Huber, K.M.; et al. Acamprosate in a Mouse Model of Fragile X Syndrome: Modulation of Spontaneous Cortical Activity, ERK1/2 Activation, Locomotor Behavior, and Anxiety. *J. Neurodev. Disord.* **2017**, *9*, 6. [CrossRef]
107. Erickson, C.A.; Wink, L.K.; Ray, B.; Early, M.C.; Stiegelmeyer, E.; Mathieu-Frasier, L.; Patrick, V.; Lahiri, D.K.; McDougle, C.J. Impact of acamprosate on behavior and brain-derived neurotrophic factor: An open-label study in youth with fragile X syndrome. *Psychopharmacology* **2013**, *228*, 75–84. [CrossRef] [PubMed]
108. Ligsay, A.; Van Dijck, A.; Nguyen, D.V.; Lozano, R.; Chen, Y.; Bickel, E.S.; Hessl, D.; Schneider, A.; Angkustsiri, K.; Tassone, F.; et al. A randomized double-blind, placebo-controlled trial of ganaxolone in children and adolescents with fragile X syndrome. *J. Neurodev. Disord.* **2017**, *9*, 26. [CrossRef] [PubMed]
109. Jung, H.; Yoon, B.C.; Holt, C.E. Axonal mRNA Localization and Local Protein Synthesis in Nervous System Assembly, Maintenance and Repair. *Nat. Rev. Neurosci.* **2012**, *13*, 308–324. [CrossRef] [PubMed]
110. Qin, M.; Kang, J.; Burlin, T.V.; Jiang, C.; Smith, C.B. Postadolescent Changes in Regional Cerebral Protein Synthesis: An In Vivo Study in the *Fmr1* Null Mouse. *J. Neurosci.* **2005**, *25*, 5087–5095. [CrossRef] [PubMed]
111. Kumari, D.; Bhattacharya, A.; Nadel, J.; Moulton, K.; Zeak, N.M.; Glicksman, A.; Dobkin, C.; Brick, D.J.; Schwartz, P.H.; Smith, C.B.; et al. Identification of Fragile X Syndrome-Specific Molecular Markers in Human Fibroblasts: A Useful Model to Test the Efficacy of Therapeutic Drugs. *Hum. Mutat.* **2014**, *35*, 1485–1494. [CrossRef]
112. Jacquemont, S.; Pacini, L.; E Jønch, A.; Cencelli, G.; Rozenberg, I.; He, Y.; D'Andrea, L.; Pedini, G.; Eldeeb, M.; Willemsen, R.; et al. Protein synthesis levels are increased in a subset of individuals with fragile X syndrome. *Hum. Mol. Genet.* **2018**, *27*, 2039–2051. [CrossRef]
113. Qin, M.; Schmidt, K.C.; Zametkin, A.J.; Bishu, S.; Horowitz, L.M.; Burlin, T.V.; Xia, Z.; Huang, T.; Quezado, Z.M.; Smith, C.B. Altered Cerebral Protein Synthesis in Fragile X Syndrome: Studies in Human Subjects and Knockout Mice. *J. Cereb. Blood Flow Metab.* **2013**, *33*, 499–507. [CrossRef]
114. Liu, Z.-H.; Huang, T.; Smith, C.B. Lithium Reverses Increased Rates of Cerebral Protein Synthesis in a Mouse Model of Fragile X Syndrome. *Neurobiol. Dis.* **2012**, *45*, 1145. [CrossRef]
115. Gross, C.; Hoffmann, A.; Bassell, G.J.; Berry-Kravis, E.M. Therapeutic Strategies in Fragile X Syndrome: From Bench to Bedside and Back. *Neurotherapeutics* **2015**, *12*, 584–608. [CrossRef]
116. Jacquemont, S.; Berry-Kravis, E.; Hagerman, R.; von Raison, F.; Gasparini, F.; Apostol, G.; Ufer, M.; Des Portes, V.; Gomez-Mancilla, B. The Challenges of Clinical Trials in Fragile X Syndrome. *Psychopharmacology* **2014**, *231*, 1237–1250. [CrossRef]
117. Pop, A.S.; Gomez-Mancilla, B.; Neri, G.; Willemsen, R.; Gasparini, F. Fragile X Syndrome: A Preclinical Review on Metabotropic Glutamate Receptor 5 (mGluR5) Antagonists and Drug Development. *Psychopharmacology* **2014**, *231*, 1217–1226. [CrossRef] [PubMed]
118. Osterweil, E.K.; Chuang, S.-C.; Chubykin, A.A.; Sidorov, M.; Bianchi, R.; Wong, R.K.S.; Bear, M.F. Lovastatin corrects excess protein synthesis and prevents epileptogenesis in a mouse model of fragile X syndrome. *Neuron* **2013**, *77*, 243–250. [CrossRef]
119. Bhattacharya, A.; Kaphzan, H.; Alvarez-Dieppa, A.C.; Murphy, J.P.; Pierre, P.; Klann, E. Genetic Removal of p70 S6 Kinase 1 Corrects Molecular, Synaptic, and Behavioral Phenotypes in Fragile X Syndrome Mice. *Neuron* **2012**, *76*, 325–337. [CrossRef]
120. Reversal of Activity-Mediated Spine Dynamics and Learning Impairment in a Mouse Model of Fragile X Syndrome. Available online: https://www.ncbi.nlm.nih.gov/pubmed/24712992 (accessed on 11 January 2019).
121. Dolan, B.M.; Duron, S.G.; Campbell, D.A.; Vollrath, B.; Rao, B.S.S.; Ko, H.-Y.; Lin, G.G.; Govindarajan, A.; Choi, S.-Y.; Tonegawa, S.; et al. Rescue of fragile X syndrome phenotypes in *Fmr1* KO mice by the small-molecule PAK inhibitor FRAX486. *Proc. Natl. Acad. Sci. USA* **2013**, *110*, 5671–5676. [CrossRef]
122. Bhattacharya, A.; Mamcarz, M.; Mullins, C.; Choudhury, A.; Boyle, R.G.; Smith, D.G.; Walker, D.W.; Klann, E. Targeting Translation Control with p70 S6 Kinase 1 Inhibitors to Reverse Phenotypes in Fragile X Syndrome Mice. *Neuropsychopharmacology* **2016**, *41*, 1991–2000. [CrossRef]

123. Gross, C.; Nakamoto, M.; Yao, X.; Chan, C.-B.; Yim, S.Y.; Ye, K.; Warren, S.T.; Bassell, G.J. Excess Phosphoinositide 3-Kinase Subunit Synthesis and Activity as a Novel Therapeutic Target in Fragile X Syndrome. *J. Neurosci.* **2010**, *30*, 10624–10638. [CrossRef]

124. Hayashi, M.L.; Rao, B.S.S.; Seo, J.-S.; Choi, H.-S.; Dolan, B.M.; Choi, S.-Y.; Chattarji, S.; Tonegawa, S.; Rao, B.S.S. Inhibition of p21-activated kinase rescues symptoms of fragile X syndrome in mice. *Proc. Natl. Acad. Sci. USA* **2007**, *104*, 11489–11494. [CrossRef]

125. Tian, M.; Zeng, Y.; Hu, Y.; Yuan, X.; Liu, S.; Li, J.; Lu, P.; Sun, Y.; Gao, L.; Fu, D.; et al. 7, 8-Dihydroxyflavone induces synapse expression of AMPA GluA1 and ameliorates cognitive and spine abnormalities in a mouse model of fragile X syndrome. *Neuropharmacology* **2015**, *89*, 43–53. [CrossRef]

126. Sawicka, K.; Pyronneau, A.; Chao, M.; Bennett, M.V.L.; Zukin, R.S. Elevated ERK/p90 ribosomal S6 kinase activity underlies audiogenic seizure susceptibility in fragile X mice. *Proc. Natl. Acad. Sci. USA* **2016**, *113*, E6290–E6297. [CrossRef]

127. Pasciuto, E.; Ahmed, T.; Wahle, T.; Gardoni, F.; D'Andrea, L.; Pacini, L.; Jacquemont, S.; Tassone, F.; Balschun, D.; Dotti, C.G.; et al. Dysregulated ADAM10-Mediated Processing of APP during a Critical Time Window Leads to Synaptic Deficits in Fragile X Syndrome. *Neuron* **2015**, *87*, 382–398. [CrossRef] [PubMed]

128. Gross, C.; Bassell, G.J. Neuron-specific regulation of class I PI3K catalytic subunits and their dysfunction in brain disorders. *Front. Neurosci.* **2014**, *7*, 7. [CrossRef] [PubMed]

129. Gross, C.; Bassell, G.J. Excess Protein Synthesis in FXS Patient Lymphoblastoid Cells Can Be Rescued with a p110β-Selective Inhibitor. *Mol. Med.* **2012**, *18*, 336–345. [CrossRef] [PubMed]

130. Gross, C.; Raj, N.; Molinaro, G.; Allen, A.G.; Whyte, A.J.; Gibson, J.R.; Huber, K.M.; Gourley, S.L.; Bassell, G.J. Selective role of the catalytic PI3K subunit p110β in impaired higher-order cognition in Fragile X syndrome. *Cell Rep.* **2015**, *11*, 681–688. [CrossRef] [PubMed]

131. Gross, C.; Chang, C.-W.; Kelly, S.M.; Bhattacharya, A.; McBride, S.M.; Danielson, S.W.; Jiang, M.Q.; Chan, C.B.; Ye, K.; Gibson, J.R.; et al. Increased expression of the PI3K enhancer PIKE mediates deficits in synaptic plasticity and behavior in Fragile X syndrome. *Cell Rep.* **2015**, *11*, 727–736. [CrossRef] [PubMed]

132. Klann, E.; Dever, T.E. Biochemical mechanisms for translational regulation in synaptic plasticity. *Nat. Rev. Neurosci.* **2004**, *5*, 931–942. [CrossRef]

133. Richter, J.D.; Sonenberg, N. Regulation of cap-dependent translation by eIF4E inhibitory proteins. *Nat. Cell Boil.* **2005**, *433*, 477–480. [CrossRef] [PubMed]

134. Brems, H.; Legius, E.; Bagni, C.; Borrie, S.C. Cognitive Dysfunctions in Intellectual Disabilities: The Contributions of the Ras-MAPK and PI3K-AKT-mTOR Pathways. *Annu. Genom. Hum. Genet.* **2017**, *18*, 115–142.

135. Zhu, P.J.; Chen, C.-J.; Mays, J.; Stoica, L.; Costa-Mattioli, M. mTORC2, but not mTORC1, is required for hippocampal mGluR-LTD and associated behaviors. *Nat. Neurosci.* **2018**, *21*, 799–802. [CrossRef]

136. Hoeffer, C.A.; Sanchez, E.; Hagerman, R.J.; Mu, Y.; Nguyen, D.V.; Wong, H.; Whelan, A.M.; Zukin, R.S.; Klann, E.; Tassone, F. Altered mTOR signaling and enhanced CYFIP2 expression levels in subjects with Fragile X syndrome. *Genes Brain* **2012**, *11*, 332–341. [CrossRef]

137. Sharma, A.; Hoeffer, C.A.; Takayasu, Y.; Miyawaki, T.; McBride, S.M.; Klann, E.; Zukin, R.S. Dysregulation of mTOR Signaling in Fragile X Syndrome. *J. Neurosci.* **2010**, *30*, 694–702. [CrossRef]

138. Monyak, R.E.; Emerson, D.; Schoenfeld, B.P.; Zheng, X.; Chambers, D.B.; Rosenfelt, C.; Langer, S.; Hinchey, P.; Choi, C.H.; McDonald, T.V.; et al. Insulin Signaling Misregulation Underlies Circadian and Cognitive Deficits in a Drosophila Fragile X Model. *Mol. Psychiatry* **2017**, *22*, 1140–1148. [CrossRef]

139. Gantois, I.; Khoutorsky, A.; Popic, J.; Aguilar-Valles, A.; Freemantle, E.; Cao, R.; Sharma, V.; Pooters, T.; Nagpal, A.; Skalecka, A.; et al. Metformin ameliorates core deficits in a mouse model of fragile X syndrome. *Nat. Med.* **2017**, *23*, 674–677. [CrossRef]

140. Kim, S.H.; Markham, J.A.; Weiler, I.J.; Greenough, W.T. Aberrant early-phase ERK inactivation impedes neuronal function in fragile X syndrome. *Proc. Natl. Acad. Sci. USA* **2008**, *105*, 4429–4434. [CrossRef]

141. Hou, L.; Antion, M.D.; Hu, D.; Spencer, C.M.; Paylor, R.; Klann, E. Dynamic Translational and Proteasomal Regulation of Fragile X Mental Retardation Protein Controls mGluR-Dependent Long-Term Depression. *Neuron* **2006**, *51*, 441–454. [CrossRef]

142. Weng, N.; Weiler, I.J.; Sumis, A.; Greenough, W.T.; Berry-Kravis, E.; Berry-Kravis, E. Early-phase ERK activation as a biomarker for metabolic status in fragile X syndrome. *Am. J. Med. Genet. Part B Neuropsychiatr. Genet.* **2008**, *147*, 1253–1257. [CrossRef]

143. Berry-Kravis, E.; Sumis, A.; Hervey, C.; Nelson, M.; Porges, S.W.; Weng, N.; Weiler, I.J.; Greenough, W.T. Open-Label Treatment Trial of Lithium to Target the Underlying Defect in Fragile X Syndrome. *J. Dev. Behav. Pediatr.* **2008**, *29*, 293–302. [CrossRef]

144. Erickson, C.A.; Weng, N.; Weiler, I.J.; Greenough, W.T.; Stigler, K.A.; Wink, L.K.; McDougle, C.J. Open-label riluzole in fragile X syndrome. *Brain Res.* **2011**, *1380*, 264–270. [CrossRef]

145. Pellerin, D.; Çaku, A.; Fradet, M.; Bouvier, P.; Dubé, J.; Corbin, F. Lovastatin corrects ERK pathway hyperactivation in fragile X syndrome: Potential of platelet's signaling cascades as new outcome measures in clinical trials. *Biomarkers* **2016**, *21*, 1–12. [CrossRef]

146. Zamzow, R. Drug Duo Delivers Brain, Behavioral Benefits for Fragile X Syndrome | Spectrum | Autism Research News. Available online: https://www.spectrumnews.org/news/drug-duo-delivers-brain-behavioral-benefits-fragile-x-syndrome/ (accessed on 11 January 2019).

147. Reinhard, S.M.; Razak, K.; Ethell, I.M. A delicate balance: Role of MMP-9 in brain development and pathophysiology of neurodevelopmental disorders. *Front. Cell. Neurosci.* **2015**, *9*, 280. [CrossRef]

148. Janusz, A.; Miłek, J.; Perycz, M.; Pacini, L.; Bagni, C.; Kaczmarek, L.; Dziembowska, M. The Fragile X Mental Retardation Protein Regulates Matrix Metalloproteinase 9 mRNA at Synapses. *J. Neurosci.* **2013**, *33*, 18234–18241. [CrossRef]

149. Taylor, A.K.; Tassone, F.; Dyer, P.N.; Hersch, S.M.; Harris, J.B.; Greenough, W.T.; Hagerman, R.J. Tissue heterogeneity of the *FMR1* mutation in a high-functioning male with fragile X syndrome. *Am. J. Med Genet.* **1999**, *84*, 233–239. [CrossRef]

150. Sidhu, H.; Dansie, L.E.; Hickmott, P.W.; Ethell, D.W.; Ethell, I.M. Genetic Removal of Matrix Metalloproteinase 9 Rescues the Symptoms of Fragile X Syndrome in a Mouse Model. *J. Neurosci.* **2014**, *34*, 9867–9879. [CrossRef]

151. Lovelace, J.W.; Wen, T.H.; Reinhard, S.; Hsu, M.S.; Sidhu, H.; Ethell, I.M.; Binder, D.K.; Razak, K.A. Matrix metalloproteinase-9 deletion rescues auditory evoked potential habituation deficit in a mouse model of Fragile X Syndrome. *Neurobiol. Dis.* **2016**, *89*, 126–135. [CrossRef]

152. Bilousova, T.V.; Dansie, L.; Ngo, M.; Aye, J.; Charles, J.R.; Ethell, D.W.; Ethell, I.M. Minocycline Promotes Dendritic Spine Maturation and Improves Behavioural Performance in the Fragile X Mouse Model. *J. Med. Genet.* **2009**, *46*, 94–102. [CrossRef]

153. Utari, A.; Chonchaiya, W.; Rivera, S.M.; Schneider, A.; Hagerman, R.J.; Faradz, S.M.H.; Ethell, I.M.; Nguyen, D.V. Side Effects of Minocycline Treatment in Patients with Fragile X Syndrome and Exploration of Outcome Measures. *Am. J. Intellect. Dev. Disabil.* **2010**, *115*, 433–443. [CrossRef]

154. Paribello, C.; Tao, L.; Folino, A.; Berry-Kravis, E.; Tranfaglia, M.; Ethell, I.M.; Ethell, D.W. Open-label add-on treatment trial of minocycline in fragile X syndrome. *BMC Neurol.* **2010**, *10*, 91. [CrossRef]

155. Leigh, M.J.S.; Nguyen, D.V.; Mu, Y.; Winarni, T.I.; Schneider, A.; Chechi, T.; Polussa, J.; Doucet, P.; Tassone, F.; Rivera, S.M.; et al. A Randomized Double-Blind, Placebo-Controlled Trial of Minocycline in Children and Adolescents with Fragile X Syndrome. *J. Dev. Behav. Pediatr.* **2013**, *34*, 147–155. [CrossRef]

156. Dziembowska, M.; Pretto, D.I.; Janusz, A.; Kaczmarek, L.; Leigh, M.J.; Gabriel, N.; Durbin-Johnson, B.; Hagerman, R.J.; Tassone, F. High MMP-9 activity levels in fragile X syndrome are lowered by minocycline. *Am. J. Med. Genet. Part A* **2013**, *161*, 1897–1903. [CrossRef]

157. AlOlaby, R.R.; Sweha, S.R.; Silva, M.; Durbin-Johnson, B.; Yrigollen, C.M.; Pretto, D.; Hagerman, R.J.; Tassone, F. Molecular biomarkers predictive of sertraline treatment response in young children with fragile X syndrome. *Brain Dev.* **2017**, *39*, 483–492. [CrossRef]

158. Kim, S.W.; Cho, K.J. Activity-dependent alterations in the sensitivity to BDNF-TrkB signaling may promote excessive dendritic arborization and spinogenesis in fragile X syndrome in order to compensate for compromised postsynaptic activity. *Med Hypotheses* **2014**, *83*, 429–435. [CrossRef]

159. Louhivuori, V.; Vicario, A.; Uutela, M.; Rantamäki, T.; Louhivuori, L.M.; Castrén, E.; Tongiorgi, E.; Åkerman, K.E.; Castrén, M.L. BDNF and TrkB in neuronal differentiation of *Fmr1*-knockout mouse. *Neurobiol. Dis.* **2011**, *41*, 469–480. [CrossRef]

160. Uutela, M.; Lindholm, J.; Louhivuori, V.; Wei, H.; Louhivuori, L.M.; Pertovaara, A.; Akerman, K.; Castrén, E.; Castrén, M.L. Reduction of BDNF expression in *Fmr1* knockout mice worsens cognitive deficits but improves hyperactivity and sensorimotor deficits. *Genes Brain* **2012**, *11*, 513–523. [CrossRef]

161. Castrén, M.L.; Castrén, E. BDNF in Fragile X Syndrome. *Neuropharmacology* **2014**, *76*, 729–736. [CrossRef]

162. Louhivuori, V.; Arvio, M.; Soronen, P.; Oksanen, V.; Paunio, T.; Castrén, M.L. The Val66Met polymorphism in the BDNF gene is associated with epilepsy in fragile X syndrome. *Epilepsy Res.* **2009**, *85*, 114–117. [CrossRef]

163. Westmark, C.J.; Westmark, P.R.; O'Riordan, K.J.; Ray, B.C.; Hervey, C.M.; Salamat, M.S.; Abozeid, S.H.; Stein, K.M.; Stodola, L.A.; Tranfaglia, M.; et al. Reversal of Fragile X Phenotypes by Manipulation of AβPP/Aβ Levels in *Fmr1* KO Mice. *PLoS ONE* **2011**, *6*, e26549. [CrossRef]

164. Erickson, C.A.; Ray, B.; Maloney, B.; Wink, L.K.; Bowers, K.; Schaefer, T.L.; McDougle, C.J.; Sokol, D.K.; Lahiri, D.K. Impact of Acamprosate on Plasma Amyloid-β Precursor Protein in Youth: A Pilot Analysis in Fragile X Syndrome-Associated and Idiopathic Autism Spectrum Disorder Suggests a Pharmacodynamic Protein Marker. *J. Psychiatr.* **2014**, *59*, 220–228. [CrossRef]

165. Lee, H.Y.; Jan, L.Y. Fragile X syndrome: Mechanistic insights and therapeutic avenues regarding the role of potassium channels. *Curr. Opin. Neurobiol.* **2012**, *22*, 887–894. [CrossRef]

166. Ferron, L.; Nieto-Rostro, M.; Cassidy, J.S.; Dolphin, A.C. Fragile X mental retardation protein controls synaptic vesicle exocytosis by modulating N-type calcium channel density. *Nat. Commun.* **2014**, *5*, 3628. [CrossRef]

167. Castagnola, S.; Delhaye, S.; Folci, A.; Paquet, A.; Brau, F.; Duprat, F.; Jarjat, M.; Grossi, M.; Béal, M.; Martin, S.; et al. New Insights Into the Role of Cav2 Protein Family in Calcium Flux Deregulation in *Fmr1*-KO Neurons. *Front. Mol. Neurosci.* **2018**, *11*, 11. [CrossRef]

168. Castaño, Z.; Gordon-Weeks, P.R.; Kypta, R.M.; Gordon-Weeks, P.R.; Gordon-Weeks, P.R. The neuron-specific isoform of glycogen synthase kinase-3β is required for axon growth. *J. Neurochem.* **2010**, *113*, 117–130. [CrossRef]

169. Min, W.W.; Yuskaitis, C.J.; Yan, Q.; Sikorski, C.; Chen, S.; Jope, R.S.; Bauchwitz, R.P. Elevated Glycogen Synthase Kinase-3 Activity in Fragile X Mice: Key Metabolic Regulator with Evidence for Treatment Potential. *Neuropharmacology* **2009**, *56*, 463–472. [CrossRef]

170. Portis, S.; Giunta, B.; Obregon, D.; Tan, J. The role of glycogen synthase kinase-3 signaling in neurodevelopment and fragile X syndrome. *Int. J. Physiol. Pathophysiol. Pharmacol.* **2012**, *4*, 140–148.

171. Yuskaitis, C.J.; Mines, M.A.; King, M.K.; Sweatt, J.D.; Miller, C.A.; Jope, R.S. Lithium ameliorates altered glycogen synthase kinase-3 and behavior in a mouse model of Fragile X syndrome. *Biochem. Pharmacol.* **2010**, *79*, 632–646. [CrossRef]

172. Mines, M.A.; Yuskaitis, C.J.; King, M.K.; Beurel, E.; Jope, R.S. GSK3 Influences Social Preference and Anxiety-Related Behaviors during Social Interaction in a Mouse Model of Fragile X Syndrome and Autism. *PLoS ONE* **2010**, *5*, e9706. [CrossRef]

173. Guo, W.; Murthy, A.C.; Zhang, L.; Johnson, E.B.; Schaller, E.G.; Allan, A.M.; Zhao, X. Inhibition of GSK3β improves hippocampus-dependent learning and rescues neurogenesis in a mouse model of fragile X syndrome. *Hum. Mol. Genet.* **2011**, *21*, 681–691. [CrossRef]

174. Franklin, A.V.; King, M.K.; Palomo, V.; Martinez, A.; McMahon, L.L.; Jope, R.S. Glycogen Synthase Kinase-3 Inhibitors Reverse Deficits in Long-Term Potentiation and Cognition in Fragile X Mice. *Biol. Psychiatry* **2014**, *75*, 198–206. [CrossRef]

175. Yuskaitis, C.J.; Beurel, E.; Jope, R.S. Evidence of Reactive Astrocytes but Not Peripheral Immune System Activation in a Mouse Model of Fragile X Syndrome. *Biochim. Biophys. Acta.* **2010**, *1802*, 1006–1012. [CrossRef]

176. Mines, M.A.; Jope, R.S.; Mines, M.M. Glycogen Synthase Kinase-3: A Promising Therapeutic Target for Fragile X Syndrome. *Front. Neurosci.* **2011**, *4*, 35. [CrossRef]

177. Goebel-Goody, S.M.; Lombroso, P.J. Taking STEPs Forward to Understand Fragile X Syndrome. *Results Probl. Cell Differ.* **2012**, *54*, 223–241.

178. Goebel-Goody, S.M.; Baum, M.; Paspalas, C.D.; Fernandez, S.M.; Carty, N.C.; Kurup, P.; Lombroso, P.J. Therapeutic Implications for Striatal-Enriched Protein Tyrosine Phosphatase (STEP) in Neuropsychiatric Disorders. *Pharmacol. Rev.* **2012**, *64*, 65–87. [CrossRef]

179. Johnson, M.A.; Lombroso, P.J. A Common STEP in the Synaptic Pathology of Diverse Neuropsychiatric Disorders. *Yale J. Boil. Med.* **2012**, *85*, 481–490.

180. Darnell, J.C.; Van Driesche, S.J.; Zhang, C.; Hung, K.Y.S.; Mele, A.; Fraser, C.E.; Stone, E.F.; Chen, C.; Fak, J.J.; Chi, S.W.; et al. FMRP stalls ribosomal translocation on mRNAs linked to synaptic function and autism. *Cell* **2011**, *146*, 247–261. [CrossRef] [PubMed]

181. Royston, S.; Tagliatela, S.M.; Naegele, J.R.; Lombroso, P.J.; Tagliatela, S.; Goebel-Goody, S.M.; Wilson-Wallis, E.D.; Goebel-Goody, S.M.; Wilson-Wallis, E.D.; Goebel-Goody, S.M.; et al. Genetic manipulation of STEP reverses behavioral abnormalities in a fragile X syndrome mouse model. *Genes Brain* **2012**, *11*, 586–600.

182. Xu, J.; Chatterjee, M.; Baguley, T.D.; Brouillette, J.; Kurup, P.; Ghosh, D.; Kanyo, J.; Zhang, Y.; Seyb, K.; Ononenyi, C.; et al. Inhibitor of the Tyrosine Phosphatase STEP Reverses Cognitive Deficits in a Mouse Model of Alzheimer's Disease. *PLoS Biol.* **2014**, *12*, e1001923. [CrossRef]

183. Chatterjee, M.; Kurup, P.K.; Lundbye, C.J.; Hugger Toft, A.K.; Kwon, J.; Benedict, J.; Kamceva, M.; Banke, T.G.; Lombroso, P.J. STEP Inhibition Reverses Behavioral, Electrophysiologic, and Synaptic Abnormalities in *Fmr1* KO Mice. *Neuropharmacology* **2018**, *128*, 43–53. [CrossRef]

184. Ashwood, P.; Nguyen, D.V.; Hessl, D.; Hagerman, R.J.; Tassone, F. Plasma Cytokine Profiles in Fragile X Subjects: Is There a Role for Cytokines in the Pathogenesis? *Brain Behav. Immun.* **2010**, *24*, 898–902. [CrossRef] [PubMed]

185. Tabet, R.; Moutin, E.; Becker, J.A.J.; Heintz, D.; Fouillen, L.; Flatter, E.; Kręzel, W.; Alunni, V.; Koebel, P.; Dembélé, D.; et al. Fragile X Mental Retardation Protein (FMRP) Controls Diacylglycerol Kinase Activity in Neurons. *Proc. Natl. Acad. Sci. USA.* **2016**, *113*, E3619–E3628. [CrossRef]

186. Lin, S.-L. microRNAs and Fragile X Syndrome. *Adv. Exp. Med. Biol.* **2015**, *888*, 107–121. [PubMed]

187. Lin, S.-L.; Chang, S.-J.; Ying, S.-Y. First in vivo evidence of microRNA-induced fragile X mental retardation syndrome. *Mol. Psychiatry* **2006**, *11*, 616–617. [CrossRef]

188. Chang, S.-J.E.; Chang-Lin, S.; Chang, D.C.; Chang, C.P.; Lin, S.-L.; Ying, S.-Y. Repeat-Associated MicroRNAs Trigger Fragile X Mental Retardation-Like Syndrome in Zebrafish~!2008-10-29~!2008-11-12~!2008-11-26~! *Open Neuropsychopharmacol. J.* **2008**, *1*, 6–18. [CrossRef]

189. Lin, S.-L.; Ying, S.-Y. Role of Repeat-Associated MicroRNA (ramRNA) in Fragile X Syndrome (FXS). In *Current Perspectives in microRNAs (miRNA)*; Springer: Dordrecht, The Netherlands, 2008; pp. 245–266.

190. Muddashetty, R.S.; Nalavadi, V.C.; Gross, C.; Yao, X.; Xing, L.; Laur, O.; Warren, S.T.; Bassell, G.J. Reversible Inhibition of PSD-95 mRNA Translation by miR-125a, FMRP Phosphorylation, and mGluR Signaling. *Mol. Cell* **2011**, *42*, 673–688. [CrossRef] [PubMed]

191. Edbauer, D.; Neilson, J.R.; Foster, K.A.; Wang, C.-F.; Seeburg, D.P.; Batterton, M.N.; Tada, T.; Dolan, B.M.; Sharp, P.A.; Sheng, M. Regulation of synaptic structure and function by FMRP-associated microRNAs miR-125b and miR-132. *Neuron* **2010**, *65*, 373–384. [CrossRef]

192. Liu, T.; Wan, R.-P.; Tang, L.-J.; Liu, S.-J.; Li, H.-J.; Zhao, Q.-H.; Liao, W.-P.; Sun, X.-F.; Yi, Y.-H.; Long, Y.-S. A MicroRNA Profile in *Fmr1* Knockout Mice Reveals MicroRNA Expression Alterations with Possible Roles in Fragile X Syndrome. *Mol. Neurobiol.* **2015**, *51*, 1053–1063. [CrossRef] [PubMed]

193. Fazeli, Z.; Ghaderian, S.M.H.; Najmabadi, H.; Omrani, M.D. High expression of miR-510 was associated with CGG expansion located at upstream of *FMR1* into full mutation. *J. Cell. Biochem.* **2018**, *120*, 1916–1923. [CrossRef]

Review

Modelling Protein Synthesis as A Biomarker in Fragile X Syndrome Patient-Derived Cells

Rakhi Pal and Aditi Bhattacharya *

Centre for Brain Development and Repair, Institute for Stem Cell Biology and Regenerative Medicine, GKVK Post, Bellary Road, Bengaluru 560065, India; rakhip@instem.res.in
* Correspondence: aditi@instem.res.in; Tel.: +91 80 67176218

Received: 3 December 2018; Accepted: 6 March 2019; Published: 11 March 2019

Abstract: The most conserved molecular phenotype of Fragile X Syndrome (FXS) is aberrant protein synthesis. This has been validated in a variety of experimental model systems from zebrafish to rats, patient-derived lymphoblasts and fibroblasts. With the advent of personalized medicine paradigms, patient-derived cells and their derivatives are gaining more translational importance, not only to model disease in a dish, but also for biomarker discovery. Here we review past and current practices of measuring protein synthesis in FXS, studies in patient derived cells and the inherent challenges in measuring protein synthesis in them to offer usable avenues of modeling this important metabolic metric for further biomarker development.

Keywords: protein synthesis; Fragile X Syndrome; biomarker; iPSC; fibroblast; lymphoblast

---

1. Fragile X Syndrome as a Translational Control Disorder—Early Studies Leading to General Consensus

Fragile X Syndrome (FXS) is a triplet repeat disorder, where runaway expansion of CGG repeats at the promoter region of *FMR1* gene causes promoter hypermethylation leading to failure of gene transcription and subsequently loss of protein expression of Fragile X mental retardation protein (FMRP) [1]. FMRP is a versatile protein important in a variety of cellular processes which include ribonucleoparticle packaging and transport [2], mediating micro RNA interactions with target mRNA [3,4], translation elongation [5] and interaction with large conductance ion channels [6]. It is, therefore, not unexpected that the major role of FMRP is in determining timely mRNA translation and by extension, loss of FMRP in FXS, has been noted to cause an imbalance in de novo protein synthesis.

The above states the general consensus that has emerged in the FXS field, through multiple progressive contributions from varied groups [7]. In the 1990s, FMRP as a constituent of ribonucleoparticle (RNP) and mRNA binding protein was uncovered, following which its role in sensing synaptic activity and coupling it to trafficking of RNPs in dendrites and spines was established [8–10]. The association of FMRP with ribosomes was established by multiple studies [11–16], while the evidence that FMRP suppresses the translation of its target mRNAs was provided by Laggerbauer et al. [17] and Li et al. [18] in rabbit reticulocyte lysates. Two seminal studies in 2011 [19] and 2014 [20] showed that FMRP actively engages in ribosome stalling and does so by structurally occupying the cleft between ribosomal subunits and mRNA respectively. These findings firmly established the idea of a 'translational brake', which received ancillary support from studies that showed FRMP-Cyfip1 association occludes eIF4E from interacting with a productive cap-binding complex and initiating translation [21].

This mechanism of FMRP action led to a natural investigation into the net rates of proteins accumulation in cells lacking FMRP, hence modeling FXS. It was actually in rabbit reticulocytes that loss of FMRP was first reported to cause enhanced protein synthesis [18]. However, almost all work that can be found on this topic is done in neurons and brain slice preparations from FXS mouse models, since it was believed that FMRP expression was largely limited to neurons. Expression profiling in the brain was undertaken by Devys et al., [22] which registered the highest expression of FMRP in cerebellar grey matter in human sections, while Bakker et al., [23] did so in mice tissues. Both studies observed little to no glial staining and focused primarily on neurons. Taminini et al. [24] testing a variety of FMRP antibodies, concluded that majority of FMRP staining was neuronal, giving rise to the notion that FMRP expression is limited to neurons in CNS. It took Wang et al., [25] followed by Pacey and Doering [26], to show the developmental stage specific expression of FMRP in glial cells. It is noteworthy that the initial characterization of the FXS mice model [27] included spine morphology and behavioral experiments. Beebe-Smith and colleagues [28] provided the first metabolic readout in FXS mice, showing elevated cerebral glucose metabolism rates across 38 brain regions. The highest elevation was seen in areas that were active in behaviors such as open field activity and passive avoidance or areas with highest FMRP expression in control animals. This group followed up this work with the first comprehensive study of protein synthesis rates in 2005 [29] using in vivo L-[1-14C] leucine intra-arterial introduction to label nascent proteins within a 60 min time window in four- and six-month-old mice. A total of 75 brain areas were monitored, and overall a 10% increase in basal protein synthesis rates was reported compared to wild type littermates, with differences ranging from 18% in the paraventricular nucleus to 4% in the periaqueductal grey. This study forms an important resource for any proteostatic study in FXS, but it is relevant to note that these were averaged across a brain region and did not distinguish between specific cell types.

Experiments tracking protein synthesis in FXS model mice have largely concentrated on the hippocampus and that too in the context of activation of metabotropic glutamate receptor 5 (mGluR5), [30–32]. This focus on one specific brain area arose from data that showed that mGluR activation in area CA1 of the hippocampus induced a protein-synthesis dependent form of long-term depression (LTD) [33] and that this LTD was exaggerated in FXS [34]. Additionally, FMRP was shown to be a conduit of this mGluR5-protein synthesis connection by Muddashetty et al. [35] using mouse synaptoneurosomes and S^{35} methionine labelling of nascent proteins. Hence hippocampal mGluR5-LTD and protein synthesis measures in slices quickly emerged as two experiments that could be used as yardsticks to measure the efficacy of pharmacologic and genetic interventions in mice models over the next several years. The biochemical readout of de novo protein synthesis was facilitated by the development of a method to measure translation in acutely-prepared intact hippocampal slices [30], marrying techniques of electrophysiological slice preparation and incubation with metabolic S^{35} methionine labeling. This technique enjoys widespread acceptance in the FXS community [31,32,36,37]. Measuring protein synthesis rates in vivo, in humans, has been fraught with experimental confounds like anesthetic administration [38]. However, the same group has recently published a method to measure cerebral translation rates in awake subjects as well [39] and hence raises the expectation of doing the same in FXS cohorts.

With the advent of non-radioactive probes for measuring protein synthesis in the past decade, it has become easier for a variety of labs and clinics to measure protein synthesis. These tools were quickly embraced for studying translation control in FXS. SUnSET or Surface sensing of translation [40] was employed by Bhattacharya et al [41], to check whether a genetic deletion of S6K1 rescued FXS mouse phenotypes. SUnSET employs sub-critical levels of puromycin as a t-RNA analog that end labels nascent proteins in the ribosomes. SUnSET was subsequently used to measure translation in FXS model mice [42–44] and human patient derived fibroblasts [45]. Cell-type specific monitoring of translation using non-canonical amino acid tagging (NCAT, [46]) was first employed to measure translation rates in FXS patient derived lymphoblastoid cells (LCLs) [47], following which this technique was adopted for use in acute brain slices [48,49]. This method has two variants BONCAT (Bio-orthogonal non-canonical amino acid tagging) and FUNCAT (fluorescent non-canonical amino acid tagging) which employs orthogonal amino acid substitutes to label newly synthesized proteins, which are then tagged using click-chemistry and then detected by western blot (BONCAT) using biotin or fluorescent-tagged alkynes or azides. It is currently unclear if there is any disruption in astrocytic protein synthesis in FXS.

Taken together, the nexus of protein synthesis, FMRP loss and FXS pathology has been a cornerstone of mechanism-based research in this field. Using multiple measurement techniques, in multiple model systems (mouse, rat, drosophila, zebrafish) it can be shown that there is a surfeit of basal translation in FXS models compared to matched controls [30,36,41,50,51]. Resetting translational homeostasis using a variety of interventions [30,37,41,43,52–55] have yielded ameliorative effects on a wide range of FXS phenotypes correlated with the human condition. Therefore, measuring protein synthesis and its regulation has become an inescapable benchmark of model validity and/or monitoring treatment outcome efficacy.

2. Patient-Derived Models of FXS that Model 'Disease in a Dish'

Studying neurologic disorders, are amongst the most challenging largely due to inaccessibility of neural tissue. The traditional source has been post mortem samples, which are perhaps more useful in studying pathological changes in ageing [56] and neurotrauma [57]. It is also easier to build a sample repository for these conditions simply due to higher numbers and lifespan ranges of the patients. In contrast it is extremely challenging to retrieve post-mortem tissue for neurodevelopmental disorders like ASDs, intellectual disabilities, spina bifida, fetal alcohol syndrome, etc., [58] owing to younger ages of the patients and far fewer fatalities. Even when post-mortem tissue is available, experimental results are strongly influenced by the manner in which tissue was harvested, preserved, tissue pH, transit time, and duration of storage. Finally, measuring live cell metrics, like metabolism and protein synthesis, is not possible in post-mortem tissues.

Human Embryonic stem cells (huESC) were first isolated by Thomson et al., [59] in 1998 from the inner cell mass of the blastocyst of pre-implantation human embryos. These cells are pluripotent and can be further differentiated to cell types of interest. Verlinsky et al. [60] isolated the first FXS huESCs followed by multiple other groups [61–63]. The *FMR1* gene is found to be expressed in pluripotent stem cells but can undergo transcriptional silencing upon differentiation [64]. Over the years, the use of hECS has been mired in ethical issues which has limited the availability of cell lines to researchers. Therefore, the bulk of precision medicine efforts in other diseases, has come to rely on the availability of induced pluripotent technology (iPSC) [65], to generate human iPSCs (hiPSCs) from fibroblast samples. In the FXS field, Urbach et al., [66], first reported the generation of hiPSC from fibroblasts of individuals carrying the FXS mutation followed by others [67,68].

One critical difference between huESC and hiPSC in FXS is that the epigenetic silencing due to hypermethylation of the CGG repeats can be retained in the reprogrammed cells. This naturally spurred researchers to use FXS hiPSCs to study DNA methylation as an in vitro human model of FXS (see Bhattacharyya and Zhao [69], 2016 for an in-depth review). Hierarchical clustering and analysis of DNA microarray show that the hiPSCs cluster together with human ESCs [70] and functionally can form embryoid bodies that develop into the three germ layers besides forming teratomas [71]. Also, the *FMR1* gene is expressed in both differentiated and undifferentiated wild type cells and is transcriptionally silenced in patients [72]. It is worth noting, that Sheridan et al., [67], first reported that reprogrammed FXS fibroblasts showed an instability in the CGG trinucleotide stretch in the 5′ UTR of the *FMR1* gene and some of the FXS hiPSC clones had repeat lengths shorter than the control fibroblast samples. Additionally, in the same study, multiple hiPSC clones were derived from mosaic individuals resulting in a set of genetically matched hiPSC lines (isogenic pairs).There have been no reports of reprogramming human lymphoblastoid cell lines (LCL) from FXS mutant patients.

Further, studies have investigated the effects of the FXS mutation on iPSC differentiation to a neural lineage. However, none of these reports have been conclusive; possibly due to the genetic differences in samples, source of human material and differentiation protocols used. Castren et al., [73] reported that differentiation of neurospheres derived from post mortem human FXS brain and control fetal brain showed differences in neurite length, number, morphology, and altered Tuj1 to GFAP ratio while Bhattacharyya et al. [74] found no significant difference in neurons differentiated from FXS and healthy samples. Contrastingly, Sheridan et al. [67] demonstrated FXS-associated morphological differences in hiPSC-derived neurons, with FXS cells having fewer and shorter neurites than control cells. Investigating beyond morphological parameters have revealed electrophysiological deficits in huES derived FXS samples. In a series of studies by Ben-Yosef's group [75–77] several functional deficits in FXS neurons differentiated in vitro huESC have been demonstrated. These include the inability to fire repetitive action potentials (AP), reduced AP amplitude and longer AP duration and immature responses to GABA [78]. Therefore, to date a variety of groups have shown different phenotypes in neurons-derived from FXS hiPSCs with more focus on discovery rather than developing a consistent cell-autonomous or network-level biomarker for modeling FXS.

3. Protein Synthesis Studies in Human Derived Cells in FXS

In contrast to the remarkable amount of work done in understanding the methylation and transcriptional silencing, structural and electrophysiological deficits in FXS huESC and hiPSCs, there are but a handful of studies that have measured protein synthesis in these cells. The first study was in two control and patient-derived LCL lines, wherein FUNCAT was used to measure steady-state and IL-2 induced translation [47]. While FXS LCLs showed increased basal protein synthesis, the translation rates decreased in IL-2 stimulated cases contrary to matched controls. A subsequent publication [79], reverted to a more traditional method to test translation in patient-derived fibroblasts using H^3 leucine autoradiography. This study utilized eight FXS and nine control fibroblast lines spanning a large age range (0–62 years for controls and 1–23 years for FXS). All FXS lines were from patients with greater than 200 repeats, including mosaics. The study reported elevated rates of leucine incorporation in five of the eight FXS lines, with the other three showing comparable translation to WT controls. Grouped data across cell lines did show an extremely significant elevation in protein synthesis rates of patients-versus controls. Curiously, protein synthesis appeared to be correlated with phospho-S6 ribosomal protein abundance in control samples, whereas it had a significant direct correlation with age for FXS lines. This runs counter to Qin (2005) study in mice, where there was a sharp drop in translation across brain areas as soon as FXS model mice attained middle-age. Furthermore, FXS lines showed the same negative correlation with age for phospho-states of mTOR, S6K1, and ERK 1/2, as has been shown in many other cases [80]. These pro-translation kinases have been found to be upregulated in FXS animal models, and targeting these molecules, either genetically or pharmacologically has been found to ameliorate several behavioral, synaptic and morphological phenotypes associated with FXS [41,52]. Therefore the conundrum of decreasing mTOR and S6K1 phosphorylation and hence activation, yet increasing protein synthesis drive remains to be reconciled.

Though the Kumari et al. [79] study had a larger cohort, it was not truly multi-center, that would better simulate a large clinical trial setting. This was addressed by a recent study that combined protein synthesis measurements across three laboratories in Europe and US [45]. The study measured rates of protein synthesis in fibroblasts from 32 individuals with FXS and compared them to 17 controls (ages 6–69 years for FXS; 10-50 years for controls). Further, they compared translation rates in FXS model mouse embryonic fibroblasts and primary neurons within the same experimental parameters. An important finding in this study is the inherent consistency in control fibroblasts across passages, which did not hold for the FXS counterparts. It is not entirely clear if control and FXS cells underwent SUnSET as yoked sets, wherein control and FXS fibroblast samples grown under similar conditions underwent SUnSET puromycin and processing at the same time, which has important implications for the metabolic variability of the samples. A noteworthy finding of this study was that almost a third of the FXS lines had puromycin labeling similar to controls. No correlation with age, mRNA and FMRP protein levels was to be found again. The study further explored the consistency of fibroblasts and neurons from FXS model mice, which has yielded far more consistent elevation in protein synthesis in previous studies [29,31,32,41]. Individual preparations of mouse embryonic fibroblasts (MEFs) and neurons had more variability for the FXS sets and the population spreads were non-normal, however the overall elevation in protein synthesis levels held in neurons and MEFs.

The disparate results from these studies gives an important insight into the heterogeneity in FXS. Contrary to simple and elegant molecular etiology, FXS is not one disease. While there are stand-alone "FXS-only" patients and there also exist large patient subsets like FXS ± ASD, FXS ± Prader-Willi etc. Therefore, sample and patient stratification is very important at the outset. Similarly, mRNA translation or protein synthesis is a metabolic process that is exquisitely sensitive, to the genetic background, culturing conditions of cells and method of detection used. It would have been useful to know if there are de novo mutations in translation control and executive factors like eEF2, in any of the cell lines that was tested. These likely have a strong bearing on the total efficiency of the translation process that is being studied (see Figure 1 for a graphical depiction). A quick comparison of the culturing conditions in these three studies yields different culturing and experimental conditions in each study. LCLs used in Gross and Bassell [47] used the obligate RPMI media supplemented with FBS, with methionine withdrawal- being the only stressor for the protein synthesis experiment. Kumari et al. [79] fibroblasts were cultured in DMEM + FBS, while the protein synthesis assay was done with cells incubated in ACSF. This incubation starved cells of amino acids and growth factors when the radioligand was added. Jaquemont et al. [45] fibroblasts were cultured in DMEM/F12 + FBS with translation assay done on cells that were serum starved for 16 hours, followed by recovering in FBS for four hours, after with puromycin was added for 30 min. Therefore, it is possible that the outcomes of this metabolic assay could be sensitive to culturing conditions and serum withdrawal states.

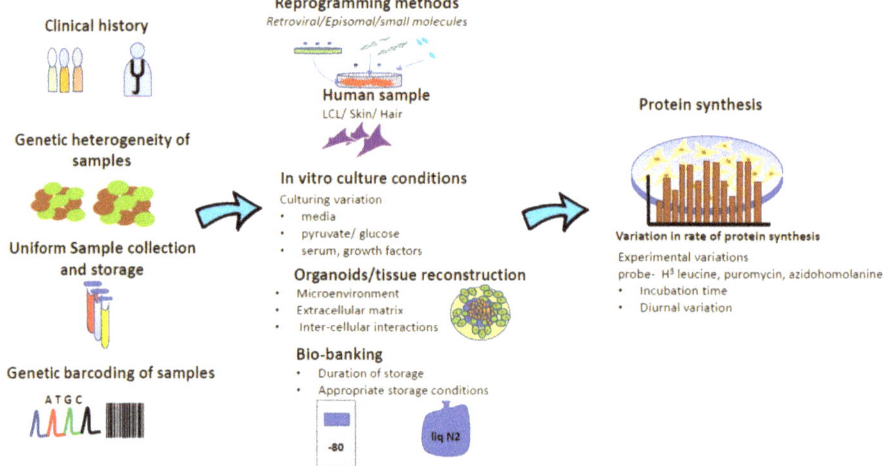

Figure 1. Schematic representation of intrinsic variation at multiple levels (tempo-spatial, tissue, experimental protocols) during development of essential patient derived substrates that influence decisive outcome measures such as protein synthesis.

4. Translation, Neural Differentiation and Loss of FMRP: Challenges for Deploying Translation as a Biomarker

A key property of any deployable biomarker is prior knowledge of its variation and natural history. Variability in translation in fibroblasts has been reported in almost all FXS studies with a modest cohort size. However, the natural consistency in protein synthesis in control fibroblasts is not well understood. In synchronized MEFs, general rates of protein synthesis undergo a diurnal variation that is dependent on circadian regulators acting on mTORC1 components [81]. Translation is a key process involved in differentiation and cell fate specification and, recently, a slew of papers has identified key proteins that mediate this process [82,83]. Indeed, patterns of activation of mTORC1 change as cells differentiate [84,85]. Hence development and stem cell differentiation are multi-layered processes where key players governing translation itself are in constant flux.

In this state of dynamic equilibrium, loss of FMRP recreating FXS, creates a perturbation that has long term effects on cell fate and translation in general. Through chorionic villus sampling of human fetus, Willemsen et al. [86] demonstrated that FMRP loss occurs between 10–13 weeks in gestation. This is approximately when neuronal migration and maturation commences from neural stem cells. There is no clear in vivo study in rodent models to show deficits in this process in FXS. Similarly, there appears to no systematic cataloging of the changes in protein synthesis flux as ES or iPSC differentiate into neurons or glia and how this course changes in FXS derived cells. Without the support of this natural history, using translation as a biomarker or treatment yardstick will always be plagued by uncertainty. This is perhaps due to the fact that understanding the fluctuations in protein synthesis as a metabolic metric has frequently been overshadowed by the motivation to find what individual proteins are misregulated and hence offer potential node points for intervention.

Recently two studies were published that delved to some degree into matter changing transcriptomic hits with differentiation in FXS. Sunamara et al. [87] generated full knockouts (KO) of *FMR1* in hiPSC cells using CRISPR/Cas9 and transcriptomically characterized the differentiating cells from their isogenic controls. Loss of FMRP did not compromise the proliferation of the neural precursor cells derived from these cells. However, FMRP ablated neural precursor cells (NPCs) aberrantly expressed glial fibrillary acidic protein (GFAP) which continued even when the cells were differentiated into neurons. KO cells displayed transcriptomic profiles that were reminiscent of a more pluripotent stage rather than being committed to a neuronal or glial fate. Intriguingly though there was an increase in mTOR and S6 phosphorylation in hiPSCs lacking FMRP, which may allude to a higher translational rate. A second paper in 2018 by Richter laboratory [88] measured translational efficiency disruption in the event of FMRP loss in murine adult neural stem cells. Using ribosomal profiles and whole RNA sequencing, they were able to show that that FMRP loss affects different transcripts at different points of gene expression including translation, ribosome stalling and mRNA abundance. This suggests a possible dysregulation of translation in stem cells (human and murine-derived) and FMRP's role in early development. Since stem cells are a population that are always in a state of dynamic equilibrium, there are likely to be effects on protein synthesis and causing variability in its measurement.

5. Patient Specific Treatment in FXS: Potential Ways to Leverage Protein Synthesis as a Diagnostic Marker

The importance of preclinical animal models in understanding the cardinal mechanisms underlying FXS and other models of autism spectrum disorders (ASD) cannot be overstated [89]. However, setbacks in the recent drug trials has perhaps had a positive outcome in highlighting the need for having interim validation of drug effects in patient-derived cells. Another transition, that has occurred, more for cancer therapeutics, is a move towards using patient-derived cells to tailor patient-specific treatments. An elegant demonstration of the power of this approach was Chia et al., [90] wherein cells isolated from cancer biopsy from the patient were used to identify the best therapy course. However, this work rested upon well-characterized and accepted yardsticks of cancer biology that are, at present, lacking for cultured patient-derived cells for FXS.

This highlights a need to not only understand disease biology in a dish, but also to generate benchmarks in such systems that can be used for diagnostic purposes. One category of this could be signatures that individually or in combination, measure a cell-autonomous feature that can be reliably measured across multiple sites in adherence to an agreed-upon, consensus protocol. Structural and physiological signatures as mentioned above in neurons from FXS samples offer one such avenue. The downside is the long time and high cost involved in generating these neurons. A faster and cost-effective readout can be protein synthesis, which can give a quantum of holistic translation in a patient-derived fibroblast or LCL provided the background mutational load is known and the culture passage, media and incubation conditions are standardized.

Imperatively, for therapeutic discovery or even therapy choice, it is prudent to appreciate that the cell types of the human brain do not function in isolation. It is a continuous orchestration of electrical and chemical signaling events organized in a spatio-temporal fashion in a tightly regulated microenvironment. It is increasingly becoming crucial to ask how each of these cell types communicate with the milieu around it. Clearly, this is lacking in the two-dimensional in vitro stem cell derived models and hence has been unable to mimic many of the diseased phenotypes [91]. Organoids are an initial step in this direction, wherein three-dimensional culturing of cells using suitable matrices, scaffolds and shearing force, has resulted in the capability to re-construct cerebral organoids [92]. These have been found to mimic neural development and incorporate structural and functional deficits in diseased samples [93,94]. Electrophysiological recordings from cerebral organoid slices [95] validates the functional properties of neurons from the different regions of the brain. Mariani et al. [96] has demonstrated that cerebral organoids derived from ASD patients show significant differences in the number of synaptic contacts made. Additionally, forebrain organoids from severe idiopathic ASD exhibit dysregulation of forkhead box G1 expression, high cell cycle progression and higher levels of GABA produced by the cells [96]. This is a simple example of how modelling using organoids can be used to identify new drugs using a candidate-based approach for studying aberrant neurodevelopmental processes. The key question would be whether protein synthesis fluxes in FXS organoids will be similar to those in 2-D cultured cells.

To summarize, the FXS research community is slowly embracing the power of patient-derived cells for biomarker discovery. However with growing appreciation, there is a need to identify robust measures of disease-states that address with patient heterogeneity to predict accurately any treatment response. Protein synthesis is a consistently aberrant molecular phenotype in FXS, however there needs to be further research devoted to mapping the natural history of cells derived from FXS patients, controlled for intrinsic genetic variation and also experimental conditions. This is a fundamental characterization effort that needs to be undertaken by any outfit aiming to utilize patient-derived cells for developing diagnostics and therapeutic interventions.

Funding: This review was made possible by support to A.B. and R.P. from funds from the Department of Biotechnology, Government of India, Centre for Neurosynaptopathies grant. APC was funded by FRAXA research foundation grant to A.B.

Conflicts of Interest: The authors declare no conflict of interest.

References

1. Santoro, M.R.; Bray, S.M.; Warren, S.T. Molecular mechanisms of fragile X syndrome: A twenty-year perspective. *Annu. Rev. Pathol.* **2012**, *7*, 219–245. [CrossRef] [PubMed]
2. Richter, J.D.; Bassell, G.J.; Klann, E. Dysregulation and restoration of translational homeostasis in fragile X syndrome. *Nat. Rev. Neurosci.* **2015**, *16*, 595–605. [CrossRef] [PubMed]
3. Caudy, A.A.; Myers, M.; Hannon, G.J.; Hammond, S.M. Fragile X-related protein and VIG associate with the RNA interference machinery. *Genes Dev.* **2002**, *1*, 16–2491.
4. Muddashetty, R.S.; Nalavadi, V.C.; Gross, C.; Yao, X.; Xing, L.; Laur, O.; Warren, S.T.; Bassell, G.J. Reversible inhibition of PSD-95 mRNA translation by miR-125a, FMRP phosphorylation, and mGluR signaling. *Mol. Cell* **2011**, *42*, 673–688. [CrossRef] [PubMed]
5. Antar, L.N.; Afroz, R.; Dictenberg, J.B.; Carroll, R.C.; Bassel, G.J. Metabotropic glutamate receptor activation regulates fragile x mental retardation protein and FMR1 mRNA localization differentially in dendrites and at synapses. *J. Neurosci.* **2004**, *24*, 2648–2655. [CrossRef] [PubMed]
6. Richter, J.D.; Coller, J. Pausing on Polyribosomes: Make Way for Elongation in Translational Control. *Cell* **2015**, *163*, 292–300. [CrossRef] [PubMed]
7. Deng, P.Y.; Rotman, Z.; Blundon, J.A.; Cho, Y.; Cui, J.; Cavalli, V.; Zakharenko, S.S.; Klyachko, V.A. FMRP regulates neurotransmitter release and synaptic information transmission by modulating action potential duration via BK channels. *Neuron* **2013**, *77*, 696–711. [CrossRef] [PubMed]
8. O'Donnell, W.T.; Warren, S.T. A decade of molecular studies of fragile X syndrome. *Annu. Rev. Neurosci.* **2002**, *25*, 315–338. [CrossRef] [PubMed]

9. De Diego Otero, Y.; Severijnen, L.A.; van Cappellen, G.; Schrier, M.; Oostra, B.; Willemsen, R. Transport of fragile X mental retardation protein via granules in neurites of PC12 cells. *Mol. Cell. Biol.* **2002**, *22*, 8332–8341.

10. Dictenberg, J.B.; Swanger, S.A.; Antar, L.N.; Singer, R.H.; Bassell, G.J. A direct role for FMRP in activity-dependent dendritic mRNA transport links filopodial-spine morphogenesis to fragile X syndrome. *Dev. Cell* **2008**, *14*, 926–939. [CrossRef] [PubMed]

11. Siomi, H.; Matunis, M.J.; Michael, W.M.; Dreyfuss, G. The pre-mRNA binding K protein contains a novel evolutionarily conserved motif. *Nucleic Acid Res.* **1993**, *21*, 1193–1198. [CrossRef] [PubMed]

12. Khandjian, E.W.; Corbin, F.; Woerly, S.; Rousseau, F. The fragile X mental retardation protein is associated with ribosomes. *Nat. Genet.* **1996**, *12*, 91–93. [CrossRef] [PubMed]

13. Eberhart, D.E.; Malter, H.E.; Feng, Y.; Warren, S.T. The fragile X mental retardation protein is a ribonucleoprotein containing both nuclear localization and nuclear export signals. *Hum. Mol. Genet.* **1996**, *5*, 1083–1091. [CrossRef] [PubMed]

14. Tamanini, F.; Meijer, N.; Verheij, C.; Willems, P.J.; Galjaard, H.; Oostra, B.A.; Hoogeveen, A.T. FMRP is associated to the ribosomes via RNA. *Hum. Mol. Genet.* **1996**, *5*, 809–813. [CrossRef] [PubMed]

15. Corbin, F.; Bouillon, M.; Fortin, A.; Morin, S.; Rousseau, F.; Khandjian, E.W. The fragile X mental retardation protein is associated with poly (A)+ mRNA in actively translating polyribosomes. *Hum. Mol Genet.* **1997**, *6*, 1465–1472. [CrossRef] [PubMed]

16. Brown, V.; Small, K.; Lakkis, L.; Feng, Y.; Gunter, C.; Wilkinson, K.D.; Warren, S.T. Purified recombinant FMRP exhibits selective RNA binding as an intrinsic property of the fragile X mental retardation protein. *J. Biol. Chem.* **1998**, *273*, 15521–15527. [CrossRef] [PubMed]

17. Laggerbauer, B.; Ostareck, D.; Keidel, E.M.; Ostareck-Lederer, A.; Fischer, U. Evidence that fragile X mental retardation protein is a negative regulator of translation. *Hum. Mol. Genet.* **2001**, *10*, 329–338. [CrossRef] [PubMed]

18. Li, Z.; Zhang, Y.; Ku, L.; Wilkinson, K.D.; Warren, S.T.; Feng, Y. The fragile X mental retardation protein inhibits translation via interacting with mRNA. *Nucleic Acids Res.* **2001**, *29*, 2276–2283. [CrossRef] [PubMed]

19. Darnell, J.C.; Van Driesche, S.J.; Zhang, C.; Hung, K.Y.; Mele, A.; Fraser, C.E.; Stone, E.F.; Chen, C.; Fak, J.J.; Chi, S.W.; et al. FMRP stalls ribosomal translocation on mRNAs linked to synaptic function and autism. *Cell* **2011**, *146*, 247–261. [CrossRef] [PubMed]

20. Chen, E.; Sharma, M.R.; Shi, X.; Agrawal, R.K.; Joseph, S. Fragile X mental retardation protein regulates translation by binding directly to the ribosome. *Mol. Cell* **2014**, *54*, 407–417. [CrossRef] [PubMed]

21. Napoli, I.; Mercaldo, V.; Boyl, P.P.; Eleuteri, B.; Zalfa, F.; De Rubeis, S.; Di Marino, D.; Mohr, E.; Massimi, M.; Falconi, M.; et al. The fragile X syndrome protein represses activity-dependent translation through CYFIP1, a new 4E-BP. *Cell* **2008**, *134*, 1042–1054. [CrossRef] [PubMed]

22. Devys, D.; Lutz, Y.; Rouyer, N.; Bellocq, J.P.; Mandel, J.L. The FMR-1 protein is cytoplasmic, most abundant in neurons and appears normal in carriers of a fragile X premutation. *Nat. Genet.* **1993**, *4*, 335–340. [CrossRef] [PubMed]

23. Bakker, C.E.; de Diego Otero, Y.; Bontekoe, C.; Raghoe, P.; Luteijn, T.; Hoogeveen, A.T.; Oostra, B.A.; Willemsen, R. Immunocytochemical and biochemical characterization of FMRP, FXR1P, and FXR2P in the mouse. *Exp. Cell Res.* **2000**, *258*, 162–170. [CrossRef] [PubMed]

24. Tamanini, F.; Willemsen, R.; van Unen, L.; Bontekoe, C.; Galjaard, H.; Oostra, B.A.; Hoogeveen, A.T. Differential expression of FMR1, FXR1 and FXR2 proteins in human brain and testis. *Hum. Mol. Genet.* **1997**, *6*, 1315–1322. [CrossRef] [PubMed]

25. Wang, H.; Ku, L.; Osterhout, D.J.; Li, W.; Ahmadian, A.; Liang, Z.; Feng, Y. Developmentally-programmed FMRP expression in oligodendrocytes: A potential role of FMRP in regulating translation in oligodendroglia progenitors. *Hum. Mol. Genet.* **2004**, *13*, 79–89. [CrossRef] [PubMed]

26. Pacey, L.K.; Doering, L.C. Developmental expression of FMRP in the astrocyte lineage: Implications for fragile X syndrome. *Glia* **2007**, *55*, 1601–1609. [CrossRef] [PubMed]

27. Bakker, C.E.; Verheij, C.; Willemsen, R.; van der Helm, R.; Oerlemans, F.; Vermey, M.; Bygrave, A.; Hoogeveen, A.; Oostra, B.A.; Reyniers, E.; et al. Fmr1 knockout mice: A model to study fragile X mental retardation. *Cell* **1994**, *78*, 23–33.

28. Qin, M.; Kang, J.; Smith, C.B. Increased rates of cerebral glucose metabolism in a mouse model of fragile X mental retardation. *Proc. Natl. Acad. Sci. USA* **2002**, *99*, 15758–15763. [CrossRef] [PubMed]

29. Qin, M.; Kang, J.; Burlin, T.V.; Jiang, C.; Smith, C.B. Post-adolescent changes in regional cerebral protein synthesis: An in vivo study in the FMR1 null mouse. *J. Neurosci.* **2005**, *25*, 5087–5095. [CrossRef] [PubMed]

30. Dölen, G.; Osterweil, E.; Rao, B.S.; Smith, G.B.; Auerbach, B.D.; Chattarji, S.; Bear, M.F. Correction of fragile X syndrome in mice. *Neuron* **2007**, *56*, 955–962. [CrossRef] [PubMed]

31. Osterweil, E.K.; Krueger, D.D.; Reinhold, K.; Bear, M.F. Hypersensitivity to mGluR5 and ERK1/2 leads to excessive protein synthesis in the hippocampus of a mouse model of fragile X syndrome. *J. Neurosci.* **2010**, *30*, 15616–15627. [CrossRef] [PubMed]

32. Michalon, A.; Sidorov, M.; Ballard, T.M.; Ozmen, L.; Spooren, W.; Wettstein, J.G.; Jaeschke, G.; Bear, M.F.; Lindemann, L. Chronic pharmacological mGlu5 inhibition corrects fragile X in adult mice. *Neuron* **2012**, *74*, 49–56. [CrossRef] [PubMed]

33. Huber, K.M.; Kayser, M.S.; Bear, M.F. Role for rapid dendritic protein synthesis in hippocampal mGluR-dependent long-term depression. *Science* **2000**, *288*, 1254–1257. [CrossRef] [PubMed]

34. Huber, K.M.; Gallagher, S.M.; Warren, S.T.; Bear, M.F. Altered synaptic plasticity in a mouse model of fragile X mental retardation. *Proc. Natl. Acad. Sci. USA* **2002**, *99*, 7746–7750. [CrossRef] [PubMed]

35. Muddashetty, R.S.; Kelić, S.; Gross, C.; Xu, M.; Bassell, G.J. Dysregulated metabotropic glutamate receptor-dependent translation of AMPA receptor and postsynaptic density-95 mRNAs at synapses in a mouse model of fragile X syndrome. *J. Neurosci.* **2007**, *27*, 5338–5348. [CrossRef] [PubMed]

36. Till, S.M.; Asiminas, A.; Jackson, A.D.; Katsanevaki, D.; Barnes, S.A.; Osterweil, E.K.; Bear, M.F.; Chattarji, S.; Wood, E.R.; Wyllie, D.J.; et al. Conserved hippocampal cellular pathophysiology but distinct behavioural deficits in a new rat model of FXS. *Hum. Mol. Genet.* **2015**, *24*, 5977–5984. [CrossRef] [PubMed]

37. Henderson, C.; Wijetunge, L.; Kinoshita, M.N.; Shumway, M.; Hammond, R.S.; Postma, F.R.; Brynczka, C.; Rush, R.; Thomas, A.; Paylor, R.; et al. Reversal of disease-related pathologies in the fragile X mouse model by selective activation of GABAB receptors with arbaclofen. *Science Transl. Med.* **2012**, *4*, 152ra128. [CrossRef] [PubMed]

38. Qin, M.; Schmidt, K.C.; Zametkin, A.J.; Bishu, S.; Horowitz, L.M.; Burlin, T.V.; Xia, Z.; Huang, T.; Quezado, Z.M.; Smith, C.B. Altered cerebral protein synthesis in fragile X syndrome: Studies in human subjects and knockout mice. *J. Cereb. Blood Flow Metab.* **2013**, *33*, 499–507. [CrossRef] [PubMed]

39. Tomasi, G.; Veronese, M.; Bertoldo, A.; Smith, C.B.; Schmidt, K.C. Effects of shortened scanning intervals on calculated regional rates of cerebral protein synthesis determined with the L-[1-11C] leucine PET method. *PLoS ONE* **2018**, *13*, e0195580. [CrossRef] [PubMed]

40. Schmidt, E.K.; Clavarino, G.; Ceppi, M.; Pierre, P. SUnSET, a nonradioactive method to monitor protein synthesis. *Nat. Methods* **2009**, *6*, 275–277. [CrossRef] [PubMed]

41. Bhattacharya, A.; Kaphzan, H.; Alvarez-Dieppa, A.C.; Murphy, J.P.; Pierre, P.; Klann, E. Genetic removal of p70 S6 kinase 1 corrects molecular, synaptic, and behavioral phenotypes in fragile X syndrome mice. *Neuron* **2012**, *76*, 325–337. [CrossRef] [PubMed]

42. Gkogkas, C.G.; Khoutorsky, A.; Ran, I.; Rampakakis, E.; Nevarko, T.; Weatherill, D.B.; Vasuta, C.; Yee, S.; Truitt, M.; Dallaire, P.; et al. Autism-related deficits via dysregulated eIF4E-dependent translational control. *Nature* **2013**, *493*, 371–377. [CrossRef] [PubMed]

43. Gantois, I.; Khoutorsky, A.; Popic, J.; Aguilar-Valles, A.; Freemantle, E.; Cao, R.; Sharma, V.; Pooters, T.; Nagpal, A.; Skalecka, A.; et al. Metformin ameliorates core deficits in a mouse model of fragile X syndrome. *Nat. Med.* **2017**, *23*, 674–677. [CrossRef] [PubMed]

44. Sethna, F.; Zhang, M.; Kaphzan, H.; Klann, E.; Autio, D.; Cox, C.L.; Wang, H. Calmodulin activity regulates group I metabotropic glutamate receptor-mediated signal transduction and synaptic depression. *J. Neurosci. Res.* **2016**, *94*, 401–408. [CrossRef] [PubMed]

45. Jacquemont, S.; Pacini, L.; Jønch, A.E.; Cencelli, G.; Rozenberg, I.; He, Y.; D'Andrea, L.; Pedini, G.; Eldeeb, M.; Willemsen, R.; et al. Protein synthesis levels are increased in a subset of individuals with fragile X syndrome. *Hum. Mol. Genet.* **2018**, *27*, 3825. [CrossRef] [PubMed]

46. Dieterich, D.C.; Lee, J.J.; Link, A.J.; Graumann, J.; Tirrell, D.A.; Schuman, E.M. Labeling, detection and identification of newly synthesized proteomes with bioorthogonal non-canonical amino-acid tagging. *Nat. Protocols* **2007**, *2*, 532–540. [CrossRef] [PubMed]

47. Gross, C.; Bassell, G.J. Excess protein synthesis in FXS patient lymphoblastoid cells can be rescued with a p110β-selective inhibitor. *Mol. Med.* **2012**, *18*, 336–345. [CrossRef] [PubMed]

48. Bowling, H.; Bhattacharya, A.; Zhang, G.; Lebowitz, J.Z.; Alam, D.; Smith, P.T.; Kirshenbaum, K.; Neubert, T.A.; Vogel, C.; Chao, M.V.; et al. BONLAC: A combinatorial proteomic technique to measure stimulus-induced translational profiles in brain slices. *Neuropharmacology* **2016**, *100*, 76–89. [CrossRef] [PubMed]

49. Bhattacharya, A.; Mamcarz, M.; Mullins, C.; Choudhury, A.; Boyle, R.G.; Smith, D.G.; Walker, W.; Klann, E. Targeting Translation Control with p70 S6 Kinase 1 Inhibitors to Reverse Phenotypes in Fragile X Syndrome Mice. *Neuropsychopharmacology* **2016**, *41*, 1991–2000. [CrossRef] [PubMed]

50. Den Broeder, M.J.; van der Linde, H.; Brouwer, J.R.; Oostra, B.A.; Willemsen, R.; Ketting, R.F. Generation and characterization of FMR1 knockout zebrafish. *PLoS ONE* **2009**, *4*, e7910. [CrossRef] [PubMed]

51. McBride, S.M.; Choi, C.H.; Wang, Y.; Liebelt, D.; Braunstein, E.; Ferreiro, D.; Sehgal, A.; Siwicki, K.K.; Dockendorff, T.C.; Nguyen, H.T.; et al. Pharmacological rescue of synaptic plasticity, courtship behavior, and mushroom body defects in a Drosophila model of fragile X syndrome. *Neuron* **2005**, *45*, 753–764. [CrossRef] [PubMed]

52. Osterweil, E.K.; Chuang, S.C.; Chubykin, A.A.; Sidorov, M.; Bianchi, R.; Wong, R.K.; Bear, M.F. Lovastatin corrects excess protein synthesis and prevents epileptogenesis in a mouse model of fragile X syndrome. *Neuron* **2013**, *77*, 243–250. [CrossRef] [PubMed]

53. Udagawa, T.; Farny, N.G.; Jakovcevski, M.; Kaphzan, H.; Alarcon, J.M.; Anilkumar, S.; Ivshina, M.; Hurt, J.A.; Nagaoka, K.; Nalavadi, V.C.; et al. Genetic and acute CPEB1 depletion ameliorate fragile X pathophysiology. *Nat. Med.* **2013**, *19*, 1473–1477. [CrossRef]

54. Gross, C.; Chang, C.W.; Kelly, S.M.; Bhattacharya, A.; McBride, S.M.; Danielson, S.W.; Jiang, M.Q.; Chan, C.B.; Ye, K.; Gibson, J.R.; et al. Increased expression of the PI3K enhancer PIKE mediates deficits in synaptic plasticity and behavior in fragile X syndrome. *Cell Rep.* **2015**, *11*, 727–736. [CrossRef] [PubMed]

55. Gross, C.; Raj, N.; Molinaro, G.; Allen, A.G.; Whyte, A.J.; Gibson, J.R.; Huber, K.M.; Gourley, S.L.; Bassell, G.J. Selective role of the catalytic PI3K subunit p110β in impaired higher order cognition in fragile X syndrome. *Cell Rep.* *11*, 681–688. [CrossRef] [PubMed]

56. Robinson, E.B.; St. Pourain, B.; Antilla, V.; Kosmickim, J.A.; Bulik-Sullivan, B.; Grove, S.J.; Ripke, S.; Martin, J.; Hollegaard, M.V.; Werge, T.; et al. Genetic risk for autism spectrum disorders and neuropsychiatric variation in the general population. *Nat. Genet.* **2016**, *48*, 552–555. [CrossRef] [PubMed]

57. Graziani, G.; Tal, S.; Adelman, A.; Kugel, C.; Bdolah-Abram, T.; Krispin, A. Usefulness of unenhanced post mortem computed tomography—Findings in post mortem non-contrast computed tomography of the head, neck and spine compared to traditional medicolegal autopsy. *J. Forensic Leg. Med.* **2018**, *55*, 105–111. [CrossRef] [PubMed]

58. Broek, J.A.; Guest, P.C.; Rahmoune, H.; Bahn, S. Proteomic analysis of post mortem brain tissue from autism patients: Evidence for opposite changes in prefrontal cortex and cerebellum in synaptic connectivity-related proteins. *Mol. Autism.* **2014**, *5*, 41. [CrossRef] [PubMed]

59. Thomson, J.A.; Itskovitz-Eldor, J.; Shapiro, S.S.; Waknitz, M.A.; Swiergiel, J.J.; Marshall, V.S.; Jones, J.M. Embryonic stem cell lines derived from human blastocysts. *Science* **1998**, *282*, 1145–1147. [CrossRef] [PubMed]

60. Verlinsky, Y.; Strelchenko, N.; Kukharenko, V.; Rechitsky, S.; Verlinsky, O.; Galat, V.; Kuliev, A. Human embryonic stem cell lines with genetic disorders. *Reprod. Biomed. Online* **2005**, *10*, 105–110. [CrossRef]

61. Pickering, S.J.; Braude, P.R.; Patel, M.; Burns, C.J.; Trussler, J.; Bolton, V.; Minger, S. Preimplantation genetic diagnosis as a novel source of embryos for stem cell research. *Reprod. Biomed. Online* **2003**, *7*, 353–364. [CrossRef]

62. Ben-Yosef, D.; Malcov, M.; Eiges, R. PGD-derived human embryonic stem cell lines as a powerful tool for the study of human genetic disorders. *Mol. Cell. Endocrinol.* **2008**, *282*, 153–158. [CrossRef] [PubMed]

63. Stephenson, E.L.; Mason, C.; Braude, P.R. Preimplantation genetic diagnosis as a source of human embryonic stem cells for disease research and drug discovery. *BJOG Int. J. Obstet. Gynaecol.* **2009**, *116*, 158–165. [CrossRef] [PubMed]

64. Eiges, R.; Urbach, A.; Malcov, M.; Frumkin, T.; Schwartz, T.; Amit, A.; Yaron, Y.; Eden, A.; Yanuka, O.; Benvenisty, N.; et al. Developmental study of fragile X syndrome using human embryonic stem cells derived from preimplantation genetically diagnosed embryos. *Cell Stem Cell* **2007**, *1*, 568–577. [CrossRef]

65. Takahashi, K.; Tanabe, K.; Ohnuki, M.; Narita, M.; Ichisaka, T.; Tomoda, K.; Yamanaka, S. Induction of pluripotent stem cells from adult human fibroblasts by defined factors. *Cell* **2007**, *131*, 861–872. [CrossRef] [PubMed]

66. Urbach, A.; Bar-Nur, O.; Daley, G.Q.; Benvenisty, N. Differential modelling of fragile X syndrome by human embryonic stem cells and induced pluripotent stem cells. *Cell Stem Cell* **2010**, *6*, 407–411. [CrossRef]

67. Sheridan, S.D.; Theriault, K.M.; Reis, S.A.; Zhou, F.; Madison, J.M.; Daheron, L.; Loring, J.F.; Haggarty, S.J. Epigenetic characterization of the FMR1 gene and aberrant neurodevelopment in human induced pluripotent stem cell models of fragile X syndrome. *PLoS ONE* **2011**, *6*, e26203. [CrossRef] [PubMed]

68. Doers, M.E.; Musser, M.T.; Nichol, R.; Berndt, E.R.; Baker, M.; Gomez, T.M.; Zhang, S.C.; Abbeduto, L.; Bhattacharyya, A. iPSC-derived forebrain neurons from FXS individuals show defects in initial neurite outgrowth. *Stem Cells Dev.* **2014**, *23*, 1777–1787. [CrossRef] [PubMed]

69. Bhattacharyya, A.; Zhao, X. Human pluripotent stem cell models of Fragile X syndrome. *Mol. Cell Neurosci.* **2016**, *73*, 43–51. [CrossRef] [PubMed]

70. Zhao, M.T.; Chen, H.; Liu, Q.; Shao, N.Y.; Sayed, N.; Wo, H.T.; Zhang, J.Z.; Ong, S.G.; Liu, C.; Kim, Y.; et al. Molecular and functional resemblance of differentiated cells derived from isogenic human iPSCsand SCNT-derived ESCs. *Proc. Natl. Acad. Sci. USA* **2017**, *114*, E11111–E11120. [CrossRef] [PubMed]

71. Hentze, H.; Soong, P.L.; Wang, S.T.; Phillips, B.W.; Putti, T.C.; Dunn, N.R. Teratoma formation by human embryonic stem cells: Evaluation of essential parameters for future safety studies. *Stem Cell Res.* **2009**, *2*, 198–210. [CrossRef] [PubMed]

72. Haenfler, J.M.; Skariah, G.; Rodriguez, C.M.; Monteiro da Rocha, A.; Parent, J.M.; Smith, G.D.; Todd, P.K. Targeted Reactivation of FMR1 Transcription in Fragile X Syndrome Embryonic Stem Cells. *Front. Mol. Neurosci.* **2018**, *11*, 282. [CrossRef] [PubMed]

73. Castrén, M.; Tervonen, T.; Kärkkäinen, V.; Heinonen, S.; Castrén, E.; Larsson, K.; Bakker, C.E.; Oostra, B.A.; Akerman, K. Altered differentiation of neural stem cells in fragile X syndrome. *Proc. Natl. Acad. Sci. USA* **2005**, *102*, 17834–17839. [CrossRef] [PubMed]

74. Bhattacharyya, A.; McMillan, E.; Wallace, K.; Tubon, T.C., Jr.; Capowski, E.E.; Svendsen, C.N. Normal Neurogenesis but Abnormal Gene Expression in Human Fragile X Cortical Progenitor Cells. *Stem Cells Dev.* **2008**, *17*, 107–117. [CrossRef] [PubMed]

75. Telias, M.; Segal, M.; Ben-Yosef, D. Neural differentiation of Fragile X human Embryonic Stem Cells reveals abnormal patterns of development despite successful neurogenesis. *Dev. Biol.* **2013**, *374*, 32–45. [CrossRef] [PubMed]

76. Telias, M.; Mayshar, Y.; Amit, A.; Ben-Yosef, D. Molecular mechanisms regulating impaired neurogenesis of fragile X syndrome human embryonic stem cells. *Stem Cells Dev.* **2015**, *24*, 2353–2365. [CrossRef] [PubMed]

77. Telias, M.; Segal, M.; Ben-Yosef, D. Immature Responses to GABA in Fragile X Neurons Derived from Human Embryonic Stem Cells. *Front. Cell Neurosci.* **2016**, *10*, 121. [CrossRef] [PubMed]

78. Boland, M.J.; Nazor, K.L.; Tran, H.T.; Szücs, A.; Lynch, C.L.; Paredes, R.; Tassone, F.; Sanna, P.P.; Hagerman, R.J.; Loring, J.F. Molecular analyses of neurogenic defects in a human pluripotent stem cell model of fragile X syndrome. *Brain* **2017**, *140*, 582–598. [CrossRef] [PubMed]

79. Kumari, D.; Bhattacharya, A.; Nadel, J.; Moulton, K.; Zeak, N.M.; Glicksman, A.; Dobkin, C.; Brick, D.J.; Schwartz, P.H.; Smith, C.B.; et al. Identification of fragile X syndrome specific molecular markers in human fibroblasts: A useful model to test the efficacy of therapeutic drugs. *Hum. Mutat.* **2014**, *35*, 1485–1494. [CrossRef] [PubMed]

80. Bhattacharya, A. Sidekick No More: Neural Translation Control by p70 ribosomal S6 kinase 1. In *The Oxford Handbook of Neuronal Protein Synthesis*; Oxford University Press: Oxford, UK, 2018.

81. Lipton, J.O.; Yuan, E.D.; Boyle, L.M.; Ebrahimi-Fakhari, D.; Kwiatkowski, E.; Nathan, A.; Güttler, T.; Davis, F.; Asara, J.M.; Sahin, M. The Circadian Protein BMAL1 Regulates Translation in Response to S6K1-Mediated Phosphorylation. *Cell* **2015**, *161*, 1138–1151. [CrossRef] [PubMed]

82. Sugiyama, H.; Takahashi, K.; Yamamoto, T.; Iwasaki, M.; Narita, M.; Nakamura, M.; Rand, T.A.; Nakagawa, M.; Watanabe, A.; Yamanaka, S. Nat1 promotes translation of specific proteins that induce differentiation of mouse embryonic stem cells. *Proc. Natl. Acad. Sci. USA* **2017**, *114*, 340–345. [CrossRef] [PubMed]

83. Yoffe, Y.; David, M.; Kalaora, R.; Povodovski, L.; Friedlander, G.; Feldmesser, E.; Ainbinder, E.; Saada, A.; Bialik, S.; Kimchi, A. Cap-independent translation by DAP5 controls cell fate decisions in human embryonic stem cells. *Genes Dev.* **2016**, *30*, 1991–2004. [CrossRef] [PubMed]

84. Crino, P.B. mTOR: A pathogenic signaling pathway in developmental brain malformations. *Trends Mol. Med.* **2011**, *17*, 734–742. [CrossRef] [PubMed]

85. Tang, G.; Gudsnuk, K.; Kuo, S.H.; Cotrina, M.L.; Rosoklija, G.; Sosunov, A.; Sonders, M.S.; Kanter Castagna, C.; Yamamoto, A.; Yue, Z.; et al. Loss of mTOR-dependent macroautophagy causes autistic-like synaptic pruning deficits. *Neuron* **2014**, *83*, 1131–1143. [CrossRef] [PubMed]

86. Willemsen, R.; Bontekoe, C.J.; Severijnen, L.A.; Oostra, B.A. Timing of the absence of FMR1 expression in full mutation chorionic villi. *Hum. Genet.* **2002**, *110*, 601–605. [CrossRef] [PubMed]

87. Sunamura, N.; Iwashita, S.; Enomoto, K.; Kadoshima, T.; Isono, F. Loss of the fragile X mental retardation protein causes aberrant differentiation in human neural progenitor cells. *Sci. Rep.* **2018**, *8*, 11585. [CrossRef] [PubMed]

88. Liu, B.; Li, Y.; Stackpole, E.E.; Novak, A.; Gao, Y.; Zhao, Y.; Zhao, X.; Richter, J.D. Regulatory discrimination of mRNAs by FMRP controls mouse adult neural stem cell differentiation. *Proc. Natl. Acad. Sci. USA* **2018**, *115*, E11397–E11405. [CrossRef] [PubMed]

89. Howe, J.R.; Bear, M.F.; Golshani, P.; Klann, E.; Lipton, S.A.; Mucke, L.; Sahin, M.; Silva, A.J. The mouse as a model for neuropsychiatric drug development. *Curr. Biol.* **2018**, *28*, R909–R914. [CrossRef] [PubMed]

90. Chia, S.; Low, J.L.; Zhang, X.; Kwang, X.L.; Chong, F.T.; Sharma, A.; Bertrand, D.; Toh, S.Y.; Leong, H.S.; Thangavelu, M.T.; et al. Phenotype-driven precision oncology as a guide for clinical decisions one patient at a time. *Nat. Commun.* **2017**, *8*, 435. [CrossRef] [PubMed]

91. Forsberg, S.L.; Ilieva, M.; Maria, M.T. Epigenetics and cerebral organoids: Promising directions in autism spectrum disorders. *Transl. Psychiatry* **2018**, *8*, 14. [CrossRef] [PubMed]

92. Lancaster, M.A.; Renner, M.; Martin, C.A.; Wenzel, D.; Bicknell, L.S.; Hurles, M.E.; Homfray, T.; Penninger, J.M.; Jackson, A.P.; Knoblich, J.A. Cerebral organoids model human brain development and microcephaly. *Nature* **2013**, *501*, 373–379. [CrossRef] [PubMed]

93. Stachowiak, E.K.; Benson, C.A.; Narla, S.T.; Dimitri, A.; Chuye, L.E.B.; Dhiman, S.; Harikrishnan, K.; Elahi, S.; Freedman, D.; Brennand, K.J.; et al. Cerebral organoids reveal early cortical maldevelopment in schizophrenia-computational anatomy and genomics, role of FGFR1. *Transl. Psychiatry* **2017**, *7*. [CrossRef] [PubMed]

94. Johnstone, M.; Vasistha, N.A.; Barbu, M.C.; Dando, O.; Burr, K.; Christopher, E.; Glen, S.; Robert, C.; Fetit, R.; Macleod, K.G.; et al. Reversal of proliferation deficits caused by chromosome 16p13.11 microduplication through targeting NFκB signaling: An integrated study of patient-derived neuronal precursor cells, cerebral organoids and in vivo brain imaging. *Mol. Psychiatry* **2018**, *24*, 294–311. [CrossRef] [PubMed]

95. Sloan, S.A.; Andersen, J.; Paşca, A.M.; Birey, F.; Paşca, S.P. Generation and assembly of human brain region-specific three-dimensional cultures. *Nat. Protocols* **2018**, *13*, 2062–2085. [CrossRef] [PubMed]

96. Mariani, J.; Coppola, G.; Zhang, P.; Abyzov, A.; Provini, L.; Tomasini, L.; Amenduni, M.; Szekely, A.; Palejev, D.; Wilson, M.; et al. FOXG1-Dependent Dysregulation of GABA/Glutamate Neuron Differentiation in Autism Spectrum Disorders. *Cell* **2015**, *162*, 375–390. [CrossRef] [PubMed]

Commentary

Best Practices in Fragile X Syndrome Treatment Development

Craig A. Erickson [1,2,*], Walter E. Kaufmann [3,4], Dejan B. Budimirovic [5,6], Ave Lachiewicz [7], Barbara Haas-Givler [8], Robert M. Miller [9], Jayne Dixon Weber [9], Leonard Abbeduto [10], David Hessl [10], Randi J. Hagerman [10] and Elizabeth Berry-Kravis [11]

[1] Division of Child & Adolescent Psychiatry, Cincinnati Children's Hospital Medical Center, Cincinnati, OH 45229, USA

[2] Department of Psychiatry and Behavioral Neuroscience, University of Cincinnati College of Medicine, Cincinnati, OH 45219, USA

[3] Department of Human Genetics, Emory University School of Medicine, Atlanta, GA 30322, USA; wekaufmann@ucdavis.edu

[4] MIND Institute, Department of Neurology, School of Medicine, University of California, Davis, CA 95817, USA

[5] Kennedy Krieger Institute, Johns Hopkins Medical Institutions, Baltimore, MD 21205, USA; budimirovic@kennedykrieger.org

[6] Behavioral Sciences-Child Psychiatry, Johns Hopkins University School of Medicine, Baltimore, MD 21287, USA

[7] Duke Department of Pediatrics, Duke University School of Medicine, Durham, NC 27710, USA; ave.lachiewicz@duke.edu

[8] Autism & Developmental Medicine Institute, Geisinger, Lewisburg, PA 17837, USA; bahaasgivler@geisinger.edu

[9] National Fragile X Foundation, McLean, VA 22102, USA; robby@fragilex.org (R.M.M.); jayne@fragilex.org (J.D.W.)

[10] MIND Institute and Department of Psychiatry and Behavioral Sciences, School of Medicine, University of California, Davis, CA 95817, USA; ljabbeduto@ucdavis.edu (L.A.); drhessl@ucdavis.edu (D.H.); rjhagerman@ucdavis.edu (R.J.H.)

[11] Departments of Pediatrics, Neurological Sciences, Biochemistry, Rush University Medical Center, Chicago, IL 60612, USA; Elizabeth_Berry-Kravis@rush.edu

[*] Correspondence: craig.erickson@cchmc.org; Tel.: +1-513-636-6265

Received: 21 November 2018; Accepted: 12 December 2018; Published: 15 December 2018

Abstract: Preclinical studies using animal models of fragile X syndrome have yielded several agents that rescue a wide variety of phenotypes. However, translation of these treatments to humans with the disorder has not yet been successful, shedding light on a variety of limitations with both animal models and human trial design. As members of the Clinical Trials Committee of the National Fragile X Foundation, we have discussed a variety of recommendations at the level of preclinical development, transition from preclinical to human projects, family involvement, and multi-site trial planning. Our recommendations are made with the vision that effective new treatment will lie at the intersection of innovation, rigorous and reproducible research, and stakeholder involvement.

Keywords: fragile X syndrome; clinical trials; treatment development; best practices

1. Introduction

Since the proposal of the "mGluR theory of fragile X syndrome (FXS)" over a decade ago, there has been an explosion of preclinical small molecule investigation in FXS leading to a number of mechanistically-targeted treatments moving into FXS human study. Despite significant pre-clinical data supporting translation of many drugs to humans, combined with some success in early-phase studies, no definitive large-scale placebo-controlled trials have met primary endpoints leading to drug use indications specific to FXS. As leaders and stakeholders in the field of FXS, we recognize the importance of issuing recommendations for methodology, including study design and strategies, which can maximize potential for success for the bench to bedside treatment development pathway. With this in mind, we review key aspects of the process of FXS treatment development with an eye towards ensuring that successful trials of new treatments, incorporating innovative research and stakeholder concerns, can be enacted.

2. Fragile X Clinical and Research Consortium Clinical Trials Committee

The National Fragile X Foundation (NFXF) has developed a Clinical Trials Committee (CTC) structure made up of Fragile X Clinical and Research Consortium (FXCRC) members, FXS clinicians, expert FXS trialists, outcome measure experts in the field, and family stakeholders to assist and support treatment developments. This committee reports and provides recommendations to the NFXF's Facilitation of Research Oversight Group (FROG). The authors of this commentary all sit on the CTC. The CTC is designed to provide a one-stop point of contact to interface with industry, academic, and other partners seeking to move forward FXS-focused new treatment development at any stage. The CTC centralizes new treatment development support and advising in an orphan disease condition of great interest to potential business and academic partners. Moving forward the FXCRC Clinics have embraced the concept that, with many potential trials in FXS proposed, it will be critical to protect patients and optimize FXS participant resources through careful discussions about proposed trials and development programs with the CTC. The CTC will review preclinical data in early stages of drug development and subsequently help with recommendations regarding trial designs, outcome measures, and development strategy for agents showing promise on appropriate rigor preclinical and/or early phase clinical studies.

Given the history of large-scale trial failures, and the large list of mechanistic targets and treatment development partners, it is critical for the FXS field to create a collaborative framework of expertise. The latter include those experts who led the first round of trials and representatives of affected families, who will collaborate in the design of trials and drug development paths that can capitalize on lessons learned and result in clear delineation of the drugs that have a positive impact in FXS and, ultimately, in the successful registration of these drugs for use by patients. Given this, it is advised that sponsors of all drug development in FXS that have reached the stage of multi-site Phase II or III trials, hold advisory meetings with and seek endorsement from the CTC, prior to use of FXCRC Clinics to carry out further development in FXS. Those seeking to engage the CTC may contact the National Fragile X Foundation CTC liaison (J.D.W.; jayne@fragilex.org) to initiate a dialogue. This recommendation is made to ensure that study sponsors receive accurate information facilitating the conduct of trials in FXS including broad stakeholder input.

3. Preclinical Development

Experience with mouse models of FXS and other neurodevelopmental disorders has demonstrated their value as experimental systems for proof-of-principle assessments of new interventions [1]. Nonetheless, this work has also shown that phenotypes that are improved in mice do not necessarily translate directly to affected individuals [2]. Therefore, our field needs to emphasize the development of preclinical animal testing that can be evaluated in a similar manner in humans [3]. As important as the limited relevance of animal models to the human disorder, is the concern about reproducibility of animal data. The NIH has published guidelines promoting standardization of experimental paradigms and "best preclinical practices" for animal model work [4] and these have begun to be applied to neurodevelopmental disorders [5]. We recommend that any new treatment should be evaluated in blinded randomized experiments performed for multiple phenotypes in different domains (e.g., electrophysiological, molecular, behavioral), with some phenotypes being directly translatable to humans, and reversal of phenotypes shown in at least two independent laboratories [2] and ideally in several species, given the fact that rat and *Drosophila* models of FXS also exist.

The presentation and publication of negative preclinical testing results is crucial to inform study design, outcome measure selection, and execution of first in human studies. Thus, this practice should be encouraged and is of importance to our field. The FXS field is challenged early on in drug development in part by a publication culture focused on "positive" preclinical data. Understanding which particular treatments may not improve murine or other animal phenotypes may be as important as understanding the aspects of success with a specific drug.

In the same manner that human trials may not result in efficacious treatments for every individual with FXS, understanding variation within animal treatment response in vivo may provide early clues to features that may predict treatment response later in affected patients. Preclinical focus to date on murine findings has been on rescue or non-rescue with drug versus placebo treatment across various behavioral, molecular, and neuroanatomical assays and not on variation within treatment response. Further investigation into potential variability in preclinical treatment response may aide future planning when treatments move into human studies.

Considering that mouse or other animal models work only provides evidence of treatment adequacy but not of specific outcomes, the possibility of skipping altogether in vivo model testing has been raised. Instead, in vitro evaluation of iPSC-derived neural cells offers the opportunity of testing directly cells from individuals with FXS and showing correction of cellular abnormalities, data that not only could be used for demonstrating treatment effectiveness but also subject selection [6]. However, this intriguing new approach does not test extension of cellular changes to neurobehavioral outcomes and will need to be systematically compared with current animal model standards prior to its full implementation [2]. In summary, the rigor, reproducibility, and specific findings of treatment development preclinical data should be evaluated in detail in planning for the scope and potentially specific aspects of the design of first in FXS human studies.

4. Transition from Preclinical Studies to Human Subject Projects

Once preclinical work has demonstrated the scientific validity of a particular treatment through in vivo and/or in vitro studies, other elements of the experimental data can inform the subsequent steps. For initial human FXS trials, a project relying on single laboratory murine data or an iPSC model alone may be best served moving first into a human Phase Ib proof-of-concept target engagement study versus a brisk move into a large-scale first in FXS Phase II study. The latter would be more adequate for treatments supported by multiple preclinical studies providing convergent evidence and/or supported by small initial target engagement focused Phase Ib work in FXS. Drugs for which there is improvement of a directly translatable animal pharmacodynamic marker (e.g., event-related potential (ERP) abnormality) can be quickly tested initially in early phase human pharmacokinetic/pharmacodynamic (PK/PD) studies applying the biomarker prior to moving to larger clinical trials to help inform a large-scale study. Elements of adaptive trial design can be considered even in the earliest stages of human projects in FXS to facilitate the establishment of a predictive model incorporating biological markers and clinical outcomes.

A process of de-risking large-scale multi-site projects in our field, utilizing well-conceived proof of concept early human studies, is quite possible. Such early work would need to include use of extensive objective and/or directly observable measures of brain function, communication, and behavior in order to determine potential clinically-relevant changes with treatment. Use of a crossover design, while potentially problematic in large-scale studies, given increased length, treatment expectation, and potential carryover effects on proxy outcome measures, may provide additional strength to Phase Ib projects that focus primarily on objective evaluations of target engagement. A rigorously designed Phase Ib trial in FXS may allow for an earlier go- or no-go decision on future larger scale multi-site studies of a drug, by requiring significantly less investment -in dollars and stakeholder commitment- than approaches with initial large-scale Phase II projects and by allowing for more informed biomarker and outcome measure selection based on a particular treatment or a particular cohort of individuals with FXS.

Drugs targeting brain mechanisms in FXS with predominant effects on molecular, anatomical, and electrophysiological parameters, resulting in reversal of multiple and diverse phenotypes in animal models, would best move into human therapy through a Phase Ib PK/PD study looking at a broad range of objective outcomes, as optimal outcome measures are too difficult to predict in humans in this kind of scenario. Drugs that have an expected or known behavioral effect deemed important in FXS that has been replicated in multiple labs, and which target behavior as supportive care, may move faster to Phase II or Phase III trials with outcome measures directed at the predicted behavioral targets. Whether human studies entered in a gradual manner or with a rapid move into regulatory grade Phase II and III projects, the transition into and through the human stages of treatment development in FXS will benefit from many practice recommendations put forth below. Although early phase trials focused on safety or PK/PD could be conducted at a single site, studies evaluating efficacy benefit from a multi-site design. Regardless of size and scope, in the case of rare diseases like FXS, expertise in the disorder at the trial sites enhances the possibility of successful implementation and stakeholder satisfaction as discussed in more detail below.

5. Best Practices in Human Subjects Projects: Detecting Treatment Response

Heterogeneity in the clinical presentation of FXS relates to several factors including *FMR1* gene variations and background genetic and epigenetic effects, environmental stimulation, as well as negative life experiences. These factors lead to variability not only in the symptoms of FXS that individuals manifest but also in their response to treatment [2,7]. Likely, not all individuals with FXS will have a uniformly positive clinical response even to the best targeted treatments for FXS. Placebo effects are also remarkable in our field, especially when parent questionnaires regarding interfering behavior are used. Animal models have helped to guide us to specific treatments for patients, such as mGluR5 antagonists, the GABA-B receptor agonist arbaclofen, minocycline, metformin and other agents. However, trials with these agents have been potentially complicated by the presence of subgroups of responders, making entire cohort efficacy difficult to demonstrate. In this context, utilization of biomarkers such as EEG findings or responses in iPSC-derived neuronal cell cultures may be able to identify potential treatment responses for specific individuals with FXS, helping with stratification or selection of subjects for future trials.

The large placebo effect and selection of outcomes have been challenges in clinical trials in FXS. For example, in a well powered study reported by Berry-Kravis and colleagues (2016) there was a remarkable change in caregiver measures in the placebo group [8]. This significant placebo effect could have obscured true treatment-related changes in the active drug group. In the related field of autism spectrum disorder (ASD), a meta-analysis of 25 studies involving 1315 subjects, investigating placebo response in medication trials, demonstrated a "moderate effect size" for overall placebo response (Hedges' $g = 0.45$, 95% confidence interval (0.34–0.56); $P < 0.001$) [9]. Reports like this raise the question whether we ought to expect a greater placebo effect when employing caregiver measures of abnormal behavior, in patients with neurodevelopmental disorders than in other populations [10]. If the answer is an affirmative one, that may reflect particularly high expectations among caregivers of individuals with developmental disabilities, including FXS, for treatments targeting key symptoms and reduction of disease burden [10], especially when widespread success of preclinical models is highly publicized. The strength of the placebo response in clinical trials for FXS supports the need to use more objective neurobehavioral and functional measures and/or to develop approaches that substantially mitigate biased reporting of treatment effects. Like all effective outcome measures, such "placebo-resistant" evaluations would need to be psychometrically sound (e.g., with known test-retest reliability), sensitive to change, and reflect improvements that are meaningful in the daily life of individuals with FXS. Ideally, the relationships between these measures and the underlying neural systems impaired in FXS should be understood so that in any given trial a measure can be chosen because it is proximal to the mechanism of action of the drug under study. It would also be useful if the clinical outcome measures have relevance to the neural cellular and circuit targets evaluated in preclinical studies [2].

Unfortunately, available outcome measures, including those in use in ongoing trials, generally do not meet all of the criteria outlined above [11]. Nonetheless, progress is being made in establishing the psychometric adequacy of several objective measures for quantifying change in important domains of general cognitive functioning [12], as well as specific areas of language [13] and executive function [14]. Preliminary results suggest that many of these promising measures are feasible for individuals of a wide range of ages and abilities, display minimal practice effects and strong test-retest reliability, and have good construct validity. At the same time, it is important to acknowledge that performance on these novel measures, as for most cognitive and behavioral measures, is likely to be affected by a number of skills and motivational factors and, thus, will reflect the functioning and complex interaction of multiple neural circuits that may result in attenuation of their sensitivity to detect drug efficacy. Also, we should acknowledge that placebo effects may occur even when using objective performance measures in individuals with intellectual disability [15].

Quantitative measures of pathophysiology are generally considered as potential continuous measures of efficacy; however, they can be equally considered as baseline predictors of treatment response. Genetic and blood-based molecular markers have already shown promise as identifiers of potential responders in drug trials in FXS [16]. While there is a high bar for official biological marker qualification as FDA-accepted surrogate biomarkers that can be used as regulatory endpoints in trials, use of biomarkers to identify target engagement, generally, and subgroups of potential responders with FXS is tractable in the near term. An approach like this may be able to parse the sometimes underappreciated heterogeneity of FXS in early trials, in order to guide subject selection or stratification in large-scale treatment trials. An example of this approach is ongoing work using single-dose placebo-controlled probe studies in adolescents and adults with FXS [17]. Another strategy, which should be incorporated as a standard procedure for all large-scale trials, is to collect blood samples for evaluation of genetic and other molecular markers. Development of brain- or blood-based markers will enrich research populations with responders to enhance treatment success and de-risk the perils faced by trials similar to unsuccessful past trials in FXS, which had broad inclusion criteria and relied on placebo-sensitive measures. For this reason, the NFXF has developed a NFXF Biobank™ program to receive biological samples and associated clinical data from persons with FXS including trial participants to provide a repository that will benefit biomarker understanding for the field regardless of final trial result.

Previous publications have examined the current status of outcome measures in FXS, providing practical recommendations and future directions [11,18]. Attributes considered critical for novel and improved outcome measures are the following: quantitative; reflecting brain circuits, function, or molecular pathways affected in FXS; relevant to experimental models of FXS; and reflecting quality of life of individuals with FXS. An example of these promising measures is event related potentials (ERPs), using repetitive auditory or visual stimuli and measuring habituation to the stimuli which is known to be abnormal because of inhibitory (GABA) deficits in FXS. Significant improvements in this measure were documented by Schneider and colleagues [19] in a controlled trial of minocycline compared to placebo and similar ERPs are being studied in the NeuroNext trial (FX-LEARN) with AFQ056 and parent-implemented language intervention (PILI, NCT02920892). Another type of promising measures are molecular markers reflecting core abnormalities in FXS, such as excessive protein production in FXS. For instance, levels of the matrix metalloproteinase 9 (MMP9) that are up-regulated in FXS have been shown to be reduced by administration of minocycline [20].

Biomarkers sensitive to change and correlated or predictive of clinical behavioral changes are of particular interest. Thus, assessment of how quantitative measures such as auditory ERPs, eye tracking and molecular biomarkers correlate with the clinical outcomes measured by behavioral ratings, cognitive and language tests, is critical. A number of recent phase 1b and 2a trials in FXS have incorporated all of these measures (NeuroNext AFQ056 (NCT02920892), metformin (NCT03479476),

AZD7325 (NCT03140813), BPN14770 (NCT03569631)), such that the outcomes of these studies will improve our understanding of the relationships between biomarkers and clinical behavioral measures. An NIH multisite initiative called the Autism Biomarkers Consortium for Clinical Trials (www.asdbiomarkers.org) aims to identify a useful set of such tools.

Lessons learned from prior unsuccessful clinical trials in FXS suggest that greater emphasis on clinician- than caregiver-based measures; training systems for caregivers, clinicians, and other raters (e.g., teachers); evaluations in multiple settings (e.g., school, home, workplace); and behavioral ratings in real time on an electronic device or by videotaping, are approaches that could complement or replace current instruments, in particular behavioral rating scales. Finally, careful attention to other study aspects such as randomization and placebo inclusion in most early- and all late-phase trials; longer trial duration (to capture cognitive and adaptive changes); approaches to minimize placebo effects (e.g., placebo run-ins, enrichment in non-placebo responders, rater training); and younger cohort age would maximize the possibility of success.

6. Family Involvement

It is important that a family stakeholder voice is heard in FXS treatment development programs. Stakeholders can optimize recruitment and retention for FXS clinical trials. To accomplish this, it is important to provide parents with the most detailed possible information about all aspects of the trial. This would include: inclusion/exclusion criteria, length of the study, location, number/length/flexibility of visits, details on what will occur at each visit, and whether there is any travel/participation reimbursement available. It is imperative that family stakeholders understand the clinical manifestation targeted by the drug or intervention under study. It should also be noted that inclusion of an open-label extension phase following a placebo-controlled treatment will inherently boost study recruitment and retention. Enhanced communication with families about trial details, such as inclusion criteria and endpoints, should include the importance of adhering to the study design and avoiding attempts to "boost" reporting of a certain feature or behavior to facilitate study participation. Education on the implications of inaccurate reporting or embellishment of symptoms on the likelihood of obtaining appropriate results, which may benefit all stakeholders, is another key issue that should be communicated to families.

Clinicians working with FXS family stakeholders often note the high level of motivation and commitment regarding treatments or clinical trials that may benefit their loved one with FXS. Given this zeal, parents/caregivers may be inadvertently drawn to a new treatment that may not be suitable for their family member with FXS. There are many reasons why a trial could not be an appropriate fit for a specific individual with FXS, including safety and efficacy concerns, inclusion or exclusion criteria, outcome measures, length of project and/or number of appointments required. From a caregiver viewpoint, it is important that drug trials have a protocol that includes an operations manual detailing the accommodations to take place, such as staff training regarding FXS, visual supports and extra time during appointments. A well-designed project that is "FXS-friendly" will ensure that families are met with the expertise and commitment necessary to maximize the opportunity for positive outcomes.

It may be useful to ask family raters (caregivers) what they have heard about the treatment/medication, to yield information about any bias that may be likely in their future assessments. Raters should be given explicit permission to not report improvement if none occurred, to emphasize observed behavior rather than make guesses or inferences or rely overly on third parties, and to remain agnostic to the extent possible about the probability of treatment benefit.

As a patient advocacy organization that has been serving the FXS community of families and professionals for nearly 35 years, the National Fragile X Foundation strongly recommends that parents have a well-defined and meaningful role in providing input regarding clinical trials and new treatment development. There has been a history of unsuccessful trials, parental confusion regarding clinical endpoints, disappointment and, at times, anger over the lack of opportunity for and/or discontinuation of open label extensions among other concerns. Some of these pitfalls can be avoided or, at least, minimized by ensuring that researchers fully understand the worries, concerns, wants, needs, and hopes of parents and other family members. Such input can for example include, among others, recommendations about targets of treatment of importance to families in terms of quality of life, preferred routes of drug administration, advice on the conduct of research procedures within a visit day, feasibility of study designs, and strategies of communication of trials to the community. By providing such an opportunity for family input during protocol development, researchers and, ultimately, those in pharmaceutical industry leadership and decision-making positions, will have access to information that will greatly increase the likelihood of successful trial outcomes and, most importantly, lead to better lives for those with FXS. Development of a family advisory committee as a resource to treatment development programs should meet these needs.

Understanding the impact of social media on FXS treatment development is also of importance to our field. Because it is not possible to completely monitor Facebook and other types of social media, trial sponsors or investigators should include a stand-alone document on the use of social media and ask parents to sign it in order to participate in the trial. It should state that it is acceptable for participants to post about their experience with the study (e.g., how accommodating the staff is, how to navigate the hospital, tricks they have found successful in helping their family member with FXS participate, and that they are glad to be supporting research). However, the document should also clearly describe what comments are inappropriate because they may bias study reporting such as those about side effects, how the patient is doing, whether they think they are on drug or placebo or how they think other families should rate scales or questionnaires to qualify for the study.

Many clinical trials have extensively evaluated children and adults with FXS. Assessments might include IQ or equivalent tests, adaptive behavior, language, and testing regarding autism spectrum disorder. Much of this information could be beneficial for families to have for school programming and for additional support services. This testing can be expensive and not readily available to all families. The CTC recommends that relevant findings be shared with parents and guardians and provided to them in a written format that would be readily understood by providers or professionals supporting the child or adult with FXS.

7. Multi-Site Trial Planning

Once FXS treatment development has moved to the stage of large-scale Phase II or Phase III investigations, opportunities exist to enhance project executions, fidelity of data gathering, and to facilitate broad human subjects' recruitment and retention. It is important that within large-scale projects an operations manual that details accommodations to individuals and families with FXS, as the one described in the preceding section, is developed. A well-designed project that is "FXS-friendly" will ensure that families are met with the expertise and commitment necessary to maximize the opportunity for positive outcomes. Rater and investigator training including presentations by FXS thought leaders and family stakeholders will ensure FXS specificity, which will overcome barriers inherent in "off the shelf" central nervous system (CNS) trial approaches grafted into a unique neurodevelopmental disorder setting.

Such multi-site work should build upon existing FXS centers and clinics with content expertise. Luckily, over a decade of multi-site trial projects in our field has laid a foundation for large-scale trials that has united clinical centers with content expertise with best trial practices and the standard operating procedures of good clinical research. While initial site qualification efforts and trainings should involve FXS content expertise, ongoing support of fidelity of study execution across sites is essential. There are also great opportunities for centralized future potential support of recruitment and retention efforts in large-scale projects given the infrastructure within the FXCRC supported by the National Fragile X Foundation.

8. Summary of Key Recommendations

Much has been learned in the burgeoning FXS treatment development field in the past decade. We have set forth to make many recommendations to enhance the success and stakeholder benefits from future treatment development in our field. From focusing on reproducible early preclinical data to moving towards quantitative markers of pathophysiology as markers of target engagement and change with treatment, clear directions have been defined to broadly support success navigating the chasm between treatment ideas and success in placebo-controlled trials. Along the way we emphasize the importance of engaging the FXS stakeholder community to ensure the meaningfulness of project results and the appropriateness of project procedures from a family and affected individual perspective. Our group remains optimistic at the near-term prospect for impactful treatment development in FXS. We believe that our advocacy and engagement with these processes will work to maximize success and we embrace the opportunity to support the best practices we have defined in this commentary in partnership with treatment developers worldwide.

Author Contributions: Conceptualization, C.A.E.; Writing—Original draft preparation, C.A.E., W.E.K., D.B.B., A.L., B.H.-G., R.M.M., J.D.W., L.A., D.H., R.J.H., E.B-K.

Funding: This commentary received no external funding.

Conflicts of Interest: C.A.E. Has received current or past funding from Confluence Pharmaceuticals, Novartis, F. Hoffmann-La Roche Ltd., Seaside Therapeutics, Riovant Sciences, Inc., Fulcrum Therapeutics, Neuren Pharmaceuticals Ltd., Alcobra Pharmaceuticals, Neurotrope, Zynerba Pharmaceuticals, Inc., and Ovid Therapeutics Inc. to consult on trial design or development strategies and/or conduct clinical trials in FXS or other neurodevelopmental disorders. C.A.E. is additionally the inventor or co-inventor on several patents held by Cincinnati Children's Hospital Medical Center or Indiana University School of Medicine describing methods of treatment in FXS or other neurodevelopmental disorders. W.E.K. has received funding from Seaside Therapeutics, Novartis, Roche, Ipsen, Cydan, Neuren, Edison, Eloxx, Newron Pharmaceuticals, EryDel, GW Pharmaceuticals, Marinus, Biohaven, AveXis, Ovid Pharmaceuticals, Anavex, Stalicla, and Zynerba to consult on trial design or development strategies and/or conduct clinical trials in Rett syndrome, FXS or other neurodevelopmental disorders. L.A. is currently serving as an advisor to Fulcrum Therapeutics and has received funding in the past to implement outcome measures in clinical trials from F. Hoffmann-La Roche Ltd., Roche TCRC, Inc., and Neuren Pharmaceuticals Ltd. D.B.B. has received support for clinical trials in FXS from Seaside Therapeutics, Ovid Therapeutics Inc., Zynerba Pharmaceuticals, Inc. and also NIH-research funding through Asuragen, Inc. Within the last three years, D.B.B. has done ad hoc consulting work (rarely) for the American Academy of Child & Adolescent Psychiatry, Ironshore Pharmaceuticals & Development, Inc., MEDACorp, Inc., Guidepoint, Ovid Therapeutics Inc. and Sunovion Pharmaceuticals Inc. A.L. has received funding from Ovid Therapeutics for consultation. D.H. provides consultation to Ovid Therapeutics Inc., Zynerba Pharmaceuticals, Inc., and Autifony Therapeutics. B.H.-G. and R.M.M. have no conflicts to report. R.J.H. has received funding from Zynerba Pharmaceuticals, Inc. for consultation regarding clinical trials for FXS and has also been an advisor for Fulcrum Therapeutics for future trials in FXS. J.D.W. has received funding from Ovid Therapeutics Inc. for consultation regarding clinical trials for FXS. E.B.-K. has received funding from Seaside Therapeutics, Novartis, F. Hoffmann-La Roche Ltd., Alcobra Pharmaceuticals, Neuren Pharmaceuticals Ltd., Cydan, Fulcrum Therapeutics, GW Pharmaceuticals plc, Neurotrope, Marinus Pharmaceuticals, Inc., Zynerba Pharmaceuticals, Inc., BioMarin Pharmaceutical, Yamo Pharmaceuticals, and Ovid Therapeutics Inc. to consult on trial design or development strategies and/or conduct clinical trials in FXS or other neurodevelopmental disorders, from Vtesse/Sucampo/Mallinkcrodt to consult on and conduct clinical trials in NP-C, and from Asuragen, Inc. to develop testing standards for FMR1 testing.

References

1. Kazdoba, T.M.; Leach, P.T.; Silverman, J.L.; Crawley, J.N. Modeling fragile X syndrome in the Fmr1 knockout mouse. *Intractable Rare Dis. Res.* **2014**, *3*, 118–133. [CrossRef] [PubMed]
2. Berry-Kravis, E.M.; Lindemann, L.; Jonch, A.E.; Apostol, G.; Bear, M.F.; Carpenter, R.L.; Crawley, J.N.; Curie, A.; Des Portes, V.; Hossain, F.; et al. Drug development for neurodevelopmental disorders: Lessons learned from fragile X syndrome. *Nat. Rev. Drug Discov.* **2018**, *17*, 280–299. [CrossRef] [PubMed]
3. Goel, A.; Cantu, D.A.; Guilfoyle, J.; Chaudhari, G.R.; Newadkar, A.; Todisco, B.; de Alba, D.; Kourdougli, N.; Schmitt, L.M.; Pedapati, E.; et al. Impaired perceptual learning in a mouse model of Fragile X syndrome is mediated by parvalbumin neuron dysfunction and is reversible. *Nat. Neurosci.* **2018**, *21*, 1404–1411. [CrossRef] [PubMed]

4. Landis, S.C.; Amara, S.G.; Asadullah, K.; Austin, C.P.; Blumenstein, R.; Bradley, E.W.; Crystal, R.G.; Darnell, R.B.; Ferrante, R.J.; Fillit, H.; et al. A call for transparent reporting to optimize the predictive value of preclinical research. *Nature* **2012**, *490*, 187–191. [CrossRef] [PubMed]

5. Katz, D.M.; Berger-Sweeney, J.E.; Eubanks, J.H.; Justice, M.J.; Neul, J.L.; Pozzo-Miller, L.; Blue, M.E.; Christian, D.; Crawley, J.N.; Giustetto, M.; et al. Preclinical research in Rett syndrome; setting the foundation for translational success. *Dis. Model Mech.* **2012**, *5*, 733–745. [CrossRef] [PubMed]

6. Faundez, V.; De Toma, I.; Bardoni, B.; Bartesaghi, R.; Nizetic, D.; de la Torre, R.; Cohen Kadosh, R.; Herault, Y.; Dierssen, M.; Potier, M.C.; et al. Translating molecular advances in Down syndrome and Fragile X syndrome into therapies. *Eur. Neuropsychopharmacol.* **2018**, *28*, 675–690. [CrossRef] [PubMed]

7. Erickson, C.A.; Davenport, M.H.; Schaefer, T.L.; Wink, L.K.; Pedapati, E.V.; Sweeney, J.A.; Fitzpatrick, S.E.; Brown, W.T.; Budimirovic, D.; Hagerman, R.J.; et al. Fragile X targeted pharmacotherapy: Lessons learned and future directions. *J. Neurodev. Disord.* **2017**, *9*, 7. [CrossRef] [PubMed]

8. Berry-Kravis, E.; Des Portes, V.; Hagerman, R.; Jacquemont, S.; Charles, P.; Visootsak, J.; Brinkman, M.; Rerat, K.; Koumaras, B.; Zhu, L.; et al. Mavoglurant in fragile X syndrome: Results of two randomized, double-blind, placebo-controlled trials. *Sci. Transl. Med.* **2016**, *8*, 321ra5. [CrossRef] [PubMed]

9. Masi, A.; Lampit, A.; Glozier, N.; Hickie, I.B.; Guastella, A.J. Predictors of placebo response in pharmacological and dietary supplement treatment trials in pediatric autism spectrum disorder: A meta-analysis. *Transl. Psychiatry* **2015**, *5*, e640. [CrossRef] [PubMed]

10. Jeste, S.S.; Geschwind, D.H. Clinical trials for neurodevelopmental disorders: At a therapeutic frontier. *Sci. Transl. Med.* **2016**, *8*, 321fs321. [CrossRef] [PubMed]

11. Budimirovic, D.B.; Berry-Kravis, E.; Erickson, C.A.; Hall, S.S.; Hessl, D.; Reiss, A.L.; King, M.K.; Abbeduto, L.; Kaufmann, W.E. Updated report on tools to measure outcomes of clinical trials in fragile X syndrome. *J. Neurodev. Disord.* **2017**, *9*, 14. [CrossRef] [PubMed]

12. Hessl, D.; Sansone, S.M.; Berry-Kravis, E.; Riley, K.; Widaman, K.F.; Abbeduto, L.; Schneider, A.; Coleman, J.; Oaklander, D.; Rhodes, K.C.; et al. The NIH Toolbox Cognitive Battery for intellectual disabilities: Three preliminary studies and future directions. *J. Neurodev. Disord.* **2016**, *8*, 35. [CrossRef] [PubMed]

13. Berry-Kravis, E.; Doll, E.; Sterling, A.; Kover, S.T.; Schroeder, S.M.; Mathur, S.; Abbeduto, L. Development of an expressive language sampling procedure in fragile X syndrome: A pilot study. *J. Dev. Behav. Pediatr.* **2013**, *34*, 245–251. [CrossRef] [PubMed]

14. Knox, A.; Schneider, A.; Abucayan, F.; Hervey, C.; Tran, C.; Hessl, D.; Berry-Kravis, E. Feasibility, reliability, and clinical validity of the Test of Attentional Performance for Children (KiTAP) in Fragile X syndrome (FXS). *J. Neurodev. Disord.* **2012**, *4*, 2. [CrossRef] [PubMed]

15. Curie, A.; Yang, K.; Kirsch, I.; Gollub, R.L.; des Portes, V.; Kaptchuk, T.J.; Jensen, K.B. Placebo Responses in Genetically Determined Intellectual Disability: A Meta-Analysis. *PLoS ONE* **2015**, *10*, e0133316. [CrossRef] [PubMed]

16. AlOlaby, R.R.; Sweha, S.R.; Silva, M.; Durbin-Johnson, B.; Yrigollen, C.M.; Pretto, D.; Hagerman, R.J.; Tassone, F. Molecular biomarkers predictive of sertraline treatment response in young children with fragile X syndrome. *Brain Dev.* **2017**, *39*, 483–492. [CrossRef] [PubMed]

17. Erickson, C.A.; Schmitt, L.M.; Pedapati, E.V.; Ethridge, L.E.; Sweeney, J.A. Development of EEG biomarkers to assist treatment development in fragile X syndrome: single dose drug human study. In Proceedings of the International Fragile X Meeting, Cincinnati, OH, USA, 13 July 2018.

18. Berry-Kravis, E.; Hessl, D.; Abbeduto, L.; Reiss, A.L.; Beckel-Mitchener, A.; Urv, T.K.; Outcome Measures Working, G. Outcome measures for clinical trials in fragile X syndrome. *J. Dev. Behav. Pediatr.* **2013**, *34*, 508–522. [CrossRef] [PubMed]

19. Schneider, A.; Leigh, M.J.; Adams, P.; Nanakul, R.; Chechi, T.; Olichney, J.; Hagerman, R.; Hessl, D. Electrocortical changes associated with minocycline treatment in fragile X syndrome. *J. Psychopharmacol.* **2013**, *27*, 956–963. [CrossRef] [PubMed]

20. Dziembowska, M.; Pretto, D.I.; Janusz, A.; Kaczmarek, L.; Leigh, M.J.; Gabriel, N.; Durbin-Johnson, B.; Hagerman, R.J.; Tassone, F. High MMP-9 activity levels in fragile X syndrome are lowered by minocycline. *Am. J. Med. Genet. A* **2013**, *161A*, 1897–1903. [CrossRef] [PubMed]

 *brain sciences*

Article

Voice of People with Fragile X Syndrome and Their Families: Reports from a Survey on Treatment Priorities

Jayne Dixon Weber [1,*], Elizabeth Smith [2,*], Elizabeth Berry-Kravis [3], Diego Cadavid [4], David Hessl [5] and Craig Erickson [6]

[1] National Fragile X Foundation, McLean, VA 22102, USA

[2] Cincinnati Children's Hospital Medical Center Division of Child & Adolescent Psychiatry, Cincinnati, OH 45229, USA

[3] Departments of Pediatrics, Neurological Sciences, Biochemistry, Rush University Medical Center, Chicago, IL 60612, USA; Elizabeth_Berry-Kravis@rush.edu

[4] Fulcrum Therapeutics, Cambridge, MA 02139, USA; dcadavid@fulcrumtx.com

[5] MIND Institute and Department of Psychiatry and Behavioral Sciences, University of California Davis School of Medicine, Sacramento, CA 95817, USA; drhessl@ucdavis.edu

[6] Cincinnati Children's Hospital Medical Center, Division of Child & Adolescent Psychiatry and the University of Cincinnati College of Medicine Department of Psychiatry and Behavioral Neuroscience, Cincinnati, OH 45229, USA; craig.erickson@cchmc.org

[*] Correspondence: jayne@fragilex.org (J.D.W.); elizabeth.smith3@cchmc.org (E.S.); Tel.: +1-303-494-2740 (J.D.W.)

Received: 14 December 2018; Accepted: 16 January 2019; Published: 23 January 2019

Abstract: To date, there has been limited research on the primary concerns and treatment priorities for individuals with fragile X syndrome (FXS) and their families. The National Fragile X Foundation in collaboration with clinical investigators from industry and academia constructed a survey to investigate the main symptoms, daily living challenges, family impact, and treatment priorities for individuals with FXS and their families, which was then distributed to a large mailing list. The survey included both structured questions focused on ranking difficulties as well as qualitative analysis of open-ended questions. It was completed by 467 participants, including 439 family members or caretakers (family members/caretakers) of someone with FXS, 20 professionals who work with a person with FXS, and 8 individuals with FXS. Respondents indicated three main general areas of concern: Anxiety, behavioral problems, and learning difficulties. Important differences were noted, based on the sex and age of the individual with FXS. The results highlight the top priorities for treatment development for family members/caretakers, as well as a small group of professionals, and an even smaller group of individuals with FXS, while demonstrating challenges with "voice of the patient" research in FXS.

Keywords: fragile X syndrome; *FMR1* gene; voice of the person; voice of the patient; characteristics that have the greatest impact; developmental disorders

---

1. Introduction

Fragile X syndrome (FXS) is a neurodevelopmental condition that is caused by the expansion of the CGG repeat in the 5' untranslated region of fragile X mental retardation 1 (*FMR1*) gene located on the X chromosome [1]. This expansion leads to methylation of the *FMR1* promoter, transcriptional silencing of the gene, and subsequent reduction or absence of fragile X mental retardation protein (FMRP). FMRP is an RNA-binding protein that regulates dendritic translation of many key synaptic proteins that influence synaptic function and plasticity [2]. FXS is the leading known inherited cause of intellectual disability. Individuals with FXS are most commonly diagnosed after presenting with language delay, and the majority of males with FXS will ultimately meet criteria for mild-to-severe intellectual disability [3]. The average IQ in men with FXS is 40–50, with a mental age of about of 5–6 years. Females with FXS are often less affected (average IQ about 80) than males, with about 25% having cognitive impairment and others frequently being diagnosed with learning disabilities [4]. There is a relatively consistent pattern of intellectual weaknesses (visuospatial skills, working memory, processing of sequential information, attention) and strengths (simultaneous processing, imitation, visual memory) characteristic of both males and females with FXS [2]. Multiple studies have shown a decrease in full-scale IQ scores with age as children with FXS become older [5,6]. Likewise, standard scores on the Vineland Adaptive Behavior Scale for overall adaptive behavior as well as subdomains have also been shown to decline with age during childhood, in males more so than in females with FXS [7]. Decline in standard scores for intelligence and adaptive function is not the result of loss of skills or regression but rather failure to keep pace with the normal rate of intellectual development. FXS is also associated with a constellation of behavioral symptoms, which can be highly problematic for functioning and family burden, including but not limited to high levels of anxiety, attention deficit/hyperactivity disorder, social communication deficits, and self-injurious and sensory-seeking behaviors [8].

Due to its known genetic and molecular underpinnings, symptoms of FXS can potentially be targeted via medical interventions. These interventions can focus on increasing expression of the missing FMRP protein and/or remediating downstream effects of neural and synaptic dysmaturation. End goals of treatment include both reducing symptom severity and improving activities of daily living (ADLs) and quality of life in affected individuals and their families. Given the array of behavioral and developmental symptoms that can exist in individuals with FXS, as well as the range of potential biochemical targets, it is useful to focus clinical research on symptoms and concerns that are experienced as most impairing to individuals with FXS and their families. Taking a patient-first perspective can be facilitated through qualitative research methods, where semistructured and free responses from patients and their families are used as primary data.

To date, there have been qualitative studies in FXS on topics including but not limited to diagnosis [9,10], communication impairments [11], physician knowledge [12], and technology use [13]. Closely related to the present work, Bailey and colleagues surveyed parents of children with FXS regarding the prevalence of developmental delay and eight other symptoms frequently associated with FXS. These included attention problems, hyperactivity, aggressive behaviors, self-injury, autism, seizures, anxiety, and depression [14]. Following this work, the same group [15] evaluated caregiver preferences for six different treatment foci (i.e., learning and applying new skills, explaining needs, controlling behavior, taking part in new social activities, caring for oneself, and paying attention). The highest priorities for treatment based on 614 responses from caregivers of males ages 5 years and older with FXS were (1) controlling behavior and (2) caring for oneself. These priorities were consistent across age groups. These results have been informative in directing treatment targets, but the present study has several features that can expand understanding of the needs of individuals with FXS and their families. These features include the addition of individuals with FXS as well as professionals to the survey, inclusion of females with FXS, inclusion of family members/caretakers of children under 5 years of age, and free response followed by coding, allowing participants to express interest in a wider range of treatment foci.

Here, we attempted to address these gaps and report on both quantitative and qualitative findings from an online survey completed by 467 respondents, including mostly family members/caretakers but also a small sample of professionals and individuals with FXS. The survey was designed around key problem areas (i.e., "concepts of interest") as well as priorities for treatment. Survey items focused on major concerns, symptom areas, daily living skills, family impact, and treatment priorities, and results are reported by age and sex of the individual affected by FXS. A free-response item was also included to allow family members/caretakers to detail concerns above and beyond those offered, as a goal of this study was to use a patient-first approach without a priori hypotheses regarding expected reporting patterns for FXS family members/caretakers and professionals.

2. Methods

Survey: Initial survey questions were created within a focus group of individuals with expertise in FXS via their involvement in the National Fragile X Foundation. The initial draft survey included both structured, forced-choice questions as well as open-ended questions, with questions focused on the following information: (1) Respondent characteristics, including individuals with FXS; (2) major concerns and symptoms experienced; (3) difficulties with daily living skills; (4) family impact; and (5) treatment priorities. The survey was divided into age groups based on standard developmental stages: Early childhood, middle childhood, adolescent/young adult, and adult. This draft survey was then presented to a focus group including parents of individuals with FXS and medical providers, with the purpose of confirming that the survey was comprehensive, clear, and respectfully worded. Minor revisions were made, incorporating the focus group feedback. Subsequently, the final survey was sent via email to four families of individuals with FXS who had volunteered to pilot the process involved in completing the survey. No further adjustments were requested based on this pilot sample. No identifying information was included in the survey. The final survey can be seen in the Appendix A.

Sample: A link to the survey (via SurveyMonkey) was sent to 10,000+ emails subscribed to receive general emails from the National Fragile X Foundation. Recipients of the email were eligible to participate if they were an individual with a full FXS mutation or were a family member or caretaker of an individual with a full FXS mutation or a professional who works with an individual with a full FXS mutation. For individuals associated with multiple children with FXS, these individuals were eligible to complete the survey once per affected child.

Analyses: For items where responses required ranking concerns, we included data from participants who did not use all ranks (e.g., only assigning ranks of 1 and 2 and leaving all others blank) but excluded rank data higher than those instructed (e.g., assigning a rank of "6" when instructed to rank top 5). To compare ranked items, a weighted mean rank score was calculated as follows. First, the number 1 ranked item was given a 5, the number 2 ranked item was given a 4, and so on. Then, the sum was calculated for each item and then divided by the number of people who completed the survey. See the Supplementary Tables S1–S5.

For survey items concerning overall problems, problematic symptoms, daily living challenges, and family impact, we therefore report rates of endorsement as well as weighted mean rank scores across the sample and within groups by age and sex. For open-ended responses, data were first open coded for themes, and keyword lists were constructed for each theme. Then, all responses were coded in vivo for presence of any of the key words, with presence of a key word indicating endorsement of a theme.

3. Results

There were 467 individual responses to the survey, including 8 individuals with Fragile X (i.e., endorsing "I have Fragile X syndrome"), 20 professionals (i.e., endorsing "I am a professional who works with a person with FXS"), and 439 family members or caretakers (i.e., endorsing "I am a family member or caretaker of someone with FXS"). It should be noted that not everyone answered every question. Reporters described the individual with FXS on whom they were reporting as being mostly males (n = 397, 84.8%), with ages distributed across the lifespan (see Figure 1). See the data in Supplementary Table S1.

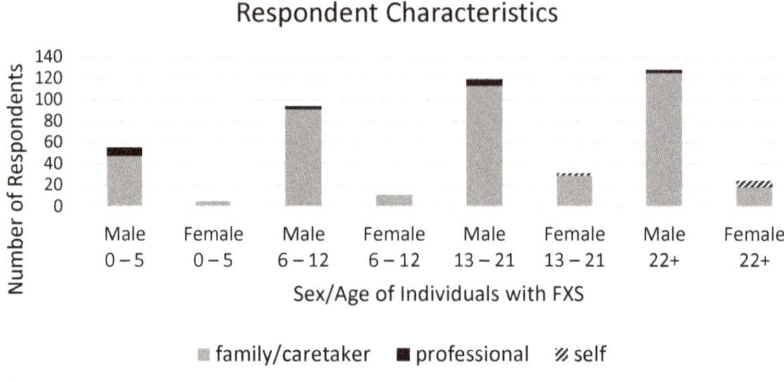

Figure 1. Survey respondent characteristics. FXS: Fragile X syndrome.

3.1. Results from Family Members/Caretakers

Major Concerns. Respondents completed the following question: "Rank these three areas (behavior, intelligence, physical abilities) from one to three to the extent it affects the person's daily life–with one having the greatest impact and three having the least impact:" No other instructions or definitions were provided for this question on the survey. Each of the three areas of impact was described as having the greatest impact by at least some family members/caretakers, with behavior endorsed most commonly as having the greatest impact for males alone. However, for females alone, intelligence was endorsed as having the greatest impact. This question was answered by 429 family members/caretakers (see Figure 2). Physical abilities were the main concern in about 15% of FXS males aged 0–5 and 10% of FXS females age 22+ but were the main concern in less than 10% of all other groups. Behavior was by far the main concern in FXS males age 12 and under; however, in each older age category, intelligence was the main concern in a larger percent of males and, by age 22+, the main concern was divided almost equally between behavior and intelligence, suggesting that intellectual deficits are perceived as increasingly limiting as FXS males (and females) become older.

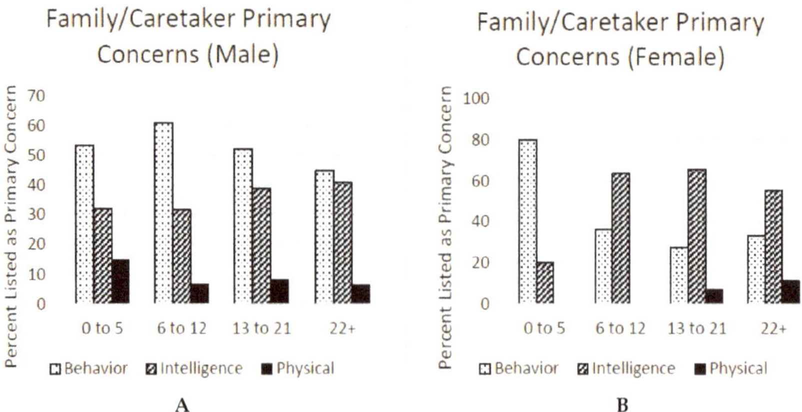

Figure 2. Primary concerns of family members/caretakers of males (**A**) and females (**B**) with FXS.

Problematic Symptoms. Family members/caretakers (438) completed the following question: "Check the five characteristics that have the greatest impact on the life of the person with FXS. Prioritize

1, 2, 3, 4, 5." Out of potential symptoms listed (please see Appendix A for a full list), the following 5 symptoms had the highest weighted mean score (see Figure 3A): (1) Anxiety–anticipatory, e.g., of new/upcoming events and or social anxiety; (2) learning or intellectual disability (problems with abstract thinking, learning); (3) speech/language delays–expressive (speaking spoken language); (4) seizures; and (5) other. Anxiety was rated highest in males beginning at age 6 and in females across all ages, although in females, anxiety was rated highest at similar rates to learning problems. In young males (ages 0–5), expressive language delays were described as being the most problematic symptom. See the data in Supplementary Table S2.

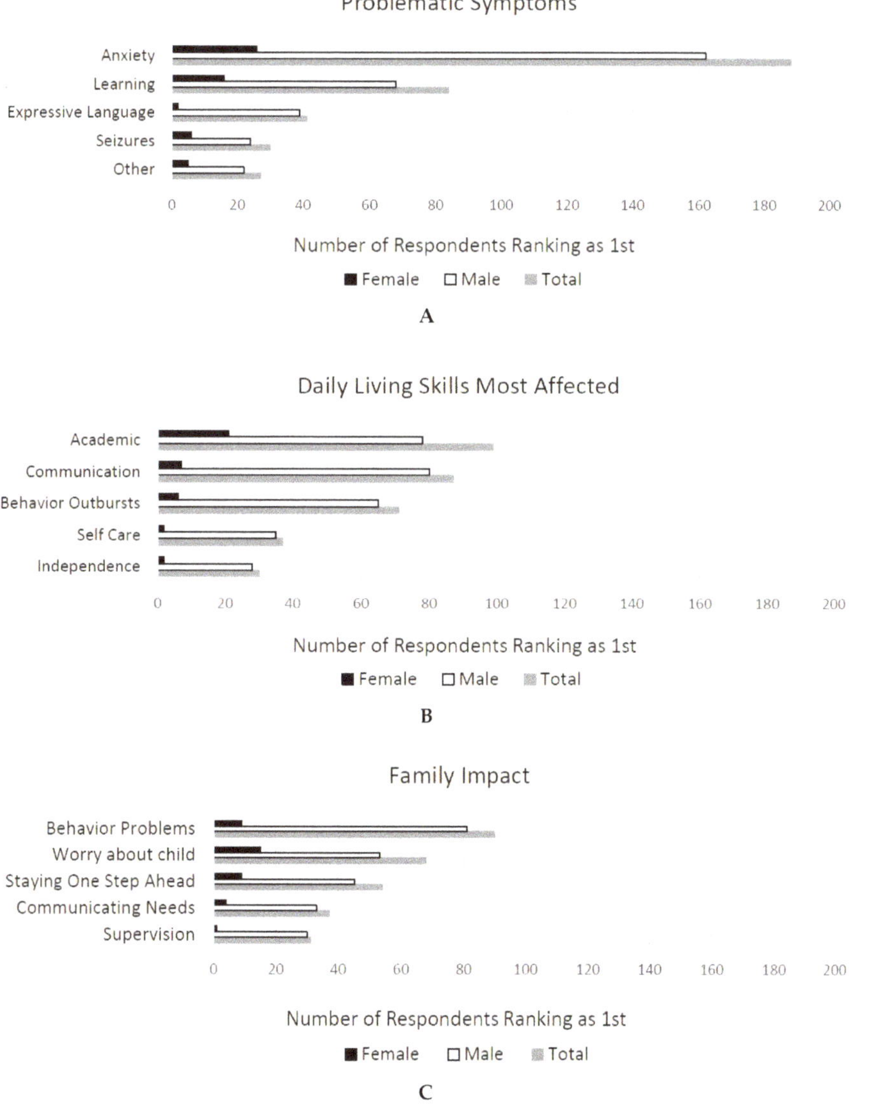

Figure 3. Family member/caretaker's 1st rank for characteristics that have the greatest impact on the life of the person with FXS (**A**), daily living skills most affected (**B**), family impact (**C**).

Daily Living Skills Most Affected. Family members/caretakers (436) answered the question "Check the top five areas of daily life the person with FXS is most affected by. Prioritize 1, 2, 3, 4, 5." Out of 15 possible areas, the five abilities that received the highest weighted rank score were (see Figure 3B): (1) Ability to learn academic skills/reading/math; (2) ability to speak/communicate; (3) ability to control behavioral outbursts; (4) ability to take care of self; and (5) independence. Again, ability to speak was a higher-rated concern in the 0–5 group than the others; however, this was one of the most highly rated items throughout childhood and adolescence, while in adulthood, ability to live independently became the highest rated daily living skill concern. In general, ablity to speak, learn academics, control behavior, and perform self-care were fairly evenenly rated as the skills weighted as most problematic across all age groups of males, while highest rated daily living problems for females reflected more social issues. See the data in Supplementary Table S3.

Family Impact. Family members/caretakers (431) answered the question "Which five specific aspects of daily living with FXS are the most challenging? Prioritize 1, 2, 3, 4, 5." Out of twenty potential options, the following five aspects were ranked as the most challenging: (1) Handling behaviors (negative)—tantrums, aggression, spitting, cussing; (2) worry about the future; (3) always thinking—how are things going, what do I need to do next? Needing to always be 'one step' ahead; (4) person is unable to tell you what he/she wants/needs; and (5) supervision (see Figure 3C). See the data in Supplementary Table S4.

Treatment Priorities. Participants responded to the question "What are the top three aspects of Fragile X syndrome that you would like to see a drug treatment address, list in order of preference, with the most important one first." After open coding of all three listed responses for all participants, the following themes emerged: Anxiety (e.g., "anxiety", "social anxiety", "reduce anxiety"), learning (e.g., "intellect", "cognitive abilities", "learning issues"), behaviors (e.g., "behavioral outbursts", "tantrums", "aggression"), Attention Deficit/Hyperactivity Disorder (ADHD) (e.g., "attention span", "focus", "impulse control"), communication (e.g., "speech delay", "communication delay", "language"), sensory (e.g., "sensory processing", "hyperarousal", "hand biting"), social (e.g., "social skills", "social behaviors", "connecting with others"), perseveration ("perseverative behaviors", "saying the same thing", "perseveration"), sleep (e.g., "sleep", "sleeping", "not sleeping"), mood (e.g., "depression", "mood stability", "mood swings"), motor (e.g., "fine motor skills", "coordination", "low muscle tone"), autism (e.g., "autism", "autistic behavior", "autistic tendencies"), eating (e.g., "weight", "hunger", "curb appetite"), and seizures (e.g., "seizures", "seizure reduction", "seizure control"). Notably, two family members/caretakers indicated that they were not interested in development of pharmaceutical treatments for symptoms of FXS. This question was answered by 439 family members/caretakers (see Figure 4A). See the data in Supplementary Table S5.

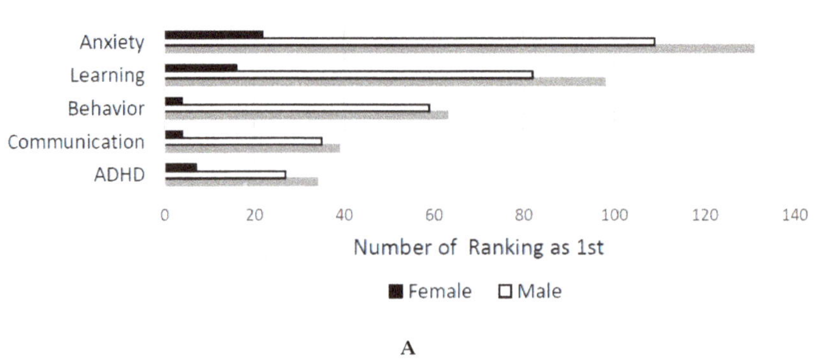

A

Figure 4. *Cont.*

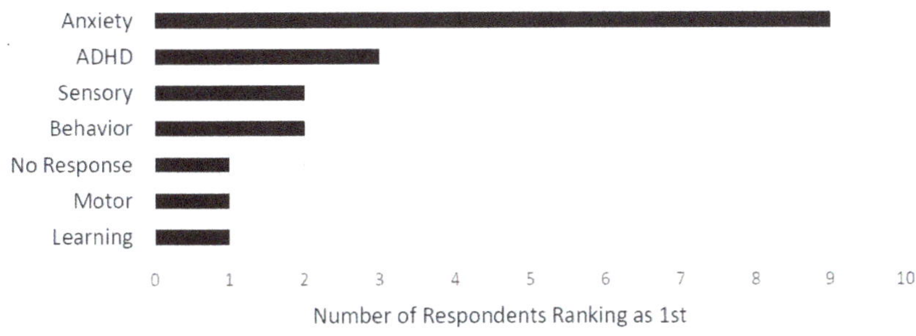

B

Figure 4. Family/caretaker (**A**) and professional drug (**B**) treatment priorities.

3.2. Voice of Professionals

Twenty professionals completed the survey. While this sample size prohibited investigations of mean response frequencies by age and sex for forced-choice questions, themes from free responses to the question about medication priorities are summarized below (see Figure 4B). Anxiety was most frequently described as the 1st priority for treatment, with almost half of professionals surveyed listing anxiety for their first priority.

3.3. Voice of Individuals with FXS

Only 8 individuals with FXS completed this survey, all of whom were female and ages 13 years or older. Given this small number, we focus here on individual answers to the free response question rather than group means for the forced-choice questions. In parallel with responses from families, anxiety was most commonly listed as a priority for treatment for this group of individuals with FXS (see Figure 5A). Due to the limited number of respondents, a weighted mean score was calculated as described above for all of the responses received (see Figure 5B).

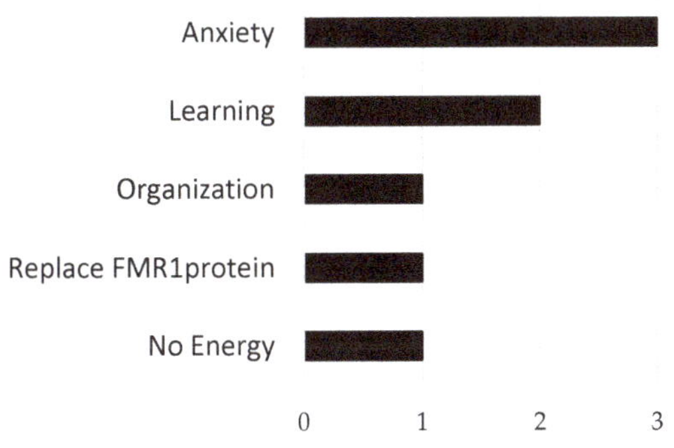

A

Figure 5. *Cont.*

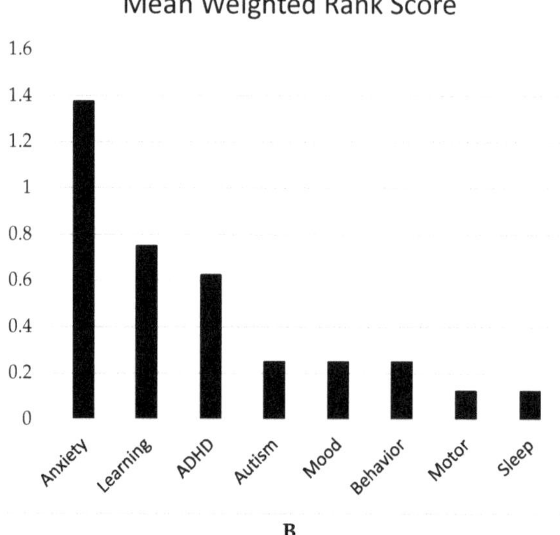

Figure 5. Drug treatment priorities as reported by 8 females with FXS. **A**: Drug treatment priorities ranked first as reported by 8 females with FXS. **B**: Drug treatment priorities, as a mean weighted rank score of all responses, as reported by 8 females with FXS.

4. Discussion

This paper presents family member/caretaker/professional/self-reported information on the characteristics of FXS that have the greatest impact on the daily lives of people living with FXS and their families/professionals and the key areas of need for treatments. Responses highlight the role of anxiety as well as some other key symptoms in the lives of individuals with FXS and also demonstrate some of the challenges that can be encountered in "voice of the person" research within this population.

While one goal of this research was to obtain information directly from individuals with FXS, only eight of the respondents were individuals with FXS themselves (all female), with around 90% of responses coming from family members/caretakers. It is likely that this distribution was skewed toward females because females tend to have more typical cognitive abilities, making the survey more accessible to them. If males with FXS were able to communicate their concerns, based on severity differences alone, they would likely communicate different concerns than females with FXS, and in fact, parents of males reported a different pattern of primary concerns than those of females. Most individuals with FXS, males in particular, would need significant support to provide responses in the format presented, and for many, intellectual impairment would make the task prohibitive. Even females who are capable of carrying out a self-report task have been shown to be fairly inaccurate and erratic in their ratings. This may be due to deficits in understanding quantity, which would interfere with assignment of severity. While this is partially alleviated by the included free response questions, this format relies on significant receptive and expressive language skills. Females with FXS may also have difficulty identifying their own problems as well, or the degree to which they differ from experiences of typically developing females. Therefore, while future work could use other adaptive methods for qualitative research (e.g., transcription of patient's statements as a way to reduce demand for reading and writing skills), integration of individual's described experiences with those of family members/caretakers and professionals will be essential for a comprehensive understanding of the challenges faced by individuals with FXS.

Importantly, while survey responses varied, three main concerns emerged as consistently problematic in the lives of people with FXS: Learning/cognitive problems, anxiety, and behavior problems. The importance of each of these concerns varied depending on the focus of the question. Specifically, anxiety was described as the most problematic symptom of FXS, while behavior problems were listed as most difficult for family members/caretakers, and learning was the most consistently reported daily living problem. Anxiety was described as a top treatment priority by family members/caretakers, professionals, and individuals with FXS alike. Anxiety, however, as defined in DSM5, is a symptom that is perceived by the patient and depends on self-report. As defined this way, anxiety is very hard to assess in most patients with FXS, given the assessment is by proxy. As such, it will be important to understand the manifestations and symptoms observed by the family members/caretakers when they report an individual with FXS as having anxiety. These observable symptoms would be expected to include: Social avoidance, anticipation of upcoming events with repeated questioning and need for constant reassurance, difficulty with performance when directly requested and while being observed, inability to transition, approach–withdrawal behaviors, eye aversion, and signs of "fight or flight" when the individual is stressed, followed by aggressive outbursts. Further qualitative work should be done to elucidate and document the types and frequencies of behaviors observed and reported as a correlate to anxiety experienced in the daily lives of individuals with FXS, as these symptoms can vary widely across the phenotype.

This study is an important first step in establishing a stakeholder-first perspective on priorities for treatment in FXS. However, several limitations can be noted. Due to its online nature, all of the information was self-reported, and we could neither confirm an FXS diagnosis nor determine the functioning level of the individuals with FXS. Many options were offered in order to capture potential areas of concern, but as a result, some participants may have found the choices confusing. In addition, most respondents were family members or caretakers, and this limits our ability to understand treatment priorities for individuals with FXS themselves. While a sample of females with FXS completed this study, that sample was relatively small and likely not representative of all individuals with FXS. Future studies directed more specifically at characterizing treatment priorities in a larger group of females with FXS are needed and should be performed. Future studies would also benefit from a design that allows individuals with FXS to self-report by including response formats and wording of questions that are more accessible to those with intellectual disability. However, due to the nature of FXS, integration with family members/caretakers and professional reports will be essential for understanding the patient perspective.

The clinical implications of these survey findings include continued focus by clinicians and researchers on the three primary concerns noted by survey respondents. Specifically, while participants noted many areas of concern and difficulty, symptoms related to learning, anxiety, and behavior problems are reported to cause the most difficulty for individuals with FXS and their families. There was, however, significant variability in response patterns across age and sex, and these reported differences will dictate a focus on different clinical features and different clinical outcome assessments for therapeutic trials, depending on the FXS age/gender subgroup being targeted in the trial.

Supplementary Materials: The following are available online at http://www.mdpi.com/2076-3425/9/2/18/s1, Table S1: Percent ranked first and weighted rank score by sex and age group for this question: Rank these three areas from one to three to the extent it affects the person's daily life—with one having the greatest impact and three having the least impact. (Primary Concerns), Table S2: Percent ranked first and weighted rank score by sex and age group for Question: Check the five characteristics that have the greatest impact on the life of the person with FXS. Prioritize 1, 2, 3, 4, 5. (Problematic Symptoms), Table S3: Percent ranked first and weighted rank score by sex and age group for this question: Check the top five areas of daily life the person with FXS is most affected by. Prioritize 1, 2, 3, 4, 5. (Daily Living Skills Most Affected), Table S4: Percent ranked first and weighted rank score by sex and age group for this question: Which five specific aspects of daily living with FXS are the most challenging? Prioritize 1, 2, 3, 4, 5 (Family Impact), Table S5: Percent ranked first and weighted rank score by sex and age group for this question: What are the top three aspects of Fragile X syndrome that you would like to see a drug treatment address, list in order of preference, with the most important one first. (Drug Treatment Priorities).

Author Contributions: Conceptualization, D.C. and J.D.W.; Formal analysis, J.D.W., E.S. and D.H.; Methodology, E.S., E.B.-K. and D.H.; Project administration, J.D.W.; Writing—original draft, J.D.W. and E.S.; Writing—review and editing, J.D.W., E.S., E.B.-K., D.C., D.H. and C.E.

Funding: This work was supported by Fulcrum Therapeutics and the National Fragile X Foundation.

Acknowledgments: We are grateful to everyone who completed this survey, and the National Fragile X Foundation for their help disseminating the survey to the FXS community.

Conflicts of Interest: J.D.W. has received funding from Ovid Therapeutics Inc. for consultation regarding clinical trials for FXS. E.S has no conflicts of interest. E.B.-K. has received funding from Seaside Therapeutics, Novartis, F. Hoffmann-La Roche Ltd., Alcobra Pharmaceuticals, Neuren Pharmaceuticals Ltd., Cydan, Fulcrum Therapeutics, GW Pharmaceuticals plc, Neurotrope, Marinus Pharmaceuticals, Inc., Zynerba Pharmaceuticals, Inc., BioMarin Pharmaceutical, Yamo Pharmaceuticals, and Ovid Therapeutics Inc. to consult on trial design or development strategies and/or conduct clinical trials in FXS or other neurodevelopmental disorders, from Vtesse/Sucampo/Mallinkcrodt to consult on and conduct clinical trials in NP-C, and from Asuragen, Inc. to develop testing standards for *FMR1* testing. D.C. is a full-time paid employee of Fulcrum Therapeutics and owns restricted stock/stock options in Fulcrum Therapeutics, a privately own discovery-stage biopharmaceutical company. D.H. provides consultation to Ovid Therapeutics Inc., Zynerba Pharmaceuticals, Inc., and Autifony Therapeutics. C.E. has received current or past funding from Confluence Pharmaceuticals, Novartis, F. Hoffmann-La Roche Ltd., Seaside Therapeutics, Riovant Sciences, Inc., Fulcrum Therapeutics, Neuren Pharmaceuticals Ltd., Alcobra Pharmaceuticals, Neurotrope, Zynerba Pharmaceuticals, Inc., and Ovid Therapeutics Inc. to consult on trial design or development strategies and/or conduct clinical trials in FXS or other neurodevelopmental disorders. C.A.E. is additionally the inventor or co-inventor on several patents held by Cincinnati Children's Hospital Medical Center or Indiana University School of Medicine describing methods of treatment in FXS or other neurodevelopmental disorders.

Appendix A

Voice of people with Fragile X syndrome and their families: Reports from a survey on treatment priorities

1. Which of the following best describes you?

    __ I have Fragile X syndrome
    __ I am a family member or caretaker of someone with FXS
    __ I am a professional who works with a person with FXS

2. What is the age of the person with FXS with whom you have the connection? Or your age if you have FXS.

    __ Male: Birth to 5 years old
    __ Female: Birth to 5 years old
    __ Male: 6 to 12 years old
    __ Female: 6 to 12 years old
    __ Male: 13 to 21 years old
    __ Female: 13 to 21 years old
    __ Male: 22 years and older
    __ Female: 22 years and older

3. Check the five characteristics that have the greatest impact on the life of the person with FXS. Prioritize 1, 2, 3, 4, 5.

    __ Anxiety—anticipatory, e.g., of new/upcoming events
    __ Auditory processing difficulties—being able to listen to instructions and react to them
    __ Autism
    __ Communication delays—initiation (asking for help) and social (turn-taking)
    __ Hyperactivity
    __ Learning or Intellectual disability (problems with abstract thinking, learning)

__ Memory—short term
__ Memory—long term
__ Motor delays (e.g., low muscle tone, poor fine motor skills, poor balance)
__ Motor stereotypes (hand-flapping, spinning around)
__Perseveration—speech (repeating things over and over)
__ Seizures
__ Sensory processing difficulties
__ Short attention span
__ Social anxiety
__ Speech/Language delays—receptive (understanding spoken language)
__ Speech/Language delays—expressive (speaking spoken language)
__ Visual information processing difficulties
__ Other - Describe

4. Rank these three areas from one to three to the extent it affects the person's daily life—with one having the greatest impact and three having the least impact:

__ Behavior
__ Intelligence
__ Physical abilities

5. Check the top five areas of daily life the person with FXS is most affected by. Prioritize 1, 2, 3, 4, 5.

__ Ability to learn academic skills/reading/math
__ Ability to take care of self-care skills/hygiene/cooking
__ Ability to speak/communicate
__ Ability to be left alone/spend time alone
__ Ability to control behavioral outbursts
__ Ability to attend and perform at school
__ Ability to find/maintain job
__ Ability to make and maintain friends
__ Ability to live independently
__ Ability to establish and maintain a relationship
__ Ability to be like other people his/her age.
__ Ability to attend events where there are a lot of people/noise
__ Willingness to go to new places
__ Willingness to travel/ go on vacation
__ Other—Describe

6. Which five specific aspects of daily living with FXS are the most challenging? Prioritize 1, 2, 3, 4, 5.

__ Always thinking—how are things going, what do I need to do next? Needing to always be 'one step' ahead
__ Checking in/setting up daily programming—school, work, etc.
__ Doctor/dentist appointments—finding/attending
__ Doing activities with friends—both child's and adult's
__ Extra costs—therapies, medications, clothing, glasses, laundry
__ Extra time it takes to do everything
__ Finding respite

  __ Food—Always hungry/wants to eat out

  __ Handling behaviors (negative)—tantrums, aggression, spitting, cussing

  __ Hygiene—shower, toileting, hair cuts

  __ Impact on non-affected family members

  __ Medications—getting prescriptions/not running out/ making changes

  __ Need for constant supervision

  __ Needing to make sure everything is "set" for the day—routine, visuals

  __ Person doesn't understand directions/can only do one thing at a time

  __ Person is unable to tell you what he/she wants/needs

  __ Running errands—how many stops can I make? What environments could be hard/noisy?

  __ Sleeping

  __ Worry about the future

  __ Other—Describe

7. List three of your favorite things about the person with FXS. (Or three things you like about yourself).

8. What are the top three aspects of Fragile X syndrome that you would like to see a drug treatment address, list in order of preference, with the most important one first.

References

1. Verkerk, A.J.; Pieretti, M.; Sutcliffe, J.S.; Fu, Y.H.; Kuhl, D.P.; Pizzuti, A.; Reiner, O.; Richards, S.; Victoria, M.F.; Zhang, F.P.; et al. Identification of a gene (*FMR-1*) containing a CGG repeat coincident with a breakpoint cluster region exhibiting length variation in fragile X syndrome. *Cell* **1991**, *65*, 905–914. [CrossRef]
2. Gross, C.; Hoffmann, A.; Bassell, G.J.; Berry-Kravis, E.M. Therapeutic Strategies in Fragile X Syndrome: From Bench to Bedside and Back. *Neurotherapeutics* **2015**, *12*, 584–608. [CrossRef] [PubMed]
3. Kaufmann, W.E.; Abrams, M.T.; Chen, W.; Reiss, A.L. Genotype, molecular phenotype, and cognitive phenotype: Correlations in fragile X syndrome. *Am. J. Med. Genet.* **1999**, *83*, 286–295. [CrossRef]
4. De Vries, B.B.; Wiegers, A.M.; Smits, A.P.; Mohkamsing, S.; Duivenvoorden, H.J.; Fryns, J.P.; Curfs, L.M.; Halley, D.J.; Oostra, B.A.; van den Ouweland, A.M.; et al. Mental status of females with an *FMR1* gene full mutation. *Am. J. Med. Genet.* **1996**, *58*, 1025–1032.
5. Fisch, G.S.; Carpenter, N.; Holden, J.J.; Howard-Peebles, P.N.; Maddalena, A.; Borghgraef, M.; Steyaert, J.; Fryns, J.P. Longitudinal changes in cognitive and adaptive behavior in fragile X females: A prospective multicenter analysis. *Am. J. Med. Genet.* **1999**, *83*, 308–312. [CrossRef]
6. Fisch, G.S.; Simensen, R.; Tarleton, J.; Chalifoux, M.; Holden, J.J.A.; Carpenter, N.; Howard-Peebles, P.N.; Maddalena, A. Longitudinal study of cognitive abilities and adaptive behavior levels in fragile X males: A prospective multicenter analysis. *Am. J. Med. Genet.* **1996**, *64*, 356–361. [CrossRef]
7. Klaiman, C.; Quintin, E.-M.; Jo, B.; Lightbody, A.A.; Hazlett, H.C.; Piven, J.; Hall, S.S.; Chromik, L.C.; Reiss, A.L. Longitudinal profiles of adaptive behavior in fragile X syndrome. *Pediatrics* **2014**, *134*, 315–324. [CrossRef] [PubMed]
8. Hagerman, R.J.; Berry-Kravis, E.; Kaufmann, W.E.; Ono, M.Y.; Tartaglia, N.; Lachiewicz, A.; Kronk, R.; Delahunty, C.; Hessl, D.; Visootsak, J.; et al. Advances in the treatment of fragile X syndrome. *Pediatrics* **2009**, *123*, 378–390. [CrossRef] [PubMed]
9. Visootsak, J.; Charen, K.; Rohr, J.; Allen, E.; Sherman, S. Diagnosis of fragile X syndrome: A qualitative study of African American families. *J. Genet. Couns.* **2012**, *21*, 845–853. [CrossRef] [PubMed]
10. Anido, A.; Carlson, L.M.; Taft, L.; Sherman, S.L. Women's attitudes toward testing for fragile X carrier status: A qualitative analysis. *J. Genet. Couns.* **2005**, *14*, 295–306. [CrossRef] [PubMed]
11. Brady, N.; Skinner, D.; Roberts, J.; Hennon, E. Communication in young children with fragile X syndrome: A qualitative study of mothers' perspectives. *Am. J. Speech-Lang. Pathol.* **2006**, *15*, 353–364. [CrossRef]
12. Budimirovic, D.B.; Cvjetkovic, S.; Bukumiric, Z.; Duy, P.Q.; Protic, D. Fragile X-Associated Disorders in Serbia: Baseline Quantitative and Qualitative Survey of Knowledge, Attitudes and Practices Among Medical Professionals. *Front. Neurosci.* **2018**, *12*, 652. [CrossRef] [PubMed]

13. Raspa, M.; Fitzgerald, T.; Furberg, R.D.; Wylie, A.; Moultrie, R.; DeRamus, M.; Wheeler, A.C.; McCormack, L. Mobile technology use and skills among individuals with fragile X syndrome: Implications for healthcare decision making. *J. Intellect. Disabil. Res.: JIDR* **2018**, *62*, 821–832. [CrossRef] [PubMed]
14. Bailey, D.B., Jr.; Raspa, M.; Olmsted, M.; Holiday, D.B. Co-occurring conditions associated with *FMR1* gene variations: Findings from a national parent survey. *Am. J. Med. Genet. Part A* **2008**, *146a*, 2060–2069. [CrossRef] [PubMed]
15. Cross, J.; Yang, J.C.; Johnson, F.R.; Quiroz, J.; Dunn, J.; Raspa, M.; Bailey, D.B., Jr. Caregiver Preferences for the Treatment of Males with Fragile X Syndrome. *J. Dev. Behav. Pediatri.: JDBP* **2016**, *37*, 71–79. [CrossRef]

Review

Closing the Gender Gap in Fragile X Syndrome: Review of Females with Fragile X Syndrome and Preliminary Research Findings

Kristi L. Bartholomay [1,†], **Cindy H. Lee** [1,*,†], **Jennifer L. Bruno** [1], **Amy A. Lightbody** [1] and **Allan L. Reiss** [1,2,3]

[1] Department of Psychiatry and Behavioral Sciences, Stanford University, Stanford, CA 94305, USA; kbarthol@stanford.edu (K.L.B.); jenbruno@stanford.edu (J.L.B.) aal@stanford.edu (A.A.L.); areiss1@stanford.edu (A.L.R.)
[2] Department of Radiology, Stanford University, Stanford, CA 94305, USA
[3] Department of Pediatrics, Stanford University, Stanford, CA 94305, USA
* Correspondence: leecindy@stanford.edu; Tel.: +1-650-724-2951
† These authors contributed equally to this work.

Received: 1 December 2018; Accepted: 10 January 2019; Published: 12 January 2019

Abstract: Fragile X syndrome (FXS) is a genetic condition known to increase the risk of cognitive impairment and socio-emotional challenges in affected males and females. To date, the vast majority of research on FXS has predominantly targeted males, who usually exhibit greater cognitive impairment compared to females. Due to their typically milder phenotype, females may have more potential to attain a higher level of independence and quality of life than their male counterparts. However, the constellation of cognitive, behavioral, and, particularly, socio-emotional challenges present in many females with FXS often preclude them from achieving their full potential. It is, therefore, critical that more research specifically focuses on females with FXS to elucidate the role of genetic, environmental, and socio-emotional factors on outcome in this often-overlooked population.

Keywords: fragile X syndrome; X chromosome; females; FMR1; anxiety; avoidance; cognition; behavior; brain

---

1. Introduction to Fragile X Syndrome

Fragile X syndrome (FXS) is a genetic condition that is commonly cited as the leading heritable cause of autism and intellectual disability. Though estimates vary widely, FXS is expected to occur in one in every 2500 to 7000 males and one in 2500 to 11,000 females [1–3]. Males with FXS tend to exhibit more cognitive and behavioral problems relative to females with FXS and, therefore, tend to come to the attention of medical and mental health providers more frequently than females.

The majority of males with FXS meet criteria for severe intellectual disability. The most common behavioral features include attention deficits and hyperactivity, anxiety, and symptoms of autism spectrum disorder [4]. In contrast, the phenotype in females with FXS is generally less severe and more frequently associated with learning disabilities, socio-emotional difficulties, and mental health issues [5]. This difference in symptom presentation and severity has led, historically, to females receiving a diagnosis only after a close male relative is diagnosed, leaving many girls and women with this condition unidentified.

To date, the vast majority of clinical intervention and research funding for FXS has focused predominately on affected males. The relative scarcity of research into the unique phenotype of females with FXS represents a significant gap in the field; however, it also represents a potent target for ongoing research and intervention. Given their typically milder phenotype, females may have more potential to attain a higher level of independence and quality of life than their male counterparts, who will likely need a moderate or greater level of support throughout their lives. However, lack of specified treatments and resources specifically targeting girls with this condition results in outcomes that are often no better than those of their more affected male counterparts. Increased emphasis on research in females with FXS and the development of new treatments is, therefore, paramount.

2. Genetics

2.1. Pattern of Inheritance of Fragile X Syndrome

FXS is a trinucleotide repeat disorder resulting from an expanded CGG repeat in the untranslated region of the fragile X mental retardation 1 (*FMR1*) gene. While the unaffected gene carried by the majority of the population contains fewer than 55 repeats, individuals with full mutation FXS typically have greater than 200 repeats in this region. Individuals who fall in the intermediate range with 55 to 200 repeats are classified as having the fragile X premutation, which is associated with its own distinct phenotype [6,7]. This paper will focus exclusively on full mutation (greater than 200 repeats) FXS.

In full mutation FXS, the repeat region is hypermethylated, resulting in transcriptional silencing of the *FMR1* gene, and, therefore, reduced production of the encoded protein, the fragile X mental retardation protein (FMRP). FMRP is an RNA binding protein which plays a key role in regulating local protein synthesis as well as a number of other functions [8,9]. In the absence of FMRP, synthesis of target proteins is dysregulated, resulting in the FXS phenotype. However, even among people with fragile X full mutation, there is significant variability in the phenotypic presentation of the syndrome [10].

2.2. Genetic Foundation of Sex Differences

The *FMR1* gene is located on the X chromosome, leading to the significant sex differences observed in the FXS phenotype. Males have only one X chromosome. Thus, if they inherit the X chromosome with the *FMR1* full mutation from their mother, 100% of their X chromosomes are potentially "affected" [11]. Females, however, have two X chromosomes. If they inherit the X chromosome with the *FMR1* full mutation from their mother and the unaffected X chromosome from their father, only 50% of their X chromosomes are potentially affected. The second, "unaffected" X chromosome allows the production of some FMRP, but the dosage is generally not sufficient to restore full FMRP function in most heterozygous females [12]. This is the basis for the typically milder, but still affected, phenotype seen in females with FXS.

The variation in phenotype among females with full mutation FXS can, at least in part, be attributed to a phenomenon known as X inactivation [11,13]. X inactivation is thought to occur so that females have approximately equal X chromosome gene dosage despite having twice as many X chromosomes (known as "dosage compensation" [14]). During female embryonic development, most of the genes on one X chromosome in each cell are randomly silenced, resulting in approximately half the cells in the body expressing the genes from each X chromosome. It is believed that only a small number of embryonic progenitor cells will go on to form the brain, so the ratio of cells that have the affected X active to silenced is thought to significantly affect the level of FMRP expression in the developing central nervous system (Figure 1). The variance of this ratio likely contributes to the widely variable phenotype for females with FXS [11].

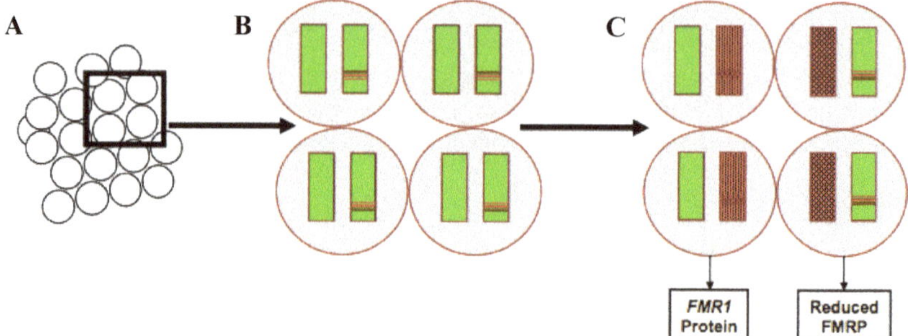

Figure 1. X-chromosome inactivation in females with fragile X syndrome and corresponding production of fragile X mental retardation 1 (*FMR1*) protein in each cell. (**A**) Group of progenitor cells. (**B**) Each cell contains two X chromosomes. The unaffected chromosomes are shown in solid green, and affected chromosomes are depicted with a red band. (**C**) One of the two chromosomes in each cell will be silenced at random indicated with red shading. In cells where the affected chromosome is silenced, there is normal production of the *FMR1* protein. In cells where the unaffected chromosome is silenced, there is reduced production of the *FMR1* protein. *FMR1*: fragile X mental retardation 1; FMRP: fragile X mental retardation protein.

3. The Fragile X Phenotype in Females

Inheriting the *FMR1* full mutation does not directly correspond to the development of the fragile X syndrome phenotype, but rather represents genetic risk for a particular set of cognitive, socio-emotional, and behavioral outcomes [15]. Phenotypic signs and symptoms will be expressed differently as a result of X inactivation, other genetic factors, and environmental influences. Many females with FXS meet criteria for one or more Diagnostic and Statistical Manual of Mental Disorders-5 (DSM-5) diagnoses—generalized anxiety disorder (GAD), social anxiety disorder (SAD), math learning disability (LD), intellectual disability (ID), autism spectrum disorder (ASD), and attention deficit hyperactivity disorder (ADHD) to name a few. These diagnoses are useful insofar as they enable the individual to receive necessary services and accommodations. DSM-5 diagnoses can also facilitate communication among mental health professionals.

The phenotypic profile for males with FXS has been relatively well established and includes moderate to severe intellectual disability by the time affected individuals reach adolescence or adulthood, high co-occurrence of autism spectrum disorder symptoms, self-injurious and aggressive behaviors, and attention deficits [4]. Very few studies have focused exclusively on females with the *FMR1* full mutation. The female phenotype is less predictable and often less severe, at least with respect to general cognitive effects. However, females with FXS do not necessarily achieve better outcomes than their male counterparts, and the socio-emotional burden on affected girls, women, and their families may still be equal to that of affected males [16,17]. It is, therefore, crucial that females be equally represented in research to establish a fundamental understanding of sex differences in FXS, the neural foundations for these differences, and potential targets for intervention to improve the quality of life and health outcomes for this often-overlooked population.

3.1. Cognitive Effects

The cognitive effect of FXS in females is variable, ranging from moderate intellectual disability to an average or above average cognitive profile [18]. Females with FXS often exhibit challenges with executive functioning and impaired spatial reasoning skills, with relative strengths in verbal skills [10,19–21]. Although females with FXS consistently show milder deficits in both cognitive and academic function than males with FXS, on average their scores fall below those of typically developing peers [10,16,22]. The discrepancy in cognitive function between girls with FXS and their typically developing peers is larger for IQ than for academic achievement scores [23].

A previous study by our group demonstrated that executive functioning scores declined over time in girls with FXS while other aspects of intelligence, such as verbal fluency and spatial ability, remained stable throughout childhood [24]. It is likely that these declines are not due to the loss of skills, but rather to a slowing rate of acquisition of new skills compared to typically developing peers, as well as the increasingly strenuous cognitive demands placed on children as they age. A recent longitudinal study on girls with FXS suggested that, while fluid intelligence was predicted only by biological or genetic factors, crystalized intelligence may be related to maternal mental health and perceived closeness of mother–child relationship [13]. Taken together, these studies suggest a vulnerable time frame in which girls with FXS are prone to decline in cognitive scores, and the importance of protective factors, such as an enriched environment, that may help mitigate this decline. It is, therefore, critical that research focus on understanding vulnerable timeframes in the cognitive development of girls with FXS, while simultaneously developing medical, psychological, and educational interventions to facilitate improved long-term outcomes.

3.2. Socio-Emotional Effects

Anxiety, avoidance, and arousal (AAA) represent three key behaviors exhibited in response to acute, potential, and sustained threat, respectively. These behaviors represent typical responses to aversive or dangerous stimuli. However, existing evidence suggests that the dysregulation of these systems can result in clinical manifestation of disorders of emotion and affect including anxiety and depression [25]. Such disorders are commonly cited as primary clinical concerns for females with FXS [4] and may be, at least in part, responsible for the often-large disparity between actual outcomes and those predicted by IQ and life skill proficiency [4,17]. Reducing AAA could improve quality of life and outcomes for girls and women with FXS. Thus, AAA represent critical targets for intervention.

In one of the first studies looking specifically at girls with FXS, Freund et al. found that females with FXS were more vulnerable to social anxiety, social avoidance, withdrawal, and depression than their IQ and age-matched peers. Furthermore, affected females showed deficits in interpersonal and social skills compared to peers who did not carry the fragile X full mutation [21]. In another study, parental report indicated that 56% of girls and women with FXS had received treatment for and/or diagnosis of an anxiety disorder while 22% had received treatment for and/or diagnosis of depression [4]. Another study found that 51.4% of girls with FXS met diagnostic criteria for specific phobia, 39.5% met criteria for social phobia, and 25.3% met criteria for selective mutism. These rates are significantly higher than those of the general population or for individuals with intellectual disability alone [26].

When completing a social challenge, girls with FXS show more gaze aversion, task avoidance, and behavioral signs of distress than their typically developing siblings, and both boys and girls with FXS exhibit aberrant cortisol reactivity, and parasympathetic and sympathetic nervous system dysregulation compared to their same-sex typically developing siblings [27,28]. Further, Hartley et al. found that the largest predictor of independence for adult women with FXS was the ability to interact appropriately in social situations and that independence was inversely correlated with the presence of co-occurring anxiety disorders or depression [17].

Together, these studies support prior findings suggesting that anxiety experienced by girls with FXS can, in part, be attributed to impairment in social skills and social communication [29,30]. The impact of atypical development of social skills and communication in girls over time may lead to a wide range of anxiety symptoms or disorders. However, there is little understanding as to how this process occurs and how biological and environmental factors interact to affect the outcome. More research into the longitudinal profile of anxiety in girls with FXS is necessary to facilitate such an understanding and for the eventual development of more effective, disorder-specific interventions to improve social and communication skills. In particular, interventions implemented before the development of anxiety symptoms and disorders in girls with FXS hold particular promise for achieving a more optimal long-term outcome.

3.3. Adaptive Behavior and Independent Living

Both males and females with FXS experience significant difficulty in the acquisition of adaptive behavior skills compared to their typically developing peers and struggle to reach independence, often continuing to live with their parents or in an assisted living environment once they reach adulthood [17]. Adaptive behaviors, or daily living skills, are a broad group of skills and abilities accumulated throughout childhood and into adulthood which allow an individual to function in their daily environment, including skills like verbal and written communication, routine self-care, and social skills.

Several studies have investigated the acquisition of adaptive behavior in children with FXS. In one such study, parents of children with FXS were interviewed about their child's adaptive behavior skills. Consistent with previous studies [31], females with FXS typically scored higher on adaptive behavior than males with FXS at all ages. However, their adaptive behavior skills appeared to decline throughout childhood. These declines were most pronounced in the domain of communication, suggesting that as girls with FXS age they fall further behind peers in their verbal and written communication skills [32]. This does not necessarily indicate that females with FXS are losing skills during childhood, but rather that they are not gaining skills at the same rate as their typically developing peers. Another study demonstrated that both IQ and quality of home environment are predictive of adaptive behavior in boys with FXS and in the unaffected siblings of children with FXS; however, only IQ appears to be predictive of adaptive behavior outcomes in girls with FXS [15]. This suggests that boys may have more adaptive behavior skills than their IQ would predict due to support and accommodations in the home and school environments allowing them to succeed, while girls with FXS may not be receiving sufficient support or intervention, or their difficulties may not be identified early enough to implement effective strategies.

Various studies have also assessed the functional and independent living skills of adults with FXS. In one study, parents of children with FXS rated their level of independence in various daily skills. By adulthood, the majority of both males and females with FXS were able to independently complete most tasks, and females acquired independent living skills significantly faster and reached a higher level of proficiency than their male counterparts [33]. In another study, parents were asked to rate the level of independence of their adult children in these types of skills as well as rate their general level of independence across various domains. In this study, for males with FXS, level of independence was best predicted by proficiency in daily living skills. However, the strongest predictor of independence for females with FXS was the ability to interact appropriately in social settings, an area of significant challenge for affected individuals. Less than half of women with FXS reached very high or high levels of independence, even though they had significantly more functional skills than their male counterparts [17].

These studies suggest that, although females with FXS have significantly higher proficiency in daily living skills both in childhood and adulthood than their male counterparts, there is not a corresponding increase in their actual levels of independence in adult life. This disparity between girls' potential, as represented by their IQ and functional skills, and their resulting level of independence represents a critical target for research and intervention. To ensure girls with FXS are reaching their maximum potential, it must first be better understood how, when, and why they are falling short of reaching independence, and interventions must directly target these timeframes and skills to substantively improve outcomes.

3.4. Barriers to Positive Outcomes

Factors other than genetics play an important role in determining outcomes for females with FXS. Such factors include severity of social-emotional symptoms, home/family environment, parental mental health, and parenting style [24], and likely account for some of the variability in outcome seen in girls with FXS. These factors also represent important potential targets of intervention.

Several studies indicate that independence and quality of life for females with FXS are heavily influenced by the DSM-5 disorders for which females show symptoms or reach diagnostic criteria [4,16,30]. The severity of autism spectrum disorder symptomology is significantly associated with independent living outcomes for individuals with FXS, and symptoms of affective disorders, such as anxiety and depression, represent a barrier to achieving independence for females with FXS specifically [17,34]. A national survey of parents of children with FXS demonstrated that ability to adapt to changing situations, thinking and reasoning skills, and perceived quality of life were all inversely correlated with the number of DSM diagnoses for which the child met criteria. The most common diagnoses among females with FXS were attention deficit, anxiety, hyperactivity, and depression [4]. FXS itself cannot be prevented or treated, so it is crucial that the symptoms of these conditions are identified early. Early detection can lead to more targeted treatment and ultimately improve the outcomes and lives of girls with FXS and their families.

The quality of the home environment also plays a role in determining the outcome for both boys and girls with FXS. One study found that the home environment influences behavioral outcomes for girls with FXS. Specifically, reported increases in parent psychopathology were correlated with an increase of child anxiety and depression, while increased efficacy of services was correlated with a decrease in thought and attention problems for the individual with FXS [35]. Another study found that increased quality of home environment corresponded with an increase in verbal IQ as well as an increase in freedom from distractibility [36], and a recent study published exclusively on girls with FXS demonstrated that lower IQ and increased social aversion could be predicted by higher levels of maternal distress and reduced perceived closeness of parent–child relationships [13]. These studies emphasize the importance of providing resources and support for families of girls with FXS to create home, family, and educational environments that promote their development and optimize behavioral, cognitive, and quality of life outcomes.

To this end, our research team is implementing a prospective longitudinal study of girls with FXS. Highlights from some preliminary findings in a small sample set are presented below.

4. Methods

4.1. Participants

The preliminary cohort presented here consists of 21 girls between the ages of 6 to 14 (mean age = 10.53). Participants were recruited through various FXS communities including regional fragile X organizations, the Fragile X Clinical and Research Consortium, and the Fragile X Online Registry With Accessible Research Database, electronic media including website and social media announcements, and with the help of the National Fragile X Foundation. All participants were diagnosed by an appropriate molecular genetic test as having more than 200 CGG repeats in the *FMR1* gene with documentation of previous testing provided by caregivers at enrollment. All participants and their caregivers were native English speakers. The data presented here represent a partial sample from the first year and a half of the current five-year longitudinal study and will constitute approximately 40% of our final overall cohort. The research team is still actively recruiting and seeing participants.

4.2. Measurements/Procedures

4.2.1. Cognition and Academics

Cognition was assessed with the Differential Ability Scales, 2nd Edition (DAS-II), which provides subscales for verbal, nonverbal, and spatial reasoning ability and an overall composite score [37]. Academic skills were assessed with The Kaufman Test of Educational Achievement, Third Edition

Brief Form (KTEA-3 Brief), which provides reading and math subscales and an overall achievement composite score [38].

4.2.2. Child Behavior and Emotion

Participants' caregivers completed the Social Responsiveness Scale-2 (SRS) [39]. The SRS addresses social awareness, social information processing, capacity for reciprocal social responses, social anxiety/avoidance, and characteristic autistic preoccupations/traits.

4.2.3. Adaptive Behavior and Functional Skills

Adaptive behavior and functional skills in this cohort were assessed utilizing the Vineland Adaptive Behavior Scales, Third Edition—Interview Form (VABS-III) which measures adaptive behavior across the domains of communication, daily living skills, and socialization and includes an overall adaptive behavior composite score [40]. Caregivers completed the VABS-III interview with a trained researcher during their research visit.

4.3. Data Analyses

Exploratory analyses were conducted to assess differences between cognitive and achievement scores and correlations between domains of cognitive, adaptive and social function. Results with a p value ≤ 0.05 were considered significant.

4.4. Preliminary Findings

4.4.1. Cognition and Academics

The distributions of the cognitive and academic assessments are presented in Figure 2. In this preliminary cohort, girls with FXS performed significantly better on the verbal domain (*Mean* = 82.25, *SD* = 11.56) than on the nonverbal domain (*Mean* = 73.65, *SD* = 16.79); $t(19) = 2.46$, $p = 0.012$ or the overall composite (*Mean* = 74.90, *SD* = 14.77); $t(19) = 3.10$, $p = 0.003$ of the DAS-II (Figure 2A). Girls in this cohort also performed significantly better on the reading domain (*Mean* = 84.85, *SD* = 13.74) than on the math domain (*Mean* = 71.20, *SD* = 14.05); $t(19) = 6.87$, $p = 0.000$ or the overall composite (*Mean* = 80.35, *SD* = 15.50); $t(19) = 3.07$, $p = 0.003$ of the KTEA-Brief (Figure 2B).

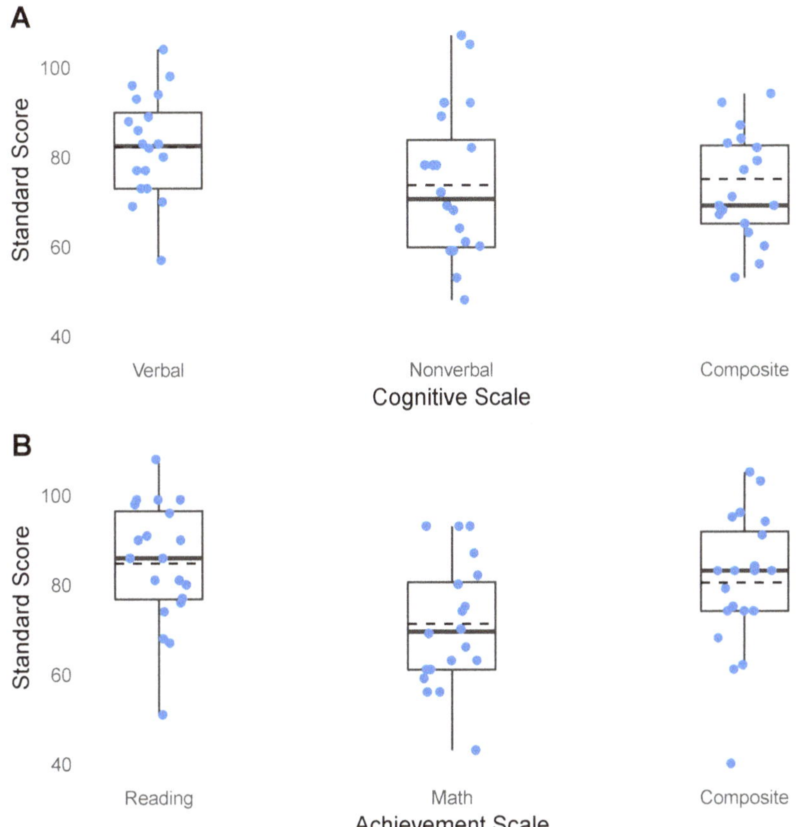

Figure 2. Cognitive and achievement score profiles. Each dot indicates a participant, solid horizontal line represents median, dashed horizontal line represents mean, box represents interquartile range, and vertical lines upper and lower extremes (excluding outliers). (**A**) Distribution of participant cognitive scores on verbal, nonverbal, and overall composite of the Differential Ability Scales, 2nd Edition (DAS-II). (**B**) Distribution of participant achievement scores on reading, math, and composite achievement on the Kaufman Test of Educational Achievement Third Edition Brief Form (KTEA-3 Brief).

4.4.2. Correlations between Cognition and Adaptive Behavior

The verbal and nonverbal domains of the DAS-II are compared with corresponding performance on each subdomain of the VABS-III in Figure 3. Significant positive (Pearson) correlations were found between nonverbal reasoning ability and VABS-III communication $r(18) = 0.58$, $p = 0.004$, daily living skills r(18) = 0.59, $p = 0.003$, and overall composite $r(18) = 0.58$, $p = 0.004$ (Figure 3B) while significant correlations were not observed between verbal reasoning ability and any subdomains of the VABS-III (Figure 3A).

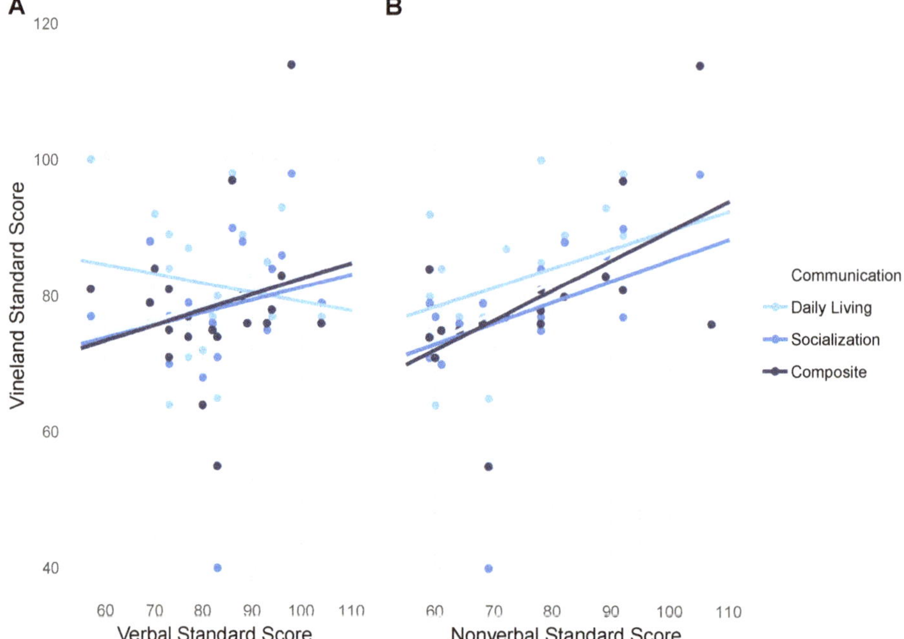

Figure 3. Correlations between cognition and adaptive behavior. (**A**) Correlations of the DAS-II verbal reasoning subscale with communication, daily living, and socialization subscales and overall adaptive behavior composite of the Vineland Adaptive Behavior Scales, Third Edition—Interview Form (VABS-III). (**B**) Correlations of the DAS-II nonverbal reasoning subscale with all domains of the VABS-III.

4.4.3. Correlations between Social Skills and Adaptive Behavior

The associations between total score on the SRS-2 and corresponding scores on each VABS-III domain are presented in Figure 4. Increasing SRS scores (representing increasingly aberrant social skills), were correlated with decreasing VABS-III scores across all subdomains: communication $r(18) = -0.62$, $p = 0.002$, daily living skills $r(18) = -0.45$ $p = 0.023$, socialization $r(18) = -0.747$, $p = 0.000$, composite $r(18) = -0.618$, $p = 0.002$.

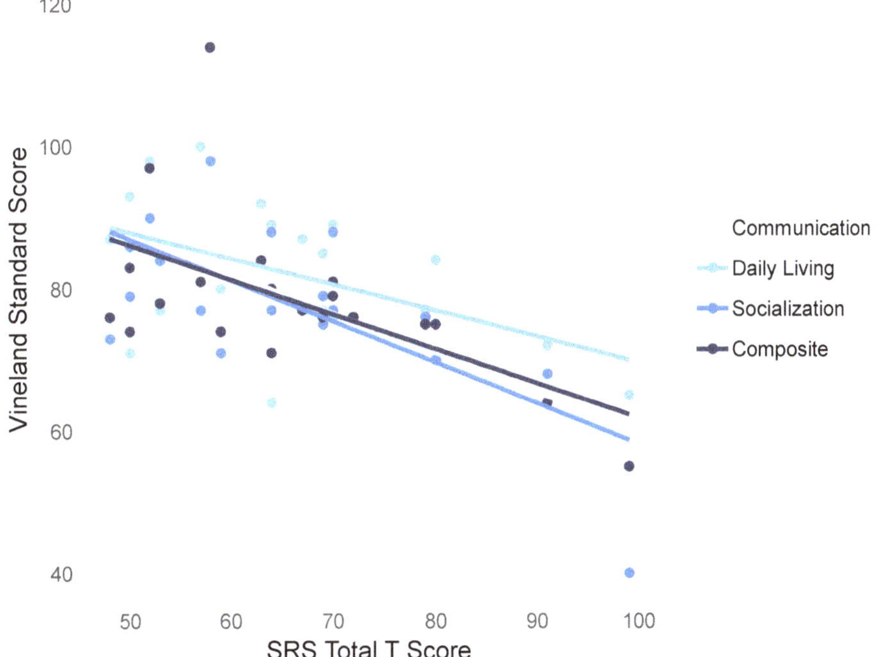

Figure 4. Correlations between social skills and adaptive behavior. Correlations between difficulties with social skills as measured by total score reported by caregivers on the Social Responsiveness Scale-2 (SRS) and the communication, daily living, and socialization subscales and overall adaptive behavior composite scores of the Vineland Adaptive Behavior Scales, Third Edition—Interview Form (VABS-III).

5. Discussion

5.1. Cognition and Academics

Consistent with prior investigations of girls with FXS, the overall cognitive profile of children from our preliminary cohort was within the borderline/low average range [41]. The preliminary cohort presented here also scored significantly higher on the verbal subdomain than on the non-verbal subdomain of the DAS-II cognitive assessment (Figure 1A) and performed significantly better on the reading domain than on the math domain of the KTEA achievement assessment (Figure 1B). These relatively lower scores on nonverbal domains of cognitive and achievement assessments are consistent with previous reports of relative strengths in the verbal ability for females with FXS. Relative strengths include acquired knowledge, long-term memory for verbal information, and simultaneous processing [10,41]. Although verbal ability represents a cognitive strength, average verbal scores for girls with FXS remain within the low average range when compared to their age-matched peers. One hypothesis put forward to explain this discrepancy is that the social demands of the language environment for children with FXS (coordination of syntax, semantics, conversational pragmatics, and eye contact) promotes AAA symptoms, which then challenge proper regulation of verbal responses [23]. Thus, it may be important to focus on the development and management of AAA symptoms in girls with FXS to optimize their cognitive as well as social-emotional outcomes.

Although the data showed no significant difference between participant scores on the cognitive assessment and the achievement assessment in our preliminary cohort (Figure 2A,B), it is often assumed that cognitive ability determines achievement. However, children with FXS frequently outperform predictions of academic achievement based on cognitive test scores [23]. One proposed explanation for the discrepancy between achievement and cognitive scores among females with FXS is a relative strength in long-term memory and the effect of repeated exposure to academic material at school [42]. Prior research has also suggested that individuals with FXS have more difficulty processing highly novel information than learning facts and school-related skills [43]. Since intelligence tests and cognitive assessments are intended to be novel and unfamiliar, deficits in flexible thinking may particularly affect performance on these types of tests. These insights may be important for facilitating successful learning outcomes for girls with FXS who exhibit a wide range of learning difficulties.

5.2. Adaptive Behavior Outcomes

Correlations between cognitive scores and adaptive behavior in our preliminary cohort suggest a significant association between nonverbal abilities and overall adaptive behavior. (Figure 3A,B). Given the relative cognitive strength in the verbal ability of individuals with FXS [10,19–22], it is particularly concerning that these strengths do not appear to be translated readily to outcomes in functional skills.

Our preliminary results also suggest a negative correlation between social skills and adaptive behavior (Figure 4), suggesting that aberrant social skills and function are associated with challenges in adaptive behavior skills. These findings are consistent with prior research, which has suggested that factors other than cognitive abilities, such as ability to interact appropriately socially [17], autism spectrum disorder type behaviors [30], and symptoms of affective disorders, such as anxiety and depression [4], have significant influence on the development of adaptive behavior skills and functional outcomes for girls with FXS.

6. Conclusions

The preliminary data presented here support prior findings that cognitive abilities do not play the only, or even necessarily the primary, role in determining functional outcomes for girls with FXS. Continued research is needed to better understand how functional outcomes are influenced by other critical factors related to academic, home and educational environments, and socio-emotional development (Figure 5).

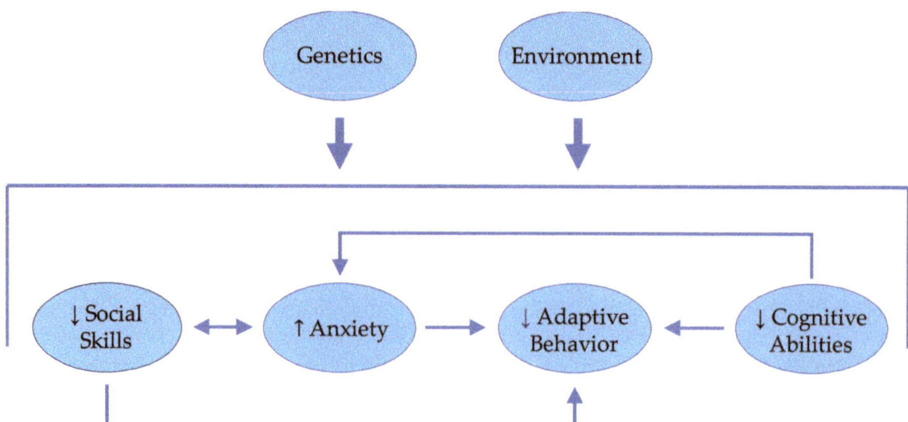

Figure 5. Genes, environment, cognitive, and socio-emotional factors all intersect to determine outcome. Anxiety may be inversely related to social skills, adaptive behavior, and cognitive abilities.

Our current study seeks to address the relative paucity of information focusing exclusively on females with the *FMR1* full mutation to elucidate the role of these factors in the development of girls and women with FXS. In particular, a fine-grained examination of gene–environment–behavior associations underlying the development and progression of social skills and symptoms of AAA will provide new information on the degree to which females with FXS experience maladaptive symptoms. We will also gain a better understanding of how and when biological and environmental factors most influence the propensity for these symptoms and the role these symptoms play in determining functional outcomes for this vulnerable population.

Author Contributions: A.L.R., A.A.L., and J.L.B. contributed to the study conception and study design. A.A.L., J.L.B., K.L.B., and C.H.L. carried out the neuropsychological assessment and collected the data. K.L.B. and C.H.L. were responsible for interpretation of the data and drafting of the manuscript. All authors critically revised and approved the manuscript.

Funding: This work and the APC were funded by the National Institute of Mental Health, Grant No. 2R01MH050047.

Acknowledgments: We thank the families who participated in this study and many members of the laboratory who assisted with this project. This work was supported by the National Institute of Mental Health Grant Nos. 2R01MH050047 (to A.L.R.).

Conflicts of Interest: The authors declare no conflict of interest. The founding sponsors had no role in the design of the study; in the collection, analyses, or interpretation of data; in the writing of the manuscript, and in the decision to publish the results.

References

1. Hunter, J.; Rivero-Arias, O.; Angelov, A.; Kim, E.; Fotheringham, I.; Leal, J. Epidemiology of fragile X syndrome: A systematic review and meta-analysis. *Am. J. Med. Genet. Part A* **2014**. [CrossRef] [PubMed]
2. Hagerman, P.J. The fragile X prevalence paradox. *J. Med. Genet.* **2008**, *5*, 498–499. [CrossRef] [PubMed]
3. Crawford, D.C.; Acuña, J.M.; Sherman, S.L. FMR1 and the fragile X syndrome: Human genome epidemiology review. *Genet. Med.* **2001**, *3*, 359–371. [CrossRef] [PubMed]
4. Bailey, D.B.; Raspa, M.; Olmsted, M.; Holiday, D.B. Co-occurring conditions associated with FMR1 gene variations: Findings from a national parent survey. *Am. J. Med. Genet. Part A* **2008**, *146*, 2060–2069. [CrossRef] [PubMed]
5. Hagerman, R.J.; Berry-Kravis, E.; Hazlett, H.C.; Bailey, D.B.; Moine, H.; Kooy, R.F.; Tassone, F.; Gantois, I.; Sonenberg, N.; Mandel, J.L.; et al. Fragile X syndrome. *Nat. Rev. Dis. Prim.* **2017**, *3*, 17065. [CrossRef] [PubMed]
6. Gallagher, A.; Hallahan, B. Fragile X-associated disorders: A clinical overview. *J. Neurol.* **2012**, *259*, 401–413. [CrossRef]
7. Bambang, K.; Metcalfe, K.; Newman, W.; McFarlane, T. Fragile X syndrome: An overview. *Obstet. Gynaecol.* **2011**, *13*, 92–97. [CrossRef]
8. Garber, K.B.; Visootsak, J.; Warren, S.T. Fragile X syndrome. *Eur. J. Hum. Genet.* **2008**. [CrossRef]
9. Bagni, C.; Oostra, B.A. Fragile X syndrome: From protein function to therapy. *Am. J. Med. Genet. Part A* **2013**, *161*, 2809–2821. [CrossRef]
10. Reiss, A.L.; Hall, S.S. Fragile X Syndrome: Assessment and Treatment Implications. *Child Adolesc. Psychiatr. Clin. N. Am.* **2007**, *16*, 663–675. [CrossRef]
11. Loesch, D.Z.; Huggins, R.M.; Hagerman, R.J. Phenotypic Variation and FMRP Levels in Fragile X. *Ment. Retard. Dev. Disabil. Res. Rev.* **2004**, *10*, 31–41. [CrossRef] [PubMed]
12. Tassone, F.; Hagerman, R.J.; Iklé, D.N.; Dyer, P.N.; Lampe, M.; Willemsen, R.; Oostra, B.A.; Taylor, A.K. FMRP expression as a potential prognostic indicator in fragile X syndrome. *Am. J. Med. Genet.* **1999**. [CrossRef]
13. Del Hoyo Soriano, L.; Thurman, A.J.; Harvey, D.J.; Ted Brown, W.; Abbeduto, L. Genetic and maternal predictors of cognitive and behavioral trajectories in females with fragile X syndrome. *J. Neurodev. Disord.* **2018**. [CrossRef] [PubMed]
14. Di, K.N.; Disteche, C.M. Dosage compensation of the active X chromosome in mammals. *Nat. Genet.* **2006**. [CrossRef]
15. Glaser, B.; Hessl, D.; Dyer-Friedman, J.; Johnston, C.; Wisbeck, J.; Taylor, A.; Reiss, A. Biological and environmental contributions to adaptive behavior in fragile X syndrome. *Am. J. Med. Genet.* **2003**. [CrossRef] [PubMed]

16. Reiss, A.L.; Dant, C.C. The behavioral neurogenetics of fragile X syndrome: Analyzing gene-brain-behavior relationships in child developmental psychopathologies. *Dev. Psychopathol.* **2003**. [CrossRef]

17. Hartley, S.L.; Seltzer, M.M.; Raspa, M.; Olmstead, M.; Bishop, E.; Bailey, D.B. Exploring the adult life of men and women with fragile X syndrome: Results from a national survey. *Am. J. Intellect. Dev. Disabil.* **2011**, *116*, 16–35. [CrossRef]

18. Hagerman, R.J.; Jackson, C.; Amiri, K.; Silverman, A.C.; O'Connor, R.; Sobesky, W. Girls with fragile X syndrome: Physical and neurocognitive status and outcome. *Pediatrics* **1992**, *89*, 395–400.

19. Kogan, C.S.; Boutet, I.; Cornish, K.; Graham, G.E.; Berry-Kravis, E.; Drouin, A.; Milgram, N.W. A comparative neuropsychological test battery differentiates cognitive signatures of Fragile X and Down syndrome. *J. Intell. Disabil. Res.* **2009**. [CrossRef]

20. Bennetto, L.; Pennington, B.F.; Porter, D.; Taylor, A.K.; Hagerman, R.J. Profile of cognitive functioning in women with the fragile X mutation. *Neuropsychology* **2001**. [CrossRef]

21. Freund, L.S.; Reiss, A.L.; Abrams, M.T. Psychiatric disorders associated with fragile X in the young female. *Pediatrics* **1993**, *91*, 321–329. [PubMed]

22. Schwarte, A.R. Fragile X Syndrome. *Sch. Psychol. Q.* **2008**. [CrossRef]

23. Cornish, K. Cognitive strengths and difficulties. In *Educating Children with Fragile X: A Multi-Professional Handbook*; Dew-Hughes, D., Ed.; RoutledgeFalmer: London, UK, 2004; pp. 20–24. ISBN 0203561538.

24. Lightbody, A.A.; Hall, S.S.; Reiss, A.L. Chronological age, but not FMRP levels, predicts neuropsychological performance in girls with fragile X syndrome. *Am. J. Med. Genet. Part B Neuropsychiatr. Genet.* **2006**. [CrossRef] [PubMed]

25. NIMH National Institute of Mental Health Research Domain Criteria (RDoC) Constucts. Available online: https://www.nimh.nih.gov/research-priorities/rdoc/constructs/negative-valence-systems.shtml (accessed on 30 November 2018).

26. Cordeiro, L.; Ballinger, E.; Hagerman, R.; Hessl, D. Clinical assessment of DSM-IV anxiety disorders in fragile X syndrome: Prevalence and characterization. *J. Neurodev. Disord.* **2011**, *3*, 57–67. [CrossRef] [PubMed]

27. Hessl, D.; Glaser, B.; Dyer-Friedman, J.; Reiss, A.L. Social behavior and cortisol reactivity in children with fragile X syndrome. *J. Child Psychol. Psychiatry Allied Discip.* **2006**, *47*, 602–610. [CrossRef]

28. Hall, S.S.; Lightbody, A.A.; Huffman, L.C.; Lazzeroni, L.C.; Reiss, A.L. Physiological correlates of social avoidance behavior in children and adolescents with fragile x syndrome. *J. Am. Acad. Child Adolesc. Psychiatry* **2009**, *48*, 320–329. [CrossRef] [PubMed]

29. Mazzocco, M.M.M.; Kates, W.R.; Baumgardner, T.L.; Freund, L.S.; Reiss, A.L. Autistic behaviors among girls with fragile X syndrome. *J. Autism Dev. Disord.* **1997**, *27*, 415–435. [CrossRef]

30. Lesniak-Karpiak, K.; Mazzocco, M.M.M.; Ross, J.L. Behavioral assessment of social anxiety in females with turner or fragile X syndrome. *J. Autism Dev. Disord.* **2003**. [CrossRef]

31. Hatton, D.D.; Wheeler, A.C.; Skinner, M.L.; Bailey, D.B.; Sullivan, K.M.; Roberts, J.E.; Mirrett, P.; Clark, R.D. Adaptive behavior in children with fragile X syndrome. *Am. J. Ment. Retard.* **2003**. [CrossRef]

32. Klaiman, C.; Quintin, E.-M.; Jo, B.; Lightbody, A.A.; Hazlett, H.C.; Piven, J.; Hall, S.S.; Chromik, L.C.; Reiss, A.L. Longitudinal profiles of adaptive behavior in fragile X syndrome. *Pediatrics* **2014**, *134*, 315–324. [CrossRef]

33. Bailey, D.B.; Raspa, M.; Holiday, D.; Bishop, E.; Olmsted, M. Functional skills of individuals with fragile X syndrome: A lifespan cross-sectional analysis. *Am. J. Intellect. Dev. Disabil.* **2009**. [CrossRef] [PubMed]

34. Hustyi, K.M.; Hall, S.S.; Quintin, E.M.; Chromik, L.C.; Lightbody, A.A.; Reiss, A.L. The Relationship Between Autistic Symptomatology and Independent Living Skills in Adolescents and Young Adults with Fragile X Syndrome. *J. Autism Dev. Disord.* **2015**, *45*, 1836–1844. [CrossRef] [PubMed]

35. Hessl, D.; Dyer-Friedman, J.; Glaser, B.; Wisbeck, J.; Barajas, R.G.; Taylor, A.; Reiss, A.L. The Influence of Environmental and Genetic Factors on Behavior Problems and Autistic Symptoms in Boys and Girls with Fragile X Syndrome. *Pediatrics* **2001**. [CrossRef]

36. Dyer-Friedman, J.; Glaser, B.; Hessl, D.; Johnston, C.; Huffman, L.C.; Taylor, A.; Wisbeck, J.; Reiss, A.L. Genetic and Environmental Influences on the Cognitive Outcomes of Children with Fragile X Syndrome. *J. Am. Acad. Child Adolesc. Psychiatry* **2002**. [CrossRef]

37. Elliott, C.D. *Differential Ability Scales*, 2nd ed.; NCS Pearson, Inc.: Minneapolis, MN, USA, 2007.

38. Kaufman, A.S.; Kaufman, N.L. *Kaufman Test of Educational Achievement, Third Edition Brief Form (KTEA-3 Brief)*; NCS Pearson, Inc.: Minneapolis, MN, USA, 2015.

39. Constantino, J.N.; Gruber, C.P. *Social Responsiveness Scale (SRS)*, 2nd ed.; Western Psychological Services: Los Angeles, CA, USA, 2012.

40. Sparrow, S.S.; Cicchetti, D.V.; Balla, D.A. *Vineland-II Adaptive Behavior Scales: Survey Forms Manual*; NCS Pearson, Inc.: Minneapolis, MN, USA, 2005.

41. Freund, L.S.; Reiss, A.L. Cognitive profiles associated with the fra(X) syndrome in males and females. *Am. J. Med. Genet.* **1991**, *38*, 542–547. [CrossRef] [PubMed]

42. Braden, M. The effect of fragile X syndrome on learning. In *Educating Children with Fragile X: A Multi-Professional Handbook*; Dew-Hughes, D., Ed.; RoutledgeFalmer: London, UK, 2003; pp. 43–47, ISBN 0203561538.

43. Kemper, M.B.; Hagerman, R.J.; Altshul-Stark, D. Cognitive profiles of boys with the fragile X syndrome. *Am. J. Med. Genet.* **1988**, *30*, 191–200. [CrossRef] [PubMed]

Article

A Pilot Quantitative Evaluation of Early Life Language Development in Fragile X Syndrome

Debra L. Reisinger [1], Rebecca C. Shaffer [1,2], Ernest V. Pedapati [3,4,5], Kelli C. Dominick [3,5] and Craig A. Erickson [3,5,*]

[1] Division of Developmental and Behavioral Pediatrics, Cincinnati Children's Hospital Medical Center, Cincinnati, OH 45229, USA; debra.reisinger@cchmc.org (D.L.R.); rebecca.shaffer@cchmc.org (R.C.S.)

[2] Department of Pediatrics, University of Cincinnati College of Medicine, Cincinnati, OH 45267, USA

[3] Division of Child and Adolescent Psychiatry, Cincinnati Children's Hospital Medical Center, Cincinnati, OH 45229, USA; ernest.pedapati@cchmc.org (E.V.P.); kelli.dominick@cchmc.org (K.C.D.)

[4] Division of Child Neurology, Cincinnati Children's Hospital Medical Center, Cincinnati, OH 45229, USA

[5] Department of Psychiatry and Behavioral Neuroscience, University of Cincinnati College of Medicine, Cincinnati, OH 45267, USA

* Correspondence: craig.erickson@cchmc.org; Tel.: +1-513-636-6265

Received: 15 December 2018; Accepted: 24 January 2019; Published: 29 January 2019

Abstract: Language delay and communication deficits are a core characteristic of the fragile X syndrome (FXS) phenotype. To date, the literature examining early language development in FXS is limited potentially due to barriers in language assessment in very young children. The present study is one of the first to examine early language development through vocal production and the language learning environment in infants and toddlers with FXS utilizing an automated vocal analysis system. Child vocalizations, conversational turns, and adult word counts in the home environment were collected and analyzed in a group of nine infants and toddlers with FXS and compared to a typically developing (TD) normative sample. Results suggest infants and toddlers with FXS are exhibiting deficits in their early language skills when compared to their chronological expectations. Despite this, when accounting for overall developmental level, their early language skills appear to be on track. Additionally, FXS caregivers utilize less vocalizations around infants and toddlers with FXS; however, additional research is needed to understand the true gap between FXS caregivers and TD caregivers. These findings provide preliminary information about the early language learning environment and support for the feasibility of utilizing an automated vocal analysis system within the FXS population that could ease data collection and further our understanding of the emergence of language development.

Keywords: fragile X syndrome; language development; automated vocal analysis

1. Introduction

Fragile X Syndrome (FXS) is the leading inherited cause of intellectual disability (ID) associated with a mutation on an unstable trinucleotide (CCG) repeat expansion on the fragile X mental retardation 1 (*FMR1*) gene [1]. FXS impacts 1 in 4,000 males and 1 in 6,000 females and, as an X linked disorder, has a more severe presentation in males. FXS is characterized by mild to severe ID with a series of other features including: anxiety, social deficits, communication deficits, gaze aversion, inattention, impulsivity, aggression and hyperactivity [2]. Within communication deficits, it is evident in the current literature that FXS is associated with significant language delay, above that expected by given cognitive deficits, with relevant strengths in receptive communication and relative weaknesses in expressive communication [3,4]. Unfortunately, it can be quite challenging to accurately assess early language acquisition in infants and young children due to the natural development of language.

This can be particularly difficult in clinical populations with known speech delays (e.g., FXS, autism spectrum disorder, Down syndrome) potentially impacting early diagnostic and treatment efforts.

Within the typically developing population, infants can perceive and attend to speech in comparison to silence or other sounds prior to speaking their first word [5,6]. The progression of expressive language development has universally been identified as cooing (between 1 and 4 months), to babbling (between 5 and 10 months), to meaningful speech (between 10 and 18 months) [7]. The social environment and interactions with caregivers throughout infancy and toddlerhood provide key building blocks for language development [8,9]. Specifically, the amount of language in a child's environment prior to the age of three is significantly correlated with language acquisition and cognitive development [10,11]. Furthermore, differences in early language development (e.g., use of babbling, frequency of vocalizations) have been found to differentiate infants with atypical development and typical development including infants with autism spectrum disorder (ASD) [12,13], Williams Syndrome [14], and FXS [15].

Prospectively in the ASD literature, infants with an older sibling diagnosed with ASD who later went on to have their own diagnosis of ASD demonstrated significant declines in their trajectories of receptive and expressive communication across 6 to 36 months of age [13]. Retroactively through home videos, infants later diagnosed with ASD have been shown to exhibit reduced canonical babbling and fewer vocalizations deemed relevant for the development of speech across 9 to 15 months of age [12]. Unfortunately, neither of these studies took into consideration the impact of cognitive development on their language development. Research examining language development in toddlers with ASD have shown a discrepancy between language abilities and their nonverbal cognitive level suggesting that these language deficits exist in this population despite their cognitive abilities [16]. Similar findings have also been observed in infants with Williams syndrome, suggesting overall delays in first word production and canonical babbling [14] despite their relative strengths in language in adolescence and adulthood.

Communication deficits in school-aged children and adolescents with FXS have been investigated extensively in the literature [3,4,17–19]. Individuals with FXS have reported deficits across all aspects of language (e.g., comprehension, pragmatics, expressive and receptive skills) with these deficits remaining throughout life into adulthood. Unfortunately, the literature assessing language development in infancy and toddlerhood is limited. Roberts, Hatton, and Bailey (2001) [20] reported the age in which infants with FXS spoke their first word was delayed by approximately 17 months; however, considerable variability was noted in their sample with 30% of the infants with FXS speaking their first word within age-expected limits. Similar findings were observed by Hinton et al. (2013) [21] where infants with FXS spoke their first word around 26.2 months. Two studies have utilized retrospective home videos to examine communication abilities of infants with FXS between the ages of 9 and 12 months [15,22]. Marschik et al. (2014) [22] utilized the Inventory of Potential Communicative Acts (IPCA) [23] with seven children with FXS to assess social-communicative forms and functions where specific deficits were identified in requesting, imitating, and decision making. Belardi et al. (2017) [15] utilized a naturalistic listening approach to identify deficits in canonical babbling (e.g., producing adult-like syllables) and the frequency of vocalizations in infants with FXS. Utilizing standardized assessments and parental report for language development to assess how visual attention at 12 and 18 months impacts language outcomes, Kover et al. (2015) [24] found that infants with FXS were significantly delayed based on both chronological and developmental expectations of language ability. Furthermore, the infants with FXS were found to acquire language at a slower rate than their chronological expectations and are likely to fall further behind over time. Overall, infants with FXS are reportedly exhibiting notable delays in their language abilities early on in development; however, the current literature lacks prospective, quantitative yet naturalistic methodologies to assess the emergence and development of these language deficits during the earliest periods of development.

Examining the language learning environment of young children, in particular their social interactions with caregivers, also provides insight into their language development. [8,9]. Within the ASD literature, Warren et al. (2010) [25] found that young children with ASD engaged in fewer caregiver interactions and vocalizations than typically developing children. They also demonstrated that their vocal productions increase as the number of words that are addressed to them increases. Within the FXS literature, little research exists examining their social or language environment and how this impacts language development. Drawing on the recent work examining maternal responsivity and language development in young children with FXS, low levels of maternal responsivity have been found to be related to deficits in receptive and expressive communication abilities along with vocabulary development in FXS [26,27]. Interestingly, the rate of child communication has been found to significantly negatively impact maternal responsivity [28] suggesting a disrupted cycle of both children with FXS and their caregivers communicating less. Further, the literature examining maternal responsivity in FXS has primarily utilized short structured activities and brief naturalistic observation to assess child language development through effortful, behavioral coding procedures. The potential ability to assess the language environment, child language abilities, and caregiver vocalizations in their natural environment through an efficient manner for longer time periods utilizing a noninvasive approach would further our current understanding of early language development in FXS.

The present study aims to build on our current knowledge of early language development in FXS while addressing some of the challenges to assessment in very young children. Utilizing a pilot sample of infants and toddlers with FXS, the present study examines child and caregiver vocalizations in their home environment utilizing an automated vocal analysis system. Consistent with the literature described above, we hypothesize that the infants with FXS will be below their chronological and developmental age expectations for vocalization use in comparison to age-matched typically developing peers. Furthermore, we hypothesize that the caregivers of the infants and toddlers with FXS will also utilize less vocalizations in comparison to other caregivers with typically developing children. Additional exploratory analyses were assessed for potential relationships between parent vocalizations and child vocalizations in the FXS sample. This preliminary study is the first to assess the utility of a noninvasive automated vocal analysis system in individuals with FXS.

2. Method

2.1. Participants

Eleven males with a confirmed molecular diagnosis of full mutation FXS between the ages of 17 to 64 months of age ($M = 41.58$, $SD = 13.43$) participated in the present pilot study. Data were drawn from a longitudinal study at Cincinnati Children's Hospital Medical Center as a subcomponent of a larger, multi-site study developing a nationwide research database in FXS. Cincinnati Children's Hospital Medical Center Institutional Review Board (IRB) approved the study protocol (IRB #: 2012-2445) and caregivers signed informed consent for their children to participate. Data were extracted from the LENA Foundation Natural Language Study [29] to derive a typically developing (TD) normative dataset to compare to the performance of the children with FXS. Comprehensive results for that study are reported in Gilkerson and Richards (2008) [29]. Two samples of TD normative data were utilized to match the FXS sample by both chronological and developmental age. The developmental age of the FXS sample ranged between 6 and 22 months of age ($M = 14.67$, $SD = 5.10$). The final FXS sample resulted in nine males between the ages of 17 and 58 months ($M = 38.33$, $SD = 13.05$) after excluding two participants (see details below under "LENA").

2.2. Measures

2.2.1. LENA

The LENA system includes a digital language processor (DLP) that is worn by the participant and a language analysis software. The DLP is a small digital recorder that is worn in a specially designed child's shirt. The device continually records the child's vocalizations and the language environment within a four to six foot radius around the child for up to 16+ hours. Once the recording is completed, an audio file from the DLP is transferred to a computer and processed by the LENA language analysis software. The software provides data for three main variables: child vocalizations (frequency and duration), adult word count, and conversational turns. The device also provides data for other variables in the environment including: TV/Electronics, Noise (e.g., bumps/rattles), Distant Sounds, Silence/Background Noise, and Overlapping Speech. Each participants' first LENA analysis data point was extracted for the current analysis. One participant was removed from the dataset due to extreme outlier findings across all variables reported. This participant was 42 months of age with a developmental age of 36 months with more than double the amount of vocalizations and conversational turns in comparison to the rest of the sample, causing the FXS sample to be skewed. Another participant was removed for only having one hour of data collected.

For the purpose of this study, we chose to focus on the three main variables provided by the LENA system. Child vocalizations (CV) included words, babbles, and pre-speech communicative sounds. Adult word count (AWC) is an estimate of the number of words spoken near the child. A normative value for average AWC was derived from the LENA Foundation Natural Language Study [29] in order to compare the FXS AWC sample to the normal population. Specifically, we utilized the AWC at the 50th percentile. Conversational turns (CT) in the LENA output occur when a child vocalizes and an adult responds, or an adult speaks and the child responds. The reliability and validity of the LENA automated vocal analysis system has been extensively researched in the literature to examine the automated vocalization systems ability to accurately label the recorded vocalizations correctly. Xu et al. (2008) [30] reported in comparison to the transcribers' labeling, the automated system correctly identified 82% of the segments transcribers labeled as Adult Speech and 76% of the segments labeled as Child Vocalizations. Further, adult word count estimates were on average 98% accurate compared to human transcribers' word counts over a 12 hour recording day. Other groups have also found adequate correlations between human coders and the LENA system ranging between 0.71 and 0.85 [31] providing additional support for the accuracy of the LENA automated vocal analysis system.

A recording was considered valid if it contained at least two hours of data. As mentioned above, one participant was dropped due to having only one hour of data. The two hour criteria was established as an attainable goal for our families given the sensory challenges in the FXS population and whether or not wearing the device would be tolerable. The amount of data collected in the remaining nine participants ranged from 7 to 18 hours (M = 12.56, SD = 3.81). Since the TD normative data was derived from the LENA Foundation Natural Language Study [29] and the chronological age range for their typical sample was between 2 and 48 months, developmentally age-matched norms were extracted for all of the FXS participants; however, two participants were outside of the 48 month window chronologically and therefore not included in the chronological age-matched analyses.

2.2.2. LENA Developmental Snapshot (LDS)

The LENA Developmental Snapshot [32] is a caregiver-report questionnaire that assesses both receptive and expressive language skills for children ages 2 to 36 months of age. The LDS consist of 52 items answered with a "yes" or "not yet" about the child's behavior (e.g., "Does your child vocalize while gesturing to let you know what he/she wants?"). Domains within the questionnaire focus on vocal behavior and preverbal communication for infants under 12 months; responsiveness to instruction, spontaneous speech production and vocabulary development for 1-year-olds; and conceptual and grammatical development for children over 24 months [33]. The LDS is scored automatically online through the LENA Online system. The number of "yes" answers reported are added up to create a total raw score which is then transformed into a Developmental Age. The LDS was found to be highly correlated (0.81–0.97) with other widely used standardized language development assessments [32]. The Developmental Age was extracted from the questionnaire for the FXS participants and used to create a developmentally age-matched TD comparison sample based on the TD data provided in LENA Foundation Natural Language Study [29]. For example, if a child with

FXS had a Developmental Age of 14 months, the 14 month data points for CV and CT from the TD sample through the Natural Language Study were extracted.

2.3. Procedures

Following completion of guardian informed consent, participants' caregivers were given in-person or mailed the LENA device along with the appropriate LENA-specific clothing to hold the device. Instructions were included on how to turn on the device and start the recording. Caregivers were instructed to have the participant wear the device during a normal day for them (e.g., avoid when they are sick or attending loud events). Additionally, they were instructed to have the participant wear the LENA clothing with the device for the entire day with the exception of taking a bath or naps; however, the device should still be nearby during these activities. The LDS was also included with the LENA device and the caregivers were asked to complete the form prior to the return of the device. Once completing the form and recording, the families were provided with materials to mail the device and questionnaire back. Once returned, the audio file was downloaded from the LENA device and uploaded to the LENA language analysis software to extract the data and the LDS was entered into the LENA online scoring system.

2.4. Data Analysis

Analyses were conducted in R version 3.5.1 (R Foundation for Statistical Computing, Vienna, Austria). First, data were examined for outliers, nonnormality, and homoscedasticity. One participant with FXS was found to be a significant outlier across all variables and was removed from the analyses. Data collected on the LENA device were all converted to hourly values in order to account for the variability in duration of data collection within and across groups. In order to analyze the differences between the FXS sample and TD infants in regard to their early language development, independent-sample *t*-tests and one sample *t*-tests were conducted. The first set of independent *t*-tests examined the FXS sample in comparison to their chronologically age-matched TD peers for CV and CT. The second set of independent *t*-tests examined the FXS sample in comparison to their developmentally age-matched TD peers for CV and CT. Next, a one sample *t*-test was conducted to compare the AWC of the FXS sample to the TD average AWC at the 50th percentile. Lastly, exploratory correlational analyses were conducted to examine relationships between AWC and the other LENA variables (CV and CT) within the FXS sample.

3. Results

3.1. Child Vocalizations

3.1.1. Chronological Age Comparisons

An independent samples *t*-test was conducted in order to determine if infants and toddlers with FXS differed significantly in the frequency of their vocalizations in comparison to chronologically age-matched TD peers. Significant group differences were found, $t(12) = -3.26$, $p = 0.007$, $d = 1.74$. Specifically, infants and toddlers with FXS ($M = 106.00$, $SD = 45.56$) had significantly less vocalizations on average per hour than their chronologically age-matched TD peers ($M = 169.31$, $SD = 23.73$). In Figure 1A, each participant with FXS's frequency of vocalizations are graphed in comparison to their chronologically age-matched TD peers.

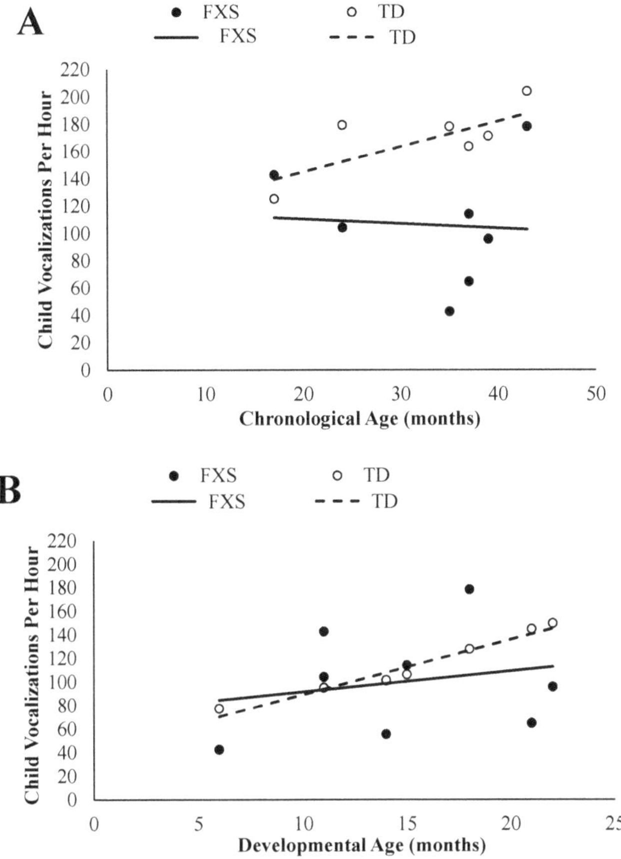

Figure 1. Infants and toddlers with fragile X syndrome (FXS) vocalizations per hour plotted in comparison to their chronologically age-matched (**A**) and developmentally age-matched (**B**) typically developing (TD) peers with trend lines.

3.1.2. Developmental Age Comparisons

An independent samples *t*-test was conducted in order to determine if infants and toddlers with FXS differed significantly in the frequency of their vocalizations in comparison to developmentally age-matched TD peers. No significant group differences emerged, $t(16) = -0.68$, $p = 0.507$, $d = 0.32$. The infants and toddlers with FXS ($M = 99.84$, $SD = 42.86$) had similar average vocalization frequencies to their developmentally age-matched TD peers ($M = 110.99$, $SD = 24.32$) per hour. In Figure 1B, each participant with FXS's frequency of vocalizations are graphed in comparison to their developmentally age-matched TD peers.

3.2. Conversational Turns

3.2.1. Chronological Age Comparisons

An independent samples *t*-test was conducted in order to determine if infants and toddlers with FXS differed significantly in the frequency of their conversational turns with caregivers in comparison to chronologically age-matched TD peers. Levene's test for equality of variances indicated unequal variances between groups ($F = 20.98$, $p = 0.001$), so the degrees of freedom were adjusted from 12 to 7. Marginally significant group differences were found, $t(7) = -1.93$, $p = 0.094$, $d = 1.03$. The infants and toddlers with FXS (M = 25.63, SD = 15.71) had less conversational turns per hour with their caregivers than their chronologically age-matched TD peers (M = 37.63, SD = 4.77). In Figure 2A, each participant with FXS's frequency of conversational turns are graphed in comparison to their chronologically age-matched TD peers.

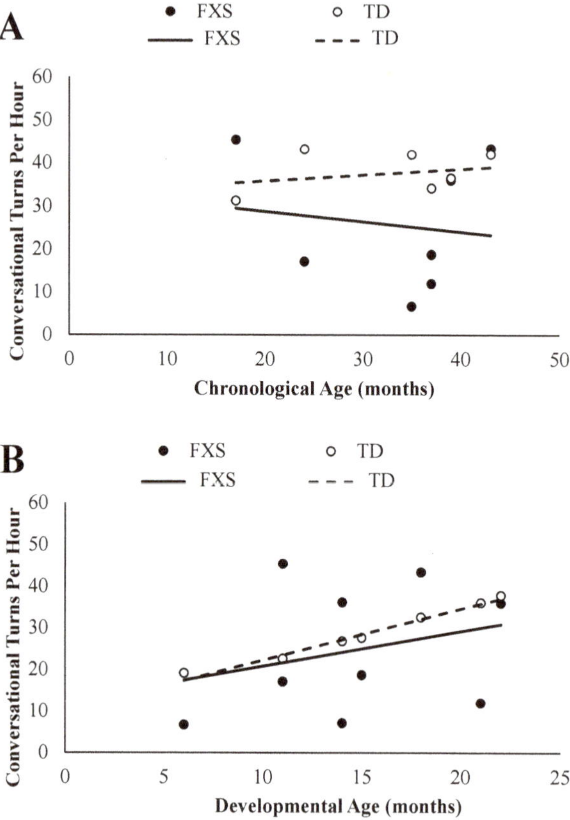

Figure 2. Infants and toddlers with fragile X syndrome (FXS) conversational turns per hour plotted in comparison to their chronologically age-matched (**A**) and developmentally age-matched (**B**) typically developing (TD) peers with trend lines.

3.2.2. Developmental Age Comparisons

An independent samples *t*-test was conducted in order to determine if infants and toddlers with FXS differed significantly in the frequency of their conversational turns with caregivers in comparison to developmentally age-matched TD peers. Levene's test for equality of variances indicated unequal variances between groups (F = 17.77, p = 0.001), so the degrees of freedom were adjusted from 16 to 11. No significant group differences emerged, $t(11)$ = -0.59, p = 0.568, d = 0.28. The infants and toddlers with FXS (M = 24.75, SD = 15.52) had similar average rates of conversational turns per hour with their caregivers to their developmentally age-matched TD peers (M = 28.04, SD = 6.33). In Figure 2B, each participant with FXS's frequency of conversational turns are graphed in comparison to their developmentally age-matched TD peers.

3.3. Adult Word Count

A single sample *t*-test was conducted to determine if a statistically significant difference existed between the FXS caregivers' and the TD caregivers' word count per hour. Results suggest that FXS caregivers (*M* = 772.04, *SD* = 405.75) had marginally significantly different word counts per hour in comparison to the TD caregivers (*M* = 1024.75, *t*(8) = −1.87, *p* = 0.099, *d* = 0.60. In Figure 3, each caregiver's word count is graphed in comparison to the average adult word count for TD caregivers.

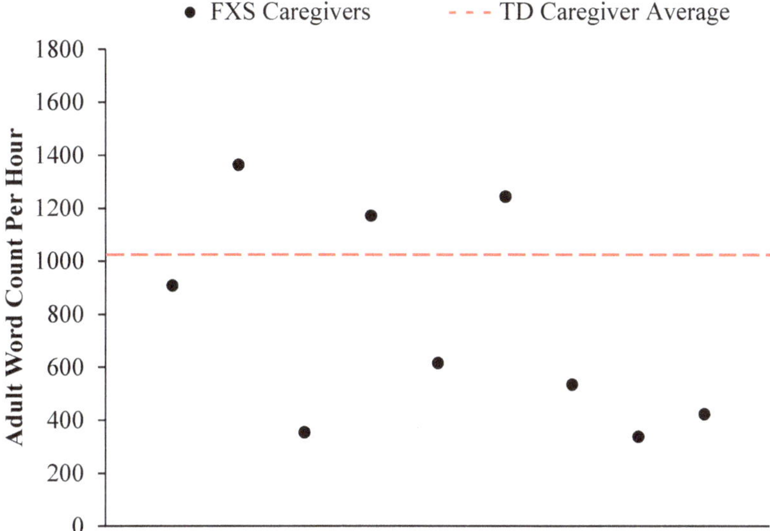

Figure 3. Caregivers' of infants and toddlers with FXS adult word count per hour plotted in comparison to the average adult word count per hour in TD caregivers (*M* = 1024.75).

Exploratory correlations were utilized to examine relationships between the AWC and the other LENA variables (CV and CT). Significant associations between AWC and CV (*r* = 0.82, *p* = 0.025) as well as AWC and CT (*r* = 0.98, *p* = 0.000) emerged.

4. Discussion

4.1. Summary of Findings

Language deficits are core characteristics of the FXS phenotype. Previous literature has identified deficits in receptive and expressive communication abilities as early as 9 months of age in FXS [15,22] with their first word being vocalized between 26 and 28 months [20,21] and a reciprocal negative relationship between child vocalizations and maternal responsivity on language development and acquisition [26,27]. The present preliminary study examined early language development through frequency of child vocalizations, conversational turns between caregivers and child, and adult vocalizations in infants and toddlers with FXS in comparison to a chronologically and developmentally age-matched typically developing sample. This pilot study is one of the first to utilize an automated vocal analysis program within the FXS population.

Partially aligning with our hypotheses and previous literature [15,22,24], infants and toddlers with FXS were found to vocalize less and engage in fewer conversational turns with their caregivers in comparison to chronologically age-matched TD peers. Despite previous research suggesting otherwise [4,24], differences in the frequency of vocalizations and conversational turns were not observed when compared to a developmentally age-matched TD group in our pilot sample. This could be explained by the differences in language assessment methodology. Specifically, majority of the literature assessing language development in infants and toddlers with FXS have utilized standardized measures obtaining norm referenced scores rather than the actual frequency of vocalizations in their normal environment. Given the known cognitive delays and receptive deficits in FXS, it can be complicated to obtain an accurate score representing their true language abilities utilizing standardized measures in the youngest children with FXS. Utilizing the LENA device allowed the present study to automatically and noninvasively obtain a naturalistic language/vocalization sample in the child's normal environment without limitations imparted by cognitive level or receptive communication deficits, which are known to impact language abilities in FXS [34]. In sum, our preliminary results support the current body of literature across the FXS lifespan suggesting deficits in verbal communication development; however, these deficits may be accounted for by their developmental level and additional research is needed to support these findings. Further, the LENA device may be a potential new mechanism for assessment of not only language but also the language environment in FXS.

Consistent with our hypotheses and previous literature [26,27], our results suggest caregivers of infants and toddlers with FXS produced fewer vocalizations around their children in comparison to caregivers with TD infants and toddlers. Despite the small sample size of this preliminary study, moderate effect sizes were still reported. These findings of reduced adult vocalizations coupled with reduced conversational turns between caregivers and their infants and toddlers with FXS are the first to provide insight into the language environment within FXS. Furthermore, a strong positive association between caregiver vocalizations and child vocalizations emerged suggesting that for the FXS caregivers who vocalized more, their children also vocalized more. These findings align with concerns for a potentially disrupted cycle of communication to evolve between caregiver and child in regard to the frequency of vocalizations in their language learning environment. Specifically, difficulty could arise for caregivers to maintain their frequency of vocalizations when their children are less responsive, which can unfortunately create a cycle of reduced communication across both groups potentially impacting language development for the child. Further work is needed to assess the impact of caregiver vocalizations and conversational turns between caregivers and their children with FXS on child vocalizations to delineate this hypothesis and to determine whether a true gap exists for FXS caregiver vocalizations in comparison to TD caregivers. Nevertheless, there is an existing body of literature demonstrating that caregivers can learn to be more responsive resulting in positive language outcomes for their children [35–37]. As for the FXS literature, there is a promising emerging body of intervention research demonstrating increases in maternal verbal responses and child prompted communication [38]. Therefore, there is hope in changing caregiver behavior through appropriate and effective interventions that can potentially close this communication gap early on, while positively impacting their child's language outcomes and potentially their overall quality of life.

4.2. Limitations

A primary limitation of this preliminary pilot study is its small sample size. Despite this, the sample is similar to those of other studies examining language development in infants and toddlers with FXS [15,21,23]. Additionally, the present study did not contain its own typically developing matched control sample to compare the FXS population too; however, utilizing the LENA Natural Language Study [29] was also a strength by allowing for the present study to utilize a more accurate, large-scale normed TD sample. Furthermore, the present study utilized a questionnaire to assess developmental level, which relies on parental report, rather than a standardized test administered by a clinician. These analyses were also limited to utilizing a cross-sectional design with one day of language data per child. Since language production can vary from day to day in infants and toddlers, it would be ideal to have more than one day of language data available. Lastly, reliability of the LENA device could be assessed through pairing human coding and the automated vocal analysis to determine the accuracy of the system specific to FXS.

5. Conclusions and Future Directions

The present study aimed to build on our current understanding of early language development in FXS utilizing new methodology. Our results suggest that communication deficits, particularly vocalization production deficits, are apparent very early on in development in comparison to chronological age expectations. However, language profiles in FXS as measured by LENA appear to potentially be in line with their developmental expectations. Additional work is needed to replicate these findings using the same methodology with a larger sample and wider age range. Utilizing a longitudinal design to obtain a more accurate assessment of language development would be ideal to further our understanding of the rate of growth in language across development. Furthermore, initial details about the language learning environment for infants and toddlers with FXS were examined with additional evidence emerging for a potentially disrupted cycle of communication between FXS caregivers and their children with reduced caregiver vocalizations being associated with reduced child vocalizations. Future studies should continue to assess the effectiveness of interventions for FXS caregivers to increase their responsiveness and vocalizations on child language outcomes. Lastly, the methodology utilized in the present study provided a measure of communication abilities in infants and toddlers with FXS and insight into the language learning environment that was noninvasive and easy to use for their families. This methodology may be promising for future researchers, the participants, and their families by simplifying data collection without reducing quality and accuracy. The LENA device may continue to be utilized in future FXS research to not only quantify vocal production development and the language learning environment, but also assist in collecting outcome data for future intervention studies.

Author Contributions: C.A.E., R.C.S., K.C.D., and E.V.P. contributed to study conceptualization and data collection. C.A.E., R.C.S., and D.L.R. contributed to the manuscript preparation and revisions. D.L.R. led the writing of the manuscript and analyzed the data.

Funding: This publication was supported by cooperative agreements #U01DD000231, #U19DD000753 and # U01DD001189, funded by the Centers for Disease Control and Prevention. Its contents are solely the responsibility of the authors and do not necessarily represent the official views of the Centers for Disease Control and Prevention or the Department of Health and Human Services.

Acknowledgments: We thank all our staff from Cincinnati Children's Hospital Medical Center for their assistance in the execution of the study, particularly Christina Harkins, Olivia Walter, and Hilary Rosselot. We would also like to thank the participants and their families who have contributed their time to make this research possible.

Conflicts of Interest: R.C.S. receives funding from Fulcrum Therapeutics. C.A.E. has received current or past funding from Confluence Pharmaceuticals, Novartis, F. Hoffmann-La Roche Ltd., Seaside Therapeutics, Riovant Sciences, Inc., Fulcrum Therapeutics, Neuren Pharmaceuticals Ltd., Alcobra Pharmaceuticals, Neurotrope, Zynerba Pharmaceuticals, Inc., and Ovid Therapeutics Inc. to consult on trial design or development strategies and/or conduct clinical trials in FXS or other neurodevelopmental disorders. C.A.E. is additionally the inventor or co-inventor on several patents held by Cincinnati Children's Hospital Medical Center or Indiana University School of Medicine describing methods of treatment in FXS or other neurodevelopmental disorders. E.V.P. has received research support by the National Institutes of Health (NIMH), American Academy of Child and Adolescent Psychiatry, and Cincinnati Children's Hospital Research Foundation. He is a clinical trial site investigator for the Marcus Autism Center (clinical trial, Autism). He receives compensation for consulting for Proctor & Gamble. He receives book royalties from Springer. There are no conflicts of interest with the current manuscript. K.C.D. has received research support from the National Institute of Neurological Disorders and Stroke (NINDS), American Academy of Child and Adolescent Psychiatry, and Cincinnati Children's Hospital Medical Center. She is a clinical trial site investigator for F. Hoffman-La Roche Ltd. and Ovid Therapeutics. There are no conflicts of interest for the current manuscript.

References

1. Hagerman, R.J.; Hagerman, P.J. (Eds.) *Fragile X Syndrome: Diagnosis, Treatment, and Research*, 3rd ed.; Johns Hopkins Univ. Press: Baltimore, MD, USA, 2002; p. 540.
2. Cordeiro, L.; Ballinger, E.; Hagerman, R.; Hessl, D. Clinical assessment of DSM-IV anxiety disorders in fragile X syndrome: Prevalence and characterization. *J. Neurodev. Disord.* **2011**, *3*, 57 67. [CrossRef] [PubMed]

3. Martin, G.E.; Losh, M.; Estigarribia, B.; Sideris, J.; Roberts, J. Longitudinal profiles of expressive vocabulary, syntax and pragmatic language in boys with fragile X syndrome or Down syndrome. *Int. J. Lang. Commun. Disord.* **2013**, *48*, 432–443. [CrossRef]

4. Roberts, J.E.; Mirrett, P.; Burchinal, M. Receptive and Expressive Communication Development of Young Males with Fragile X Syndrome. *Am. J. Ment. Retard.* **2001**, *106*, 216–230. [CrossRef]

5. Pena, M.; Maki, A.; Kovacic, D.; Dehaene-Lambertz, G.; Koizumi, H.; Bouquet, F.; Mehler, J. Sounds and silence: An optical topography study of language recognition at birth. *Proc. Natl. Acad. Sci. USA* **2003**, *100*, 11702–11705. [CrossRef] [PubMed]

6. Vouloumanos, A.; Werker, J.F. Listening to language at birth: Evidence for a bias for speech in neonates. *Dev. Sci.* **2007**, *10*, 159–164. [CrossRef]

7. Ferguson, C.A.; Menn, L.; Stoel-Gammon, C. *Phonological Development: Models, Research, Implications*; York Press: Timonium, MD, USA, 1992; ISBN 9780912752242.

8. Tamis-LeMonda, C.S.; Bornstein, M.H.; Baumwell, L. Maternal responsiveness and children's achievement of language milestones. *Child Dev.* **2001**, *72*, 748–767. [CrossRef] [PubMed]

9. Kuhl, P.K. Early language acquisition: Cracking the speech code. *Nat. Rev. Neurosci.* **2004**, *5*, 831–843. [CrossRef] [PubMed]

10. Hart, B.; Risley, T.R. *Meaningful Differences in the Everyday Experience of Young American Children*; Paul H Brookes Publishing: Baltimore, MD, USA, 1995; ISBN 1-55766-197-9.

11. Tamis-LeMonda, C.S.; Shannon, J.D.; Cabrera, N.J.; Lamb, M.E. Fathers and Mothers at Play with Their 2- and 3-Year-Olds: Contributions to Language and Cognitive Developmen. *Child Dev.* **2004**, *75*, 1806–1820. [CrossRef] [PubMed]

12. Patten, E.; Belardi, K.; Baranek, G.T.; Watson, L.R.; Labban, J.D.; Oller, D.K. Vocal patterns in infants with autism spectrum disorder: Canonical babbling status and vocalization frequency. *J. Autism Dev. Disord.* **2014**, *44*, 2413–2428. [CrossRef]

13. Longard, J.; Brian, J.; Zwaigenbaum, L.; Duku, E.; Moore, C.; Smith, I.M.; Garon, N.; Szatmari, P.; Vaillancourt, T.; Bryson, S. Early expressive and receptive language trajectories in high-risk infant siblings of children with autism spectrum disorder. *Autism Dev. Lang. Impair.* **2017**, *2*, 1–11. [CrossRef]

14. Masataka, N. Why early linguistic milestones are delayed in children with Williams syndrome: Late onset of hand banging as a possible rate-limiting constraint on the emergence of canonical babbling. *Dev. Sci.* **2001**, *4*, 158. [CrossRef]

15. Belardi, K.; Watson, L.R.; Faldowski, R.A.; Hazlett, H.; Crais, E.; Baranek, G.T.; McComish, C.; Patten, E.; Oller, D.K. A Retrospective Video Analysis of Canonical Babbling and Volubility in Infants with Fragile X Syndrome at 9–12 Months of Age. *J. Autism Dev. Disord.* **2017**, *47*, 1193–1206. [CrossRef] [PubMed]

16. Ellis Weismer, S.; Lord, C.; Esler, A. Early language patterns of toddlers on the autism spectrum compared to toddlers with developmental delay. *J. Autism Dev. Disord.* **2010**, *40*, 1259–1273. [CrossRef] [PubMed]

17. Roberts, J.E.; Hennon, E.A.; Price, J.R.; Dear, E.; Anderson, K.; Vandergrift, N.A. Expressive Language During Conversational Speech in Boys with Fragile X Syndrome. *Am. J. Ment. Retard.* **2007**, *112*, 1–17. [CrossRef]

18. Haebig, E.; Sterling, A.; Hoover, J. Examining the Language Phenotype in Children with Typical Development, Specific Language Impairment, and Fragile X Syndrome. *J. Speech Lang. Hear. Res.* **2016**, *59*, 1046–1058. [CrossRef] [PubMed]

19. Finestack, L.H.; Abbeduto, L. Expressive Language Profiles of Verbally Expressive Adolescents and Young Adults with Down Syndrome or Fragile X Syndrome. *J. Speech Lang. Hear. Res.* **2010**, *53*, 1334–1348. [CrossRef]

20. Roberts, J.E.; Hatton, D.D.; Bailey, D.B. Development and Behavior of Male Toddlers with Fragile X Syndrome. *J. Early Interv.* **2001**, *24*, 207–223. [CrossRef]

21. Hinton, R.; Budimirovic, D.B.; Marschik, P.B.; Talisa, V.B.; Einspieler, C.; Gipson, T.; Johnston, M.V. Parental reports on early language and motor milestones in fragile X syndrome with and without autism spectrum disorders. *Dev. Neurorehabil.* **2013**, *16*, 58–66. [CrossRef]

22. Marschik, P.B.; Bartl-Pokorny, K.D.; Sigafoos, J.; Urlesberger, L.; Pokorny, F.; Didden, R.; Einspieler, C.; Kaufmann, W.E. Development of socio-communicative skills in 9- to 12-month-old individuals with fragile X syndrome. *Res. Dev. Disabil.* **2014**, *35*, 597–602. [CrossRef]

23. Sigafoos, J.; Woodyatt, G.; Keen, D.; Tait, K.; Tucker, M.; Roberts-Pennell, D.; Pittendreigh, N. Identifying Potential Communicative Acts in Children with Developmental and Physical Disabilities. *Commun. Disord. Q.* **2000**, *21*, 77–86. [CrossRef]

24. Kover, S.T.; McCary, L.M.; Ingram, A.M.; Hatton, D.D.; Roberts, J.E. Language development in infants and toddlers with fragile X syndrome: Change over time and the role of attention. *Am. J. Intellect. Dev. Disabil.* **2015**, *120*, 125–144. [CrossRef] [PubMed]

25. Warren, S.F.; Gilkerson, J.; Richards, J.A.; Oller, D.K.; Xu, D.; Yapanel, U.; Gray, S. What automated vocal analysis reveals about the vocal production and language learning environment of young children with autism. *J. Autism Dev. Disord.* **2010**, *40*, 555–569. [CrossRef] [PubMed]

26. Warren, S.F.; Brady, N.; Sterling, A.; Fleming, K.; Marquis, J. Maternal responsivity predicts language development in young children with fragile X syndrome. *Am. J. Intellect. Dev. Disabil.* **2010**, *115*, 54–75. [CrossRef] [PubMed]

27. Brady, N.; Warren, S.F.; Fleming, K.; Keller, J.; Sterling, A. Effect of Sustained Maternal Responsivity on Later Vocabulary Development in Children with Fragile X Syndrome. *J. Speech Lang. Hear. Res.* **2014**, *57*, 212–226. [CrossRef]

28. Sterling, A.M.; Warren, S.F.; Brady, N.; Fleming, K. Influences on maternal responsivity in mothers of children with fragile X syndrome. *Am. J. Intellect. Dev. Disabil.* **2013**, *118*, 310–326. [CrossRef] [PubMed]

29. Gilkerson, J.; Richards, J. *The LENA Natural Language Study (Technical Report LTR-02-2)*; LENA Foundation: Boulder, CO, USA, 2008; pp. 1–26.

30. Xu, D.; Yapanel, U.; Gray, S. Reliability of the LENA™ language environment analysis system in young children's natural home environment. *LENA Tech. Rep.* **2009**, *5*, 1–16.

31. Oetting, J.B.; Hartfield, L.R.; Pruitt, S.L. Exploring LENA as a Tool for Researchers and Clinicians. *ASHA Lead.* **2009**, *14*, 20–22. [CrossRef]

32. Gilkerson, J.; Richards, J.A. *The LENA Developmental Snapshot*; LENA Foundation: Boulder, CO, USA, 2008; pp. 1–7.

33. Gilkerson, J.; Richards, J.A.; Greenwood, C.R.; Montgomery, J.K. Language assessment in a snap: Monitoring progress up to 36 months. *Child Lang. Teach. Ther.* **2017**, *33*, 99–115. [CrossRef]

34. Finestack, L.H.; Richmond, E.K.; Abbeduto, L. Language Development in Individuals with Fragile X Syndrome. *Top. Lang. Disord.* **2009**, *29*, 133–148. [CrossRef]

35. Siller, M.; Hutman, T.; Sigman, M. A parent-mediated intervention to increase responsive parental behaviors and child communication in children with ASD: A randomized clinical trial. *J. Autism Dev. Disord.* **2013**, *43*, 540–555. [CrossRef]

36. Landry, S.H.; Smith, K.E.; Swank, P.R.; Guttentag, C. A Responsive Parenting Intervention: The Optimal Timing Across Early Childhood for Impacting Maternal Behaviors and Child Outcomes. *Dev. Psychol.* **2008**, *44*, 1335–1353. [CrossRef] [PubMed]

37. Venker, C.E.; McDuffie, A.; Ellis Weismer, S.; Abbeduto, L. Increasing verbal responsiveness in parents of children with autism: A pilot study. *Autism* **2012**, *16*, 568–585. [CrossRef] [PubMed]

38. McDuffie, A.; Oakes, A.; Machalicek, W.; Ma, M.; Bullard, L.; Nelson, S.; Abbeduto, L. Early language intervention using distance video-teleconferencing: A pilot study of young boys with fragile X syndrome and their mothers. *Am. J. Speech Lang. Pathol.* **2016**, *25*, 46–66. [CrossRef] [PubMed]

Review

Executive Function in Fragile X Syndrome: A Systematic Review

Lauren M. Schmitt [1], Rebecca C. Shaffer [2,3], David Hessl [4] and Craig Erickson [1,5,*]

[1] Division of Child and Adolescent Psychiatry, Cincinnati Children's Hospital Medical Center, Cincinnati, OH 45229, USA; lauren.schmitt@cchmc.org
[2] Division of Developmental and Behavioral Pediatrics, Cincinnati Children's Hospital Medical Center, Cincinnati, OH 45229, USA; rebecca.shaffer@cchmc.org
[3] Department of Pediatrics, University of Cincinnati College of Medicine, Cincinnati, OH 45267, USA
[4] MIND Institute, Department of Psychiatry and Behavioral Sciences, University of California, Davis, CA 95616, USA; drhessl@ucdavis.edu
[5] Department of Psychiatry and Behavioral Neuroscience, University of Cincinnati College of Medicine, Cincinnati, OH 45267, USA
* Correspondence: craig.erickson@cchmc.org; Tel.: +1-513-636-6553

Received: 15 December 2018; Accepted: 11 January 2019; Published: 16 January 2019

Abstract: Executive function (EF) supports goal-directed behavior and includes key aspects such as working memory, inhibitory control, cognitive flexibility, attention, processing speed, and planning. Fragile X syndrome (FXS) is the leading inherited monogenic cause of intellectual disability and is phenotypically characterized by EF deficits beyond what is expected given general cognitive impairments. Yet, a systematic review of behavioral studies using performance-based measures is needed to provide a summary of EF deficits across domains in males and females with FXS, discuss clinical and biological correlates of these EF deficits, identify critical limitations in available research, and offer suggestions for future studies in this area. Ultimately, this review aims to advance our understanding of the underlying pathophysiological mechanisms contributing to EF in FXS and to inform the development of outcome measures of EF and identification of new treatment targets in FXS.

Keywords: fragile X syndrome; executive function; working memory; set-shifting; cognitive flexibility; inhibitory control; attention; planning; processing speed

---

1. Introduction

Fragile X syndrome (FXS) is the leading inherited monogenic cause of intellectual disability, resulting from >200 CGG trinucleotide repeat expansions in the 5′ untranslated region of the Fragile X Mental Retardation 1 Gene (*FMR1*). The resulting hyper-methylation and silencing of FMR protein (FMRP) production disrupts synaptic structure and function [1–5], leading to neural hyper-excitability and atypical brain development. The characteristic phenotypic features in humans with FXS, including impaired cognition, are hypothesized to be downstream effects of the altered neurodevelopment [6,7]. Because FXS is an X-linked genetic disorder, females with FXS are typically less severely affected than males with FXS due to the variability of X-chromosome inactivation in females [8]. Thus, females with FXS who demonstrate a greater degree of methylation and lower FMRP levels have a phenotype more similar to males with FXS, whereas females with less methylation and greater FMRP levels demonstrate more subtle clinical features. This is consistent with numerous documentations that the severity of cognitive impairments are associated with the degree of methylation mosaicism and FMRP expression in individuals with FXS [9–12], suggesting a progressive FMRP deficit leading to cognitive impairments. Yet, the precise mechanisms underlying specific cognitive impairments remain poorly understood in FXS.

Executive function (EF), or a group of discrete cognitive abilities that support adaptive goal-directed behavior [13], has been consistently documented as impaired in individuals with FXS, even beyond what is expected given their general cognitive impairments (for examples, see [14–23]). Whether this reflects a generalized deficit in EF or multiple deficits to specific EF domains (i.e., working memory, response inhibition, cognitive flexibility, attention, planning, and processing speed) is less clear. Previous studies using parent-report questionnaires have consistently documented high rates of inattention and hyperactivity in FXS, but these studies have seldom used questionnaires targeting a broader range of EF impairments or other questionnaires to assess additional EF domains (for examples, see [24,25]). In contrast, previous studies using traditional neuropsychological and experimental performance-based measures of EF have been able to capture deficits across all domains of EF in individuals with FXS. Behavioral performance-based measures have a distinct advantage over parent-report measures in their potential to be translated into tasks used during brain imaging studies (i.e., functional magnetic resonance imaging (fMRI) or electroencephalogram (EEG)) or into analogous versions to be used in studies of rodent models of FXS (i.e., *FMR1* KO mouse). Further, compared to parent-reports, performance-based measures are better able to objectively quantify performance, which may help provide insights into the specific brain regions affected in FXS as well as identify EF deficits that may be specific to FXS rather than those related to general cognitive impairments and/or common comorbid diagnoses, like attention deficit/hyperactivity disorder (ADHD) and autism spectrum disorder (ASD). Given previous indications of abnormalities in the frontostriatal regions supporting EF in FXS (for example, see [26]), translational studies of EF offer great promise for use as quantifiable biomarkers of brain function useful for early-phase drug development and intervention trials.

Despite over 50 studies examining EF deficits in FXS using performance-based measures, a comprehensive review of findings from males and females with FXS and their implications still is needed (Table 1). Generally, previous studies of EF in FXS documented impairments across each domain of EF relative to both chronologically age-matched (CA) controls and mentally aged-matched (MA) controls, yet closer examination reveals key differences in performances depending on domain, measure, task difficulty, sex, and/or control group studied. Understanding specific profiles of executive dysfunction in FXS whose etiology may be distinct, yet overlapping, with that of general cognitive impairments, is critical to understanding pathophysiological processes in FXS and developing novel treatments for this patient population. Though EF impairments are major cause of distress for individuals with FXS and their families [27,28] and poor EF leads to worse learning and academic achievement outcomes [29,30], the development of behavioral and pharmacological interventions aimed at improving EF have lagged behind those targeted towards other key phenotypes, like anxiety and sensory hyper-sensitivity. Thus, a comprehensive review of previous studies is needed to summarize EF deficits and their clinical and biological correlates in FXS, establish potential underlying brain mechanisms of these deficits, and address critical limitations of previous studies. As more clinical research begins to use EF as outcome measures in treatment trials, it is crucial to review previous studies to guide future research studies examining EF in individuals with FXS.

Table 1. Executive function measures and findings in individuals with FXS.

Executive Function Domain	Measure [1]	Findings [2]		References
		Males	**Females**	
Working Memory				
Verbal	WJ-III Memory for Words; NIH Tool Box List Sorting; RAKIT Learning Names; WMTB Nonword Repetition; RBMT Story Recall; WMS-R Logical Memory I; SB-IV Sentence Memory; Weschler Digit Forward, Digit Backward; WJ-III Auditory Memory; WISC Letter-Number Sequencing; KABC Verbal Number Recall; WISC, WAIS Arithmetic; TEA-Ch Elevator Counting	vs. CA ↓ vs. MA ↓ vs. iDD ↓ vs. DS ↓≈	vs. CA ↓ vs. IQ ≈ vs. Enviro ≈	[15,16,21,22,31–46]
Nonverbal/Visuospatial	Leiter Spatial Memory; CANTAB Spatial Span; KABC Spatial Span; SB5 Spatial Span; Delayed-no-matching-to-position; Toy Recall	vs. CA ↓ vs. MA ↓ vs. iDD ↓ vs. DS ↓≈	vs. CA ↓	[16,21,22,31,32,34,41,42,47–49]
Nonverbal/Visuoperceptual	Delayed-non-matching-to-sample; WMS-R Figural Memory I; Fish color; Sequence Memory Task; n-back; SB-IV Bead Memory; Card Task; Hidden Object Memory Test	vs. CA ↓ vs. MA ↓ vs. DS ↓≈	vs. MA ↓ vs. Enviro ↓≈	[15,16,34,35,37,43,49–51]
Inhibitory Control				
Prepotent Response Inhibition	KiTAP Go/No-go; Lab-TAB Snack Delay; Antisaccade; TEA-Ch Walk task	vs. CA ↓ vs. MA ↓ vs. iDD ≈ vs. DS ↓	vs. CA≈	[14,16,26,52–56]
Distractor Interference	Stroop, Day/Night Task, CNT; Visual Selection; TEA-Ch Same-Opposite task; KiTap Distractibility, NIH Toolbox Flanker	vs. CA ≈↓ vs. MA ↓ vs. iDD ↓ vs. DS ↓	vs. IQ ↓≈ vs. Enviro ↓ vs. TS ≈	[16,22,31,34,35,52,55–60]
Cognitive Flexibility	NIH Toolbox Dimensional Change Card Sort Test; CANTAB IED; Wisconsin Card Sorting; iTap FlexiContingency Naming Task Subtest 3; Kbility	vs. CA ↓ vs. MA ↓ vs. iDD vs. DS ↓≈ vs. PWS ≈ vs. Enviro ≈	vs. Enviro ↓≈	[15,22,23,31,32,34,35,46,48,55,56,61]
Attention				
Sustained	CPT; KiTap Vigilance; KiTAP Sustained Attention; WIAT Vigilance; WATT Vigilan task; Elevator Counting	vs. CA ↓ vs. MA ↓ vs. DS ↓≈	n/a	[15,16,37,50,52,56,62–64]

Table 1. *Cont.*

Executive Function Domain	Measure [1]	Findings [2]		References
Working Memory		Males	Females	
Attention				
Selective	Symbol Search, Cancelation, Map Search, WATT Visearch task; WISC Symbol Search; KiTAP Visual Scanning; TEA-ch Map Search	vs. CA ↓ vs. MA ≈ vs. DS ↑ vs. WS ↑	n/a	[15,16,18,19,52,56,65]
Divided	KiTAP Divided Attention; WATT Visearch dual task	vs. MA ↓ vs. DS ≈	n/a	[16,56]
Planning	NEPSY Tower; WJ-III Planning; CANTAB Stockings of Cambridge	vs. MA ↓	n/a	[22,31,32]
Processing Speed	WISC Coding, Cancelation; NIH Toolbox Pattern Comparison; Reaction Time from CNT; KiTAP Alertness	vs. CA ↓ vs. MA ≈	vs. CA ↓	[22,31,33,35,56]

[1] Abbreviations for cognitive measure used: CANTAB—Cambridge Neuropsychological Test Automated Battery [66], CNT—Contingency Naming Task [67], CPT—Continuous Performance Test [68], IED—Intra-/Extra-Dimensional task, KABC—Kaufman Test of Attention Performance for Children [69], Lab-TAB—Laboratory Temperament Assessment Battery [70], NEPSY—Developmental Neuropsychological Assessment, RAKIT—Revised Amsterdamse Kinder Intelligence Test, RMBT—Rivermead Behavioral Memory Test [71,72], SB-IV—Stanford-Binet Intelligence Scale-4th Edition [73], SB-5—Stanford Binet Intelligence Scale-5th Edition, TEA-Ch—Test of Everyday Attention for Children [74], WAIS—Wechsler Adult Intelligence Scale, WATT—Wilding Attention Tasks [75], WIAT—Wechsler Individual Achievement Test, WISC—Wechsler Intelligence Scale for Children [76], WJ-III—Woodcock Johnson [77], WMS-R—Wechsler Memory Scale-Revised [78], WMTB—Working Memory Test Battery, Third Edition [79]; [2] ↓—indicates participants with FXS performed worse than specified control group, ↑—indicates participants with FXS performed better than specified control group, ≈—indicates participants with FXS performed similar to specified control group, ≈↓—indicates mixed findings; Participant type: CA—chronologically-aged match controls, MA—mentally-aged matched control, iDD—control with idiopathic developmental delay, DS—control with Down Syndrome, WS—control with William Syndrome control, PWS—control with Prader–Willi Syndrome, Enviro—environmental control (unaffected in-house relative).

2. Methods

2.1. Search Strategy

To identify studies for inclusion in the review, computerized databases, including PubMed and PsychINFO, were used to conduct searches. Keywords such as "Fragile X" AND "executive function", "working memory", "inhibitory control", "cognitive flexibility", "set-shifting", "attention", "processing speed", or "planning" as well as variants on these terms were used. After collecting all available peer-reviewed published articles, their reference sections were scanned to identify additional articles that may have been previously missed. Authors of articles that were not available were contacted.

2.2. Selection Criteria

Broad inclusion criteria were used in order to provide the most comprehensive review of the literature. The following criteria were used to determine whether an article could be included in the review:

1. The paper was published in a peer-reviewed journal,
2. At least one participant sample had full mutation FXS,
3. Performance of individuals with FXS was compared against either a control population or normative sample data OR performance of individuals with FXS is documented in the context of a feasibility study,
4. At least one measure of EF was used,
5. The study reported quantitative scores (e.g., raw score, T-score, Standard score, etc.) beyond completion rates,
6. EF was a primary or secondary research question, and
7. The study was not a case study.

2.3. Study Organization and Consolidation

The studies and corresponding measures were organized into six executive domains commonly reported in the literature: working memory, response inhibition, cognitive flexibility, attention, processing speed, and planning. In the first section, we summarized primary findings within each executive function domain, separated by males and females (Table 1) and then in relation to clinical and psychological variables. Next, we discussed current knowledge regarding neurobiology of EF deficits in FXS by reviewing biological correlates, findings from studies using brain imaging and rodent models, and potential pathophysiological mechanisms underlying these deficits. Finally, we addressed critical considerations from previous studies and provide recommendations for future studies.

3. Review

3.1. Executve Function Deficits

3.1.1. Working Memory

Working memory is necessary for temporarily storing information for immediate use, like remembering the homework assignment the teacher announced, and it often requires the manipulation of that information, such as remembering to bring home the textbook needed for the assignment. Theories propose a phonological loop aids in storage of verbal information, while a visuospatial sketchpad aids in storage of visual-spatial or visual-perceptual information [80]. Here, we discuss working memory measures based on the type of information needed to be retained: verbal or nonverbal (i.e., spatial and perceptual).

Verbal working memory was the most common EF domain assessed across FXS studies, with the majority of studies reporting impairments in males with FXS from school-age to adulthood compared to both chronological age (CA) and mental age (MA) control groups [15,17,22,31–41]. Among males with FXS, the type of stimuli (i.e., words, digits) was less influential on performance than task difficulty, or cognitive load, and this pattern emerged regardless of age. Cognitive load can depend on the amount of information needing retention or the degree of manipulation of information needed. For example, verbal working memory was relatively intact when cognitive load was low, like remembering two to five words in forward order. In contrast, verbal working memory was significantly impaired in males with FXS when cognitive load was high, like remembering the first word from two different five-word lists [21,41,42]. This finding is consistent with studies documenting males with FXS had a greater likelihood of hitting floor effects on higher load verbal working memory tasks [22,31,32]. Similar findings also arose in some studies comparing males with FXS to other syndromic participant groups (e.g., individuals with Down syndrome (DS)). For example, males with FXS performed similar to males with DS when working memory load was either very low (e.g., digit span forward; [17,41] or very high (e.g., story retelling; [17]). This suggests both groups had relatively intact abilities when working memory demands were low, whereas both groups were equally taxed compared to CA controls under high working memory demand conditions. Yet, under moderate verbal working memory conditions, males with FXS performed worse compared to males with DS, suggesting verbal working memory deteriorated more rapidly with task difficulty in FXS than in DS. However, other studies do not support this finding, and instead suggest task difficulty may have been less influential within syndromic groups [15,35,47,48], warranting future studies that systematically vary difficulty to clarify these inconsistencies.

With regards to development, two studies reported verbal working memory developed slower in young males with FXS compared to CA controls even after accounting for mental age [36,50]. However, in contrast, another study documented a narrowing gap in verbal working memory performance in male and female children with FXS compared to a normative sample [33]. These contradictory findings may be accounted, in part, by the presence of females in the latter sample. Alternatively, because the narrowing gap was documented in an older sample of children (up to age 16 years), it also is possible that verbal working memory performance in males with FXS developed more slowly during early and middle childhood, then improved dramatically during adolescence. Visual inspection of growth curves demonstrating a relatively flat maturation rate from 6–12 years, followed by a rapid increase in performance beginning around 12 years confirmed this assertion [33]. Still prior studies have documented impaired verbal working memory in adult males with FXS, suggesting that though performance may become more similar to CA controls, it is nonetheless impaired. Overall, previous studies of verbal working memory in males with FXS suggest performance is highly dependent on cognitive load with more severe impairments as the amount of information is increased or when manipulation is required and, to some extent, chronological age. However, verbal working memory deficits may not be specific to FXS relative to other syndromic disorders.

With regards to nonverbal working memory, both visual–spatial (i.e., remembering the location of an item) and visual–perceptual (i.e., remembering what item was previously shown) subdomains are impaired in males with FXS relative to both CA and MA control groups [16,21,22,31,32,42,48]. As with verbal working memory, performance depended on cognitive load, with relatively intact abilities when demands were low, like remembering two locations in an array, but significantly worse abilities compared to controls when demands were high, like remembering three locations on two different arrays [21,42,49,51]. Notably, one study reported a relative strength in visual–perceptual compared to visual–spatial working memory in males with FXS [48], suggesting the ability to remember what object was shown is stronger than the ability to remember where the object was located. Young males with FXS demonstrated slower development of nonverbal working memory compared to CA controls after accounting for mental age [36,50], suggesting deficient growth of skills. Still longitudinal studies of nonverbal working memory have not yet been completed in older children and adult males with FXS, suggesting performance may improve more rapidly in later develop as seen in verbal working memory. Most studies found that males with FXS performed similarly to other syndromic participant groups on measures of nonverbal working memory, [15,35,47,48] with some evidence of worse performance than syndromic control groups during moderate level tasks [17,41] as found in studies of verbal working memory. Taken together, this suggests verbal and nonverbal working memory deficits are present in males with FXS across the lifespan that worsen with increased cognitive load, but are unlikely specific to this syndromic population.

Few studies examined verbal and nonverbal working memory in females with FXS. Verbal working memory performance was found to be impaired relative to CA controls [33,38] but relatively intact compared to age- and IQ-matched [43] as well as environmental (i.e., unaffected family member) control groups [34,39,40,44]. Though neither stimulus type nor cognitive load emerged as relevant factors in performance. One study documented verbal working memory performance became more similar to a normative sample from school age to late adolescence in females with FXS; however, since males also were included in the sample, it is difficult to determine whether these developmental findings occurred in one or both sexes [33]. Visual–spatial working memory abilities in females with FXS also were impaired compared to CA controls, and unlike verbal working memory abilities, seemed to depend on cognitive load as found in males with FXS [49,51]. For example, females with FXS performed worse than CA controls when required to match the spatial position with that of two trials ago (i.e., two-back) but similarly to controls when required to match with one trial ago (i.e., one-back) [49]. With regards to visual-perceptual working memory, studies reported females with FXS showed impairments relative to age- and IQ matched [43] and environmental control groups [34]. However, intact visual-perceptual working memory also has been documented compared to an environmental control group [44]. With fewer studies completed in FXS females, inconsistent findings make interpretations regarding verbal and nonverbal working memory less clear. Still previous research suggests working memory performance may be more variable in females with FXS consistent with their wide spectrum of general cognitive and adaptive abilities.

When comparing verbal and nonverbal working memory, some [32,36,42] but not all [41,51] studies found verbal working memory to be a relative weakness compared to nonverbal working memory in males with FXS, whereas the opposite pattern emerged in females with FXS [34,43,51]. For example, Pierpont and colleagues documented females with FXS outperformed males with FXS on verbal working memory measures regardless of cognitive load [45]. The finding supports studies documenting stronger expressive language skills in females with FXS compared to males with FXS [81] as well as more consistent findings of nonverbal, as opposed to verbal, working memory impairments in females with FXS. Though future studies are warranted to confirm these observations, as these findings may have distinct treatment implications for males and females with FXS in terms of areas targeted and strategies used.

3.1.2. Inhibitory Control

Inhibitory control refers to the ability to suppress contextually-inappropriate responses, and it is critical for adapting behavior to changing and often uncertain environmental demands. Two distinct cognitive components comprise inhibitory control: prepotent response inhibition (i.e., the ability to withhold a previously reinforced behavior) and distractor interference (i.e., the ability to ignore irrelevant information) [82], each of which have been evaluated in FXS.

Studies consistently documented reduced ability to withhold prepotent behavioral responses in young male children with FXS compared to both MA and CA controls [14,16,52,53]. Fewer studies have been conducted with older children and adolescent males with FXS, and findings have been more inconsistent, demonstrating both intact [54] or impaired performance [55]. However, intact performance only was reported relative to an age- and IQ-matched DD control group during an fMRI task, suggesting cognitive and behavioral issues may have been less severe in this sample of males with FXS since they were able to complete an fMRI session [54]. Thus, it is more probable that prepotent inhibition deficits persists into adolescence in males with FXS, as supported by findings from longitudinal and cross-sectional studies [14,52,55]. Prepotent inhibition errors also occurred at higher rates among males with FXS relative to both males with William Syndrome (WS) and males with DS during school-age and adolescence [16,52]. This suggests withholding prepotent behavioral responses may be more severely affected in FXS relative to other syndromic groups. Though whether this reflects different developmental trajectories and/or persists into adulthood is unclear.

Impairments in distractor interference were documented from preschool-age to adulthood in individuals with males with FXS compared to CA-, MA-, and IQ-matched control groups [16,22,31, 35,52,57] (but see [55] for null findings). This suggests difficulties inhibiting behavioral responses to environmental distractors are present early in childhood and persist into adulthood in males with FXS regardless of measure used (e.g., Stroop, Flanker). Findings that distractor interference improved a rate similar to MA controls during early childhood suggest attenuated maturation of underlying brain processes emerge early in development, but whether similar findings are observed during later childhood to adulthood remains unclear. Comparisons with other syndromic populations suggested distractor interference was relatively similar in males with FXS and males with WS [52] but impaired compared to males with DS [16,35]. Though, the study of males with WS was completed in preschool-age participants, whereas the studies of males with DS included school-age to adult participants, suggesting inconsistencies may have emerged regarding differences in ages studied or actual differences in the developmental trajectories of distractor interference. Taken together, since males with FXS demonstrated impaired prepotent response inhibition and distractor interference regardless of measure used (e.g., antisaccade, go/no-go), this suggests failure to inhibit behavioral responses may be less dependent on task stimuli and complexity, thus making measurement selection in future studies less constrained.

Only one study of prepotent response inhibition has been completed in females with FXS, in which behavioral performance during an fMRI task was similar to that of female CA controls [26]. This suggests inhibiting prepotent behavioral responses may be relatively intact in females with FXS, though participant recruitment may have been biased towards less affected females due to using an fMRI protocol. Studies assessing distractor interference in females with FXS also were equivocal, with some findings of higher error rates during Stroop-like measures compared to environmental controls and IQ-matched controls [34,57,58] and others of similar error rates to CA controls and IQ-matched controls [59,60]. This suggests inhibiting responses towards distractors may be more variable in females with FXS, consistent with their wider range of functioning and general cognitive impairments. Of note, Tamm and colleagues' observed females with FXS to have a similar distractor interference error rate to CA controls, but they were slower to respond [60], suggesting some females with FXS may have adopted a strategy to sacrifice speed for sake of accuracy. This cognitive strategy to slow responses in order to improve poor inhibitory control was not observed in males with FXS, suggesting this compensatory strategy may only be present in less affected individuals, like females with FXS. Still, future studies are warranted to better understand potential compensatory strategies to improve inhibitory control in males and females with FXS.

Though several studies examined both prepotent response inhibition and distractor interference performance [16,52,55], no study to date has directly compared these abilities directly in individuals with FXS. Thus, whether prepotent response inhibition or distractor interference is relatively more impaired than the other or whether they both reflect a more general impaired process of behavioral inhibition is not yet clear. Though Woodcock and colleagues documented impaired prepotent response inhibition in school-aged/adolescent males with FXS relative to CA controls, but intact distractor interference [55], it is important to note that authors used different variables for each measure. For example, error rate was used to assess performance on the prepotent response inhibition measure and reaction time was used to assess performance on the distractor interference measure. Thus, comparisons between domains are less appropriate since they are quantifying difference aspects of inhibitory processes. Furthermore, to our knowledge, no study to date has compared performance between males and females with FXS. Hessl and colleagues [35] reported impaired performance on a distractor interference measure in males and females with FXS relative to CA, DS, and iDD controls; however, results were not provided for each sex. Thus, clarifying patterns of deficits between prepotent response inhibition and distractor interference as well as between sexes is needed and will be important for underlying the discrete brain mechanisms disrupted in FXS as well as informing potential treatment targets.

3.1.3. Cognitive Flexibility

Cognitive flexibility refers to the ability to adaptively change behavior based on contextual demands. This is most commonly assessed with dimensional sorting tasks in which participants are required to switch from sorting figures based on one intra-dimensional feature (e.g., spatial location on screen) or extra-dimensional feature (e.g., color or size) to another. Difficulty switching is most commonly quantified by calculating the total number of trials to reach a specified criterion (i.e., higher number indicates worse performance) or by number of categories achieved (i.e., lower number indicates worse performance). However, several studies also calculated error rates, with a few categorizing errors as either perseverative (i.e., continuing to choose previously correct responses despite feedback that it is no longer correct), regressive (i.e., returning to the previously correct response after establishing the new correct response), or distractor (i.e., choosing a never-reinforced or distractor response).

Across studies, males with FXS performed worse on cognitive flexibility measures compared to CA and MA controls [22,31,32,35,48,55,83]. Deficits were observed from school-age to adulthood, though error rates reduced with increasing MA in males with FXS [31]. Several studies also found that impaired performance was predominantly due to increased perseveration errors [34,48,83]. This suggests individuals with FXS have difficulty shifting away from previously rewarded responses and choosing new responses even after the previous response is no longer rewarded. This finding from cognitive performance-based measures is consistent with findings from clinical-ratings that report increased perseverate speech and behavior in individuals with FXS. Interestingly, Van der Molen and colleagues [46] reported that perseverative responses were more prominent when cognitive demands were low, whereas distractor errors were more prominent as cognitive demands increased. This provides additional evidence that males with FXS have a perseverative response style, which may reflect failure to disengage attention from a previously reinforced stimulus even when it is no longer rewarded. Yet, when too many distractors were present, impaired distractor interference (an aspect of inhibitory control) contributed to inflexibility, and thus attention was more readily diverted towards irrelevant stimuli. This importantly demonstrates certain EF measures may require more than one type of EF, and that individuals with FXS may be more disadvantaged during these certain measures since multiple EF domains are disrupted.

Of note, Garner and colleagues reported absence of cognitive flexibility deficits in school-age males with FXS relative to age- and IQ-matched controls during a modified version of the Wisconsin Cart Sorting Task (WCST-M) created for individuals with cognitive impairments [61]. This suggests cognitive flexibility may be intact relative to individuals of similar general cognitive impairments. It also suggests the adapted versions of measures may be more appropriate for FXS participants, and may reflect a better estimate of true cognitive flexibility skills as it requires less recruitment of other EF domains. Inconsistent findings emerged comparing cognitive flexibility performance in males with FXS relative to other syndromic groups, with some studies documenting more impaired performance in males with FXS compared to males with idiopathic ID and males with DS [15] and others documenting intact performance compared to these groups as well as males with Prader–Willi Syndrome (PWS) [35,48,55]. Though these findings suggest deficits in cognitive flexibility may not be specific to FXS, Van der Molen and colleagues conducted a discriminant analysis to classify error types during a dimensional sorting task to differentiate groups and found that perseverative responses differentiate males with FXS from DS and MA control groups [46]. Thus, though cognitive flexibility impairments may not be specific to males with FXS, their perseverative response style may be unique to this patient population, implicating a critical area to be further explored in future studies.

Few studies of cognitive flexibility have been conducted with females with FXS. Still, cognitive flexibility deficits were documented in females with FXS relative to environmental controls [23,34], and like their male counterparts, these deficits predominantly emerged due to increased rate of perseverative errors in one [34] but not the other study [23]. Thus, it is possible that a subgroup of females with FXS may show cognitive flexibility deficits similar to those in males with FXS, though more studies are needed.

3.1.4. Attention

Attention refers the ability to concentrate awareness with a specific behavioral or cognitive goal. For example, sustained attention is needed to maintain general focus, selective attention is required when concentrating on a specific information while ignoring other, and divided attention is used when focus is required towards multiple goals. Across studies, a variety of measures were used and quantified attention performance in different ways. The most common variables were: reaction time (time to respond to target), hit rates (correct target identification), miss rates (omission of target identification), and false alarm rates (incorrect target identification). Though most studies examined visual attention towards visual stimuli, a few also examined auditory attention.

In school-aged males with FXS, sustained attention was found to be impaired relative to CA and MA controls as they demonstrated both slower reaction times [16,50,52,62] and reduced hit rates [50,62]. One study reported reduced reaction time in the absence of differences in hit rate [16], suggesting some males with FXS may attempt to slow behavioral responses in order to perform more accurately. Still this cognitive strategy to improve accuracy was not observed in other studies, suggesting the majority of males with FXS may not slow responses at all or enough to be effective [50,62]. Sustained attention was impaired across both the auditory and visual domains, though males with FXS demonstrated relatively weaker performance (fewer correct hits and fewer correct rejections) during auditory compared to visual measures [62,63], consistent with findings of strong visual-perceptual skills and disturbances in auditory processing in males with FXS [84]. School-age males with FXS showed comparable performance to MA controls in one study that used a shorter version of a sustained attention measure [63], suggesting males with FXS may be able to maintain attention for a specific behavioral goal for a short duration (<5 min), but have greater difficulty when tasks require longer durations of sustained attention. Additionally, relatively intact sustained attention also was reported in males with FXS from 11–38 years, suggesting sustained attention may be less impaired in later childhood into adulthood [37]. Longitudinal studies in young males with FXS reported sustained attention developed at a slower rate compared to CA controls, but at a similar rate when adjusted for MA [50]. Yet, whether gaps in performance relative to CA controls narrows with age remains unclear. Previous studies reported sustained attention was similar among males with FXS and males with DS [16,52], but stronger in males with FXS compared to males with WS [52] in terms of both hit rates and reaction times. Taken together, previous studies have indicated consistent findings of poor sustained attention among school-age males with FXS though performance may be less impaired when tasks are shorter and with increasing age.

During selective attention tasks, previous studies have reported similar hit rates in males with FXS relative to MA controls from toddlerhood to adulthood [15,18,50,52,65]. However, a few studies documented worse performance compared to MA [16,19,65] and CA controls [19], especially under certain conditions. For example, males with FXS had lower hit rates and increased errors when more distractors were present and/or when distractors were more similar to target stimuli. A similar finding was documented during studies of distractor interference, suggesting both selective attention and inhibitory controls skills may be highly dependent on context of the environment. For example, certain EF abilities are relatively intact in a less distracting and ambiguous setting for males with FXS, but as the cognitive and neural processes supporting these abilities become more taxed, they are more likely to make errors. Of note, multiple studies reported that males with FXS demonstrated increased rates of perseverative responding during selective attention measures, or responding to the same correct target multiple times even after the end of the trial [52,65]. This finding is consistent with studies of cognitive flexibility and distractor interference that also reported a higher number of perseveration errors relative to all control groups. This suggests males with FXS demonstrated both impaired ability to shift behavior and attention away from previously correct responses even in the absence of reinforcement. Taken together, these findings indicate that a similar underlying cognitive and neural mechanism may drive perseverative responding and/or failures to disengage. Further evidence that this response style is specific to males with FXS comes from selective attention studies demonstrating increased perseverative errors compared to both DS [18,52] and WS groups [19,52] even when overall performance was similar across groups [18,19]. Yet, not all previous selective attention studies categorized error types, suggesting additional studies are needed to confirm these findings [16,19]. Though one study reported stronger selective attention performance in adult males with FXS compared to adult males with DS [83], it is unclear whether this reflects greater improvements in selective attention from adolescence to adulthood relative to individuals with DS or differences in measures used [18]. Overall, previous studies noted relatively intact selective attention in males with FXS across the lifespan that becomes impaired as task difficulty increases. Yet, the presence of perseverative errors suggests a more subtle impairment in this area that may be related to other areas of executive dysfunction as well as specific to FXS.

Only one study comparing divided attention performance to MA controls found fewer hits but similar response times and rates of false alarms [16]. Authors also documented greater distance moved when locating stimuli [16], which may account, in part, for the lower hit rate. More studies are warranted in assessing divided attention in FXS. Studies examining attention in females with FXS are needed to determine whether their profile of deficits is similar to those in males with FXS.

3.1.5. Planning

Planning is the ability to manage current and future goals and involves formulation, selection, and execution of specific sets to reach those goals. Few studies have examined planning in individuals with FXS, which may be, in part, due to prominent floor effects [32]. This suggests available planning measures may be too difficult for individuals with FXS to complete and/or planning abilities are significantly more impaired in FXS relative to other domains of EF. During a version of the Tower of Hanoi task, school-age males with FXS were observed to achieve fewer correct trials compared to MA controls, suggesting males with FXS had an impaired ability to plan or problem solve during increasingly complex scenarios [22,31]. Still, future studies are warranted in this area, especially those that include females with FXS, and studies utilizing tests with lower floors to measure performance of a broader range of individuals. Additionally, assessment of error types may be useful in future studies as has been found in measures of other EF domains.

3.1.6. Processing Speed

There is some controversy whether processing speed is an executive function or a cognitive process that supports all executive functions (for example, see [85]). For the purpose of this review and its relevancy in FXS, we include processing speed as a component of EF, using the more conservative definition of simple reaction time [85]. Processing speed performance varied across studies but was highly dependent on control group used. For example, males with FXS from school-age to adulthood demonstrated longer reaction times compared to CA controls [33,35], but comparable reaction times to MA, iDD, and DS control groups [22,31,35]. This is consistent with findings that reaction times increased at a slower rate relative to CA [33] but at a similar rate relative to MA controls in FXS males [31]. Longer reaction times also were observed for females with FXS relative to CA controls, but it is unclear whether performance differs from MA controls [33,35]. Together, this suggests processing speed is deficient compared to typical development, but impairments are likely not specific to FXS.

3.1.7. Behavioral and Psychological Correlates

Parent-report measures used to identify clinical relationships include those that exclusively examined EF dysfunction (e.g., Behavior Rating Inventory of Executive (BRIEF)), those that contained subdomains related to EF impairment (e.g., hyperactivity subscale of the Aberrant Behavior Checklist (ABC)), and those examining other aspects of behavior and function, like daily living skills. Most studies demonstrated significant relationships between behavioral performance and parent-report measures, though associations were not limited to the specific EF domain behaviorally assessed [14,35,56]. For example, impaired prepotent response inhibition was associated with parent-reports of inattention and hyperactivity and impaired distractor interference was associated with parent-reports of inattention, hyperactivity, stereotyped speech, and reduced adaptability to change [14,56]. Likewise, cognitive flexibility related to attention problems and adaptability to change [56] and processing speed related to hyperactivity, stereotyped speech, and reduced adaptability to change [56]. This extensive overlap in parent-reported clinical presentation and behavioral performance has several implications. First, it suggests that the EF domains captured in parent-report measures may not correspond well to those domains examined using behavioral measures. For example, difficulty flexibly shifting away from a previous behavioral response during a cognitive flexibility task may be reported by parents as difficulty shifting attention. This also could mean that both or either lack discriminant validity, reflecting measurement specificity issues. Second, the extensive overlap in symptoms may reflect less differentiated brain processes underlying separate EF domains in FXS typically found in younger typically developing children. It also may reflect multiple neural system dysfunctions. Lastly, it also could suggest that behavioral measures used in studies are likely confounded by co-occurring conditions, like ADHD, ASD, or anxiety. For instance, difficulties in sustained attention and cognitive flexibility may interfere with performance during a Stroop-like task. Similarly, studies examining EF performance in relation to ASD symptom severity provided consistent findings of more severe ASD symptomalogy relating to worse EF deficits, which was relatively independent of domain [31,35,36,63,64]; (but see [38] for null finding). Of note, Cornish and colleagues reported that poorer auditory, but not visual, attention predicted more severe ASD symptoms 12 months later, suggesting auditory attention may be an important risk marker for ASD in young males with FXS [64]. Future studies are needed to clarify these relationships by directly comparing FXS participants with and without co-occurring diagnoses. Additionally, since these studies examined correlations without controlling for MA or global IQ, it also will be important for future studies to determine the extent to which some of these relationships are driven by more general cognitive deficits than those related specifically to EF.

Additionally, adaptive behavior skills were related to inhibitory control as well as verbal and nonverbal working memory in a co-ed FXS sample [35] and to inhibitory control in a female only sample [59]. For example, working memory deficits could make performing self-care routines challenging without visual reminders of necessary steps to bathing or brushing teeth. Likewise, difficulty inhibiting prepotent behaviors may manifest as maladaptive coping behaviors or violations of social norms. Yet, neither study found a relationship between cognitive flexibility and adaptive behavior [35,59]. These findings begin to demonstrate that certain disrupted EF skills may have downstream effects on adaptive skills necessary for daily living, suggesting interventions targeting EF also may improve adaptive skills. Still more studies are needed in this area.

Importantly, EF abilities were predictive of later functioning as demonstrated by multiple studies. For example, Pierpont and colleagues found that higher working memory performance in school-aged males with FXS predicted greater rate of development of vocabulary and language skills two years later [45]. This highlights that language acquisition in FXS is dependent on the ability to hold verbal representations online as seen in typically developing youths, suggesting the importance of interventions improving working memory in early childhood [86]. Though study authors did not replicate this finding in females with FXS, this may be, in part, due to stronger baseline vocabulary and language skills in females [45]. Additionally, better performance on sustained attention tasks was found to be a strong predictor of lower teacher-rated hyperactivity and problem behaviors and greater prosocial behaviors one year later in school-age males with FXS [50,62]. Likewise, better performance on selective attention tasks in both auditory and visual domains predicted lower parent-reported ADHD symptoms and higher nonverbal IQ 12 months later in school-age males with FXS [64]. This suggests intact early attention skills may be a protective factor against certain comorbid conditions in males with FXS. Interestingly, higher parent-ratings of inattention during preschool years predicted greater improvements in prepotent response inhibition at school-age in males with FXS [14]. Though this finding seems counterintuitive, it suggests the malleability of EF in early development and that weaknesses observed at one timepoint may be related to strengths at another timepoint. It also could suggest that FXS participants with attention issues were identified early and provided treatment, and thus greater improvements were seen across domains. Taken together, findings of relationships between EF and multiple areas of functioning emphasize these skills are likely connected at the brain level and develop very early in childhood. It also highlights the importance of EF in multiple areas of functioning and that early treatment targeting EF may have important, long-lasting impacts.

3.1.8. Summary of EF Domain Findings

Few studies directly compared performance between EF domains in FXS, making it difficult to determine whether a consistent profile of EF strengths and weaknesses is evident. This is critical to better understanding whether certain EF deficits emerge as "syndrome specific" to FXS as suggested by some groups [52,87]. Among child males with FXS, verbal working memory was a relative weakness relative to processing speed [33]; however, among adult males with FXS, working memory emerged as a relative strength compared to cognitive flexibility and planning [32]. These studies suggest that working memory may shift from a relative weakness to a relative strength from childhood to adulthood, consistent with findings that the observed gap between verbal working memory performances in FXS and a normative sample narrows during adolescence [33]. A similar observation also was observed in females with FXS with working memory as a relative weakness compared to processing speed in childhood but becoming a relative strength compared to inhibitory control in adulthood [33,38]. This provides important evidence that EF profiles in FXS likely change over the course of development in line with changes in brain maturation and potential compensatory mechanisms, a finding also supported by differences in developmental patterns of specific EF domains in males and females with FXS. For example, Cornish et al. [52] found that for children with FXS, selective and sustained attention increased more with age compared to their inhibition skills, whereas inhibition skills increased more than selective and sustained attention for children with Down syndrome. Additionally, whether developmental and maturation changes also may differentiate males with FXS from females with FXS across certain EF domains also remains unclear and warrants future studies.

Likewise, no study to date has compared strengths/weaknesses profiles across syndromic disorder. Still it is important to note that individuals with FXS demonstrated distractor interference errors during distractor interference and cognitive flexibility measures as well as perseveration errors during distractor interference, selective attention, and cognitive flexibility measures [15,16,18,19,46,48,52,65]. This suggests these EF deficits may reflect underlying cognitive and neural abnormality that manifests itself behaviorally in different contexts of uncertainty/difficulty. Additionally, since perseverative responding was not observed as prominently in other syndromic developmental disorders, it also suggests that perseverative errors may be specific to FXS, reflecting an inability to shift behavior or attention away from a response that has been previously reinforced and/or reflect difficulty managing multiple possible responses [88]. Thus, rather than being separate deficits in 'inhibitory control', 'cognitive flexibility', and 'selective attention', perseverative responding may represent a deficit in a selective component process that impacts multiple EF domains and areas of functioning. As this was one of the most consistent findings across studies, aside from documentation of the presence of EF deficits beyond those expected given cognitive functioning, this warrants further attention in future studies.

3.2. Neurobiology of EF Deficits in FXS

Establishing the extent to which deficits on EF measures relate to biologically-based substrates (i.e., FMRP levels, brain anatomy) may provide critical insights into the pathophysiological mechanisms underpinning these cognitive deficits in FXS. Following a review of these findings from previous studies, we will briefly address studies using translational models to assess for EF deficits in *FMR1* Knockout (KO) rodents and then finally summarize potential pathways from gene to behavior.

3.2.1. FMRP

Several studies indicated lower peripheral FMRP expression was associated with more severe EF deficits in both males and females with FXS [39,40,49,51], suggesting a progressive FMRP deficit has a dose-dependent relationship with EF skills. For example, FMRP expression contributed up to 60% variance in EF performance in these studies, suggesting protein levels largely contributed to EF dysfunction, though other neurobiological and environmental factors influence these deficits as well [40]. On the other hand, one study reported cognitive flexibility performance no longer related to FMRP expression once FSIQ was controlled for in a co-ed sample [23]. Given the well-documented relationship between FMRP levels and intellectual ability [8–11], it suggests general cognitive functioning may have a stronger relationship with FMRP than specific EF skills, like cognitive flexibility. This finding also may be due, in part, to less variation in FMRP levels in males with FXS despite a wider spectrum of cognitive abilities. Thus, the link between FMRP levels and EF deficits may be more evident in females only. Several studies of females with FXS reported that reduced FMRP expression was associated with worse processing speed [39,40] and cognitive flexibility deficits related to lower X activation ratio [34], which directly affects the amount of protein produced. In contrast, relationships between CGG repeat count and EF measures largely were absent [34,47,83], consistent with other studies documenting CGG repeat count is a less reliable biological correlate than FMRP levels. Still, FMRP levels themselves remain limited as taken from blood samples, and thus are a less objective quantification of levels in the brain. Together, these findings provided an important link between causal pathology and observed phenotype of EF deficits in individuals with FXS.

3.2.2. Structural Brain Imagining Studies

Structural MRI studies consistently have reported increased volume of the caudate nucleus in individuals with FXS [89–99]. The caudate is critical for goal-directed actions, whereby individuals can successfully execute correct behavioral responses and appropriate subgoals [100]. Of note, abnormal caudate volume has been associated with perseverative responding in a number of psychiatric and neurological conditions (for examples, see [101–103]), suggesting its role in this behavior in FXS. Caudate volumetric increases appear to occur early in development and persist throughout the lifespan in FXS [96], consistent with findings documenting EF deficits, including perseverative responding, are found from early childhood to adulthood in individuals with FXS. Furthermore, caudate enlargement also has been found to be related to reduced FMRP expression, suggesting a possible progressive dose-dependent relationship between protein and brain volume [92,95]. In conjunction with similar findings of a relationship between reduced FMRP expression and more severe EF deficits, this indicates a potential pathway from reduced FMRP leading to increased caudate to EF dysfunction.

Individuals with FXS also have demonstrated reduced volume of the cerebellar vermis [92,96,98, 104,105], a region with known involvement in working memory, cognitive flexibility, and planning [106]. On the other hand, inconsistent findings have emerged regarding cerebral volume in FXS, though regional differences largely accounted for these inconsistencies [41,91,96,99]. For example, several groups have found increased volume of parietal lobes [92,98] but reduced volume of frontal lobes [94,98,99]. Frontal lobe involvement in EF has been widely-documented (for review see [107]), thus it is not surprising FXS patients have demonstrated reduced volume in this region. The alterations in the parietal cortex previously have been associated with failures to flexibly shift behavior in ASD [108,109], suggesting it may contribute to these deficits in individuals with FXS.

Additionally, previous findings have found altered white matter tract circuitry in individuals with FXS using diffusion tensor imaging [110]. The dorsal–prefrontal circuitry, which includes the caudate, has prominent roles in working memory, set-shifting, and processing speed (for review see [111]). Haas and colleagues [112] also reported greater relative fiber density in the left ventral frontostriatal pathway in young male FXS participants compared to both typically-developing and developmentally-delayed controls. As these findings were observed as early as one year, this suggests frontostriatal white matter tract abnormalities, like increased caudate volume, appear early in life and thus likely reflect alterations in pre- or perinatal brain development. Together, this suggests abnormal development of the frontal-striatum-parietal-cerebellum networks likely is involved in executive dysfunction in FXS [26,113].

In a previous study, regional differences in caudate volume related to distinct behavioral phenotypes. For example, the ventromedial caudate of the orbitofrontal cortex (OFC) was associated with social abnormalities, whereas the dorsolateral caudate of the dorsolateral prefrontal cortex (DLPFC) was associated with repetitive speech and behavior, aspects of impaired cognitive flexibility [89]. Though behavioral findings have implicated extensive overlap in neural structures responsible for EF, this finding suggests subtle regional difference in neuroanatomical abnormalities may selectively disrupt processes involved in certain areas of EF. It also may suggest that observed behavioral deficits may arise from numerous neural abnormalities. For example, the OFC also is involved in reward processing [114], which is important in dimensional sorting tasks of cognitive flexibility. Thus, high rates of perseverative errors found in FXS may be maintained by both disruptions to reward processing in OFC and propensity towards repetitive behavior in DLPFC. Still, more studies directly examining the relationships between structural brain abnormalities and clinical phenotypes are needed.

3.2.3. Functional Brain Imaging Studies

Only a few functional brain imaging studies during EF measures have been completed in FXS [26,49,54,60]. Though these studies differed in sample sex, control group, and EF domain, findings consistently demonstrated that FXS patients show a pattern of reduced activation in frontostriatal regions critical for EF. For example, during an n-back task of nonverbal working memory, females with FXS did not exhibit expected increases in frontal activation when cognitive load increased, and this reduced activation was related to worse working memory performance as well as reduced FMRP expression [49]. A similar finding was observed in females with FXS during an inhibitory control task, such that reduced FMRP expression related to both worse inhibitory control performance and greater reductions in prefrontal cortical, basal ganglion, and hippocampal activation. Together, these findings implicate the involvement of FMRP in the disruption of frontal-striatum-parietal-cerebellum circuitry, and in turn, executive dysfunction. As this appears to be a progressive deficit with reduced FMRP expression, the circuitry may be more disrupted and less able to appropriately modulate activity when cognitive load increases. This assertion is further supported by several previous behavioral studies discussed earlier, in which EF performance deteriorated as cognitive load increased (i.e., working memory, selective attention).

Interestingly, several fMRI studies documented evidence of compensatory brain mechanisms to support EF performance in males and females with FXS [26,54,60]. For example, Hoeft and colleagues [54] found that during a go/no-go task of prepotent response inhibition, males with FXS showed reduced right frontostriatal activation, but increased left ventrolateral prefrontal cortex activation compared to both IQ-matched DD and CA controls. These findings occurred in the absence of differences in behavioral performance, suggesting compensatory bilateral activation of prefrontal regions may have improved abilities to withhold prepotent responses in males with FXS. The authors also found that males with FXS who demonstrated greater activation in left ventrolateral prefrontal cortex also had higher levels of FMRP expression. This provides further evidence of the effect of progressive FMRP deficit on EF, suggesting the ability of the brain to develop and use compensatory mechanisms only may be afforded to males with FXS with some FMRP production. Consistent with these findings, previous fMRI studies also have documented that females with FXS with higher levels of FMRP demonstrated increased compensatory activation with recruitment of bilateral (versus left-lateralized) prefrontal regions during a distractor interference task [59]. Females with FXS demonstrated comparable errors to CA females, but reduced reaction times, suggesting they also may have adopted a behavioral strategy to sacrifice speed for sake of accuracy. Though differences in activation patterns for FXS and CA participants also may have been due, in part, to differences in behavioral performance. Thus, several studies provide evidence that some individuals with FXS show compensatory bilateral activation of regions known to support specific EF domains (i.e., prepotent response inhibition, distractor interference, nonverbal working memory) that tracks not only with better behavioral performance, but also with FMRP expression. Taken together, previous structural and functional imaging studies provide critical insight into disrupted brain regions and circuitry that likely contribute to EF deficits in FXS, though the specificity of these findings to FXS and selectivity to distinct brain alterations and corresponding EF deficits remains unclear.

3.2.4. Potential Mechanisms Underlying EF Deficits in FXS

Together, this section has highlighted the potential critical links between FMRP, brain function, and EF deficits in FXS. FMRP is required for normal dendritic pruning, and its absence can lead to immature synapses, aplastic, and non-specific connections, and presumed aberrant activity within affected structures [7,115,116]. The absence of FMRP is presumed to alter structural integrity of neurons and lead to downstream aberrant neural connectivity [117]. This provides one potential mechanism linking the *FMR1* gene to executive dysfunction through disrupted development of frontal-striatum-parietal-cerebellum circuitry. Still the precise processes underlying this developmental cascade remains poorly understood. FMRP is expressed throughout the cerebral cortex, cerebellum, hippocampus, and thalamus during embryonic development [118,119], suggesting its absence could have widespread neural effects as demonstrated in behavioral findings. Regional differences in volumetric findings may provide key insights into possible divergent trajectories in neurodevelopment. For example, volumetric enlargements may indicate reduced post-natal synaptic pruning, as suggested by findings of increased volume of caudate nucleus and of increased frontostriatal white matter tract density as early as 1–3 years in males with FXS [112]. In contrast, volumetric reductions may indicate pre-natal effects of deficient FMRP leading to disrupted post-natal maturation.

Additionally, imbalance of excitatory: inhibitory neural activity, including within PFC, repeatedly has been documented in slice, rodent, and human models, yet its relation to cognitive deficits has been sparsely investigated [120–122]. Enhanced gamma frequency activity in local circuits during rest and disrupted evoked gamma oscillations have emerged as a relatively conserved and stable biomarkers of neural hyper-excitability in translational models of FXS [123,124]. Gamma oscillations are generated by local synaptic interactions of excitatory and inhibitory neurons and controlled by rhythmic firing of inhibitory interneurons, including parvalbumin positive (PV+) fast-spiking interneurons [125]. Thus, observed neurophysiological alterations in FXS humans and *FMR1* KO mice may reflect failures of PV+ neurons to mediate gamma oscillations. Phasic variation in gamma power in association cortex is known to regulate crucial cognitive functions in mice and humans, and thus alterations in gamma band neurophysiology we observe may contribute to cognitive deficits, including executive dysfunction, in FXS [126,127]. For example, high background gamma in FXS may restrict the neural system's ability to send high gain signals to alert a change in behavior is needed. Studies of individuals with schizophrenia and mouse models of this disorder have documented the association between PV+ interneuron dysfunction and altered gamma oscillations and EF deficits, including those in working memory and cognitive flexibility [125,127–131]. This provides evidence of failure to phasically increase gamma oscillations needed to perform EF functions due to saturated background gamma activity. Still future translational studies are needed determine if or how well-documented EEG abnormalities of both local and long-distance connections are related to EF deficits in individuals with FXS as well as *FMR1* KO mouse models, as this is critical for drug discovery and novel treatment development.

4. Crucial Considerations

The >50 studies of EF in FXS reviewed were completed across a wide range of IQ levels (male: 30–82; female: 46–112) and ages (1–75 years), included both single and co-ed sex samples, and used many different EF measures. Though important findings have emerged with new insights in potential component processes underlying EF deficits in FXS, inconsistences across studies still limit our ability to interpret these findings. Multiple crucial methodological issues and confounding factors that could have affected EF performance in individuals with FXS are addressed, with suggestions for future work in this area.

4.1. FXS Sample Characteristics

The inclusion of a wide spectrum of ability levels and ages in EF studies of FXS helps capture the breadth of behavioral and cognitive presentations in this patient population, but also likely confounded findings by potentially washing out effects within specific subgroups of individuals. Factors such as medication usage and co-occurring diagnoses also likely confounded EF performance in individuals with FXS. For example, stimulants may have improved certain aspects of EF like attention, processing speed, and inhibitory control, whereas atypical antipsychotics and benzodiazepines may have punitively impacted aspects of EF like processing speed, attention, working memory, and inhibitory control as previously shown [132]. Yet, the majority of studies did not provide data on medication usage for FXS participants (or control groups), and among those studies that did, medication classes were not specified. This makes interpreting findings challenging, especially as some studies showed on-medication participants performed better than those off-medication [56], while others found the opposite trend [31]. Only three studies excluded for psychotropic medication, which is reasonable given known effects on EF [16,41,50]; however, medication-naïve studies are neither representative or feasible in the FXS population. Thus, it is critical for future studies to address potential confounds of medication usage on performance as well as specifically examine performance by medication class when possible within FXS participants.

Furthermore, few studies reported or accounted for co-occurring conditions in FXS participants. Because co-occurring conditions like ASD and ADHD likely arise from the FMRP deficit, it is difficult to determine the extent to which the pathophysiological processes underlying EF deficits overlap or differ from those underlying these neurodevelopment disorders, which have their own well-documented EF deficits (for review see [133,134]). It also is possible that affected component cognitive processes clinically manifest as both EF deficits and behavioral presentations of these co-occurring conditions or that the observed EF deficits may reflect cognitive traits of other genetic liabilities superimposed upon the FXS phenotype [135,136]. Though the mechanisms remain unknown, it is not surprising that numerous studies indicated EF deficits worsened with more severe ASD symptoms in individuals with FXS [31,35,36,63,64]. Two previous studies excluded participants with DSM-IV diagnoses and one specifically excluded for ASD [32,46,61], though the majority of previous studies did not take ASD of ADHD symptomatology into account (i.e., using clinical variable as covariate) when assessing for EF deficits. Thus, this latter approach as well as comparing EF performance in FXS participants with and without ASD (or ADHD) may be important considerations for future studies.

4.2. Control Group Selection

In addition to highly variable patient groups, previous studies also were highly variable in their choice of comparison control groups. Using a CA control group was not common ($n = 12$) among studies, as it simply compares groups that by definition operate at different developmental levels. Using a CA control group still may be appropriate in initial studies characterizing how EF in FXS differs from typical development, as done in the majority of structural and functional brain imaging studies in FXS [26,49,60]. In contrast, using an MA control group was the most common approach in previous studies ($n = 21$). Yet, using a MA control group is based on the assumption that acquisition of skills and performance on target variables should be similar between groups despite not matching on CA [137]. However, previous studies demonstrated that developmental profiles of EF often differ between individuals with FXS and MA controls [31], contradicting the assumption. This is especially problematic in a case when a 30-year old FXS participant with a mental age of five years is matched to a six-year old control participant with a similar mental age. Due to possible confounds of maturation and history effects, it would be difficult to determine, for example, whether the absence of a deficit was due to compensatory processes developed over time in the individual with FXS [26,49,60] or whether the presence of a deficit only appeared at certain chronological ages as implicated by some prior findings [31,50]. Another assumption put into question is whether overall MA is representative of current functioning, as van der Molen and colleagues showed that EF performance varied based on whether verbal or nonverbal MA was used in comparison [32]. This suggests previous studies using combined MA comparison group may have over- or under-estimated EF deficits in FXS. Taken together, CA and MA control groups each have their own pitfalls, many of which are difficult to avoid. Matching on both mental and chronological age is the ideal option; however, is not always feasible from a recruitment standpoint, and it is often unclear which types of comparison disabilities should be utilized (e.g., Down syndrome, iDD, etc.).

Additionally, similar issues also arose in studies using iDD or syndromic control groups. Studies widely varied on whether these groups were matched (if at all) on chronological age, mental age, or IQ. The critical problem here is that it assumes differences on target variables are genuine difference between syndromic groups rather than confounds such as differences in developmental trajectory or brain maturation rate. Though using iDD or syndromic control groups carry many advantages, including determining the specificity of findings, additional caution should be made when interpreting findings in future studies that do not control for additional aspects. Overall, choosing appropriate control groups is extremely challenging in FXS studies as usually the most ideal group often is not feasible. Careful consideration and acknowledgment of potential confounds related to control groups is recommended in future studies as these decisions may limit implications of findings.

4.3. Measures

The majority of measures chosen to assess EF in FXS were either part of a common neurophysiological battery (e.g., Woodcock Johnson III, Wechsler tests) or adapted from these more traditional measures (e.g., day/night task adaptation of Stroop). Many studies also chose commonly used measures that do not have standardized versions (e.g., n-back, antisaccade, go/no-go), and less frequently, studies created new experimental measures [62,64]. Independent of domain or type of measure used, individuals with FXS had reduced completion rates compared to CA and MA controls, especially among males with FXS. However, a smaller percentage of FXS participants completed verbal and nonverbal working memory measures compared to measures of inhibitory control [14,16,22,26,31,34,35,52,53,56,57,59], attention [15,18,19,50,52,56,63], and processing speed [22,31,35,56]. Working memory completion rates were highly dependent on task complexity, with higher load tasks with lower completions rates than lower load tasks [22,31,32], consistent with findings demonstrating worse performance as complexity increased. A similar finding was observed in studies using selective attention measures [56]. Cognitive flexibility and planning measures had among the lowest completion rates (e.g., <30%; [22,23,31,46,48,55]). Of note, individuals who did not complete measures were more likely to have lower MA, higher autistic symptomology, and not taking psychotropic medication [14,22,23,35,63]. This suggests previous studies only captured EF performance in a smaller subset, and perhaps less representative sample, of individuals with FXS. Thus, given the high rate of failures across EF measures, this suggests development of more appropriate measures is needed for individuals with FXS.

Traditional neuropsychological measures have many benefits, including verified psychometric properties, published normative data, and standardization of administration and scoring procedures. Still the vast majority of these measures are not be suitable for the FXS population due to heightened floor effects and task complexity as well as lack of normative data for developmental delay populations [138,139]. In fact, completion rates were lowest among standardized measures compared to adapted or experimental measures. Additionally, many traditional neuropsychological measures assessed multiple domains of executive function simultaneously, making it difficult to determine component processes impaired. For example, the Wisconsin Card Sorting Task (and other dimensional sorting tasks) is primarily a measure of cognitive flexibility; however, working memory is necessary to keep the current rule online, inhibition of distractors and prepotent responding is required to limit perseverative and non-perseverative responses, and both selective and sustained attention are important in selecting responses and staying on task, respectively. Thus, poor performance on the task may be less specific to cognitive flexibility deficits in FXS, but may be due to, in part, other aspects of executive dysfunction. Lastly, many traditional measures often heavily depend on verbal instructions and sometimes even verbal responses, which increases potential confounding factors in this disorder with prominent expressive and receptive language deficits. Overall, the psychometric advantages of traditional neuropsychological measures may not outweigh the challenges associated with using these measures in FXS participants. Thus, careful consideration should be made prior to choosing standardized measures, especially in term of floor effects and specificity of findings.

Several studies examined feasibility of using standardized computer or application-based testing batteries of executive function abilities, including NIH Toolbox, Cogmed Working Memory, and Kiddie Test of Attentional Performance (KiTAP) [35,56,140]. These electronic batteries had distinct advantages over more traditional neuropsychological batteries, including increased participant familiarity with computer/tablet interface, flexibility in testing positions, button or touch response, limited verbal demands, and increased motivation based on 'game' environment. Though developmental extensions for two of the NIH Toolbox measures (e.g., dimensional change card test and flanker) were available for FXS participants and allowed for higher completion rates and lower basals, psychometric properties of the development extensions have not yet been established. This suggests the potential benefit of modifying traditional measures to be more developmentally appropriate for this population, but more studies are needed to confirm the validity of these measures. Though many of these measures did not have such modifications, there still is promise in using batteries based on initial feasibility studies based on findings of high test-retest reliability, convergent and divergent validity, and acceptability among participants [35,56,140], especially when developmental modifications are available.

Less often, groups adapted standardized or non-standardized versions of EF measures to be more child-friendly and appropriate for use in FXS participants or even rarer, developed their own experimental measures [20,50,53,62,64,65,83]. Among these previous studies, common modifications were implemented, including incorporating a simple story to increase engagement, using visually appealing stimuli, and rewarding correct responses. Given the high levels of completion and reduced floor effects among these studies, it suggests minor modifications may allow for the assessment of EF in a wider range of FXS participants, consistent with findings from the modified versions of NIH Toolbox measures. Though, test-retest reliability and other psychometric properties largely have not yet been established for these performance-based measures, which is a critical aspect to measure selection, especially within clinical trials. Taken together, modified measures appear to be the most appropriate when examining EF in FXS, though the psychometric properties and sensitivity to change over time longitudinally or in response to intervention warrant future study. Importantly, this suggests that the majority of measures used in previous studies are not ideal for individuals with FXS and additional work is needed to develop more appropriate measures of EF in this and similar patient populations.

4.4. Scoring and Analysis Method

One reason the majority of measures were not appropriate for individuals with FXS was because adequate scores often could not be obtained from FXS participants due to floor effects. Though this may be an effect of measure, it also suggests alternative scoring or analysis methods are needed when assessing EF in this developmental disorder. Floor effects are well-documented in this population [138], which become particularly problematic when trying to track change over time or in response to treatment, as a large range of low raw scores equate to the lowest standard score. Indeed, many studies instead used raw scores, which has been recommended from several groups when assessing cognitive functioning in this population [141]. Additionally, a promising method was developed to calculate deviation scores based on raw scores in order to better capture cognitive performance in FXS by lowering the floor of the Stanford-Binet fifth edition (SB-5) and Wechsler Intelligence Scale for Children-Third Edition (WISC-III) [138,139]. For example, by expanding the floor of the SB-5, individuals with FXS were noted to have significantly lower verbal working memory performance than was indicated by standard scores, and it became a clear weakness as evidenced by being approximately six standard deviations below the mean, relative to most other domains [138]. Yet, this deviation score approach only has been applied to SB-5 and WISC-III thus far, warranting exploration of its use for other neuropsychological measures of EF, especially as this may be an alternative solution to developing new or modified measures.

In addition, choice of dependent variable is an important consideration when assessing EF in FXS. The majority of studies used one or two variables (e.g., reaction times and correct response rates) to quantify performance, which could greatly simplify the complex processes assessed during EF measures. As a result, relevant factors could be overlooked and thus impede our understanding of mechanisms underlying impairments. For example, categorizing error types proved useful in multiple studies as it showed individuals with FXS had a propensity towards repetitive, or perseverative responding, during distractor interference, cognitive flexibility, and selective attention measures that was not observed in other syndromic disorders like DS and WS [18,19,46,52]. Thus, consideration of additional relevant variables that may better reflect component processes may be important for future EF studies in FXS. Additionally, choice of dependent variable (and measure) is critical to consider in the context of clinical trials in its ability to detect real change when it occurs amid other factors leading to variability/improvement. Relatedly, it also is important to consider whether certain measures and dependent variables reflect more meaningful clinically significant changes as opposed to statistically significant changes. However, our review of the literature demonstrates we remain limited in this regard and future studies helping to determine these answers are critically needed.

4.5. Lack of Analogous Paradigms in Translational Studies of Rodent Models of FXS

In order to better understand the mechanistic bridge from gene to behavior in FXS, it is important to examine EF performance in *FMR1* KO mouse models of FXS during translational behavioral measures. The development of clinically- and biologically-relevant behavioral assays comparable to those used in humans is an area that warrants further consideration. Though tests of anxiety, seizure susceptibility, sensorimotor gating, sociability, and sensory hypersensitivity have been readily implemented in FXS rodent models, few have explored executive dysfunction [126]. Moreover, no study to date has examined EF performance in both species using parallel measures. Previous studies have documented EF deficits in *FMR1* KO mice, though findings are more variable than those found in human studies. Additionally, measures used in rodent students often are not analogous to those used in human studies. For example, several studies have reported mild to absent working memory and cognitive flexibility deficits in FXS rodent models [142–144] despite the consistency of these findings in FXS humans. One possible explanation for inconsistent findings across species is the use of the Morris water maze to assess nonverbal (spatial) working memory in *FMR1* KO mice [142,145], for which there is no human equivalent. In addition, absence of findings in mice may have been due, in part, to relatively intact nonverbal working memory in FXS humans compared to verbal working memory, especially when cognitive load is low, suggesting Morris water maze may be too easy for the mice to complete.

On the other hand, during a five-choice serial reaction time task [146], *FMR1* KO mice show quicker response times, more false alarms, and more perseverative responding compared to WT mice during reversal trials [147,148]. Yet, perseverative responding normalized with successive training in *FMR1* KO mice, suggesting behavioral intervention similarly may help reduce perseverative behavior in individuals with FXS. Krueger and colleagues [149] also reported increased cognitive flexibility errors during a spatial discrimination reversal learning task in *FMR1* KO mice (though error type was not specified). Thus, difficulty extinguishing a previously rewarded stimulus is consistent with findings from FXS participants during measures of distractor interference, cognitive flexibility, and selective attention, and suggests perseverative responding may be relatively conserved across species.

Additionally, Krueger and colleagues [149] reported that cognitive flexibility errors were associated with decreased synaptic markers in orbitofrontal and medial prefrontal cortices, and that reductions in postsynaptic proteins preceded expression of cognitive flexibility impairments. These findings suggest a potential causal link between loss of FMRP expression in the PFC and cognitive dysfunction that has only previously been implicated in human imaging studies. Given evidence of the selectivity of specific cognitive flexibility errors in FXS as well as potential insights into gene to behavior pathways, development of parallel measures in individuals with FXS and mouse models may be particularly important. For example, analogous reversal learning paradigms have been used in individuals with ASD and BTBR mouse models of ASD and have identified similar cognitive flexibility deficits in both species [108,149,150]. Currently, our group is piloting the same measure in individuals with FXS, with preliminary findings suggesting increased perseverative errors compared to CA controls (unpublished). This ongoing work in collaboration with groups studying *FMR1* KO mice may be critical to the development of translational biomarkers in FXS. Though examining EF deficits in *FMR1* KO mice is still in its infancy, previous studies offer promising findings that suggest the importance of this area in future research.

5. Conclusions and Future Directions

Previous studies have begun to characterize EF deficits in FXS as well as provide evidence linking FMRP expression to frontal-striatum-parietal-cerebellum circuitry and, ultimately, executive dysfunction. Our review also highlighted that perseverative responding emerged as one the most consistent findings across measures and most specific to FXS. Yet, critical gaps in our mechanistic understanding of these deficits remain. A focused effort on developing translational measures that can be used across species and methods (i.e., behavioral, EEG), is selective towards one EF domain or cognitive process, and minimizes undue burden on the FXS participant is critical to bridging this gap. By advancing our understanding of the pathophysiological processes underlying EF deficits in FXS, as a field we will be better-suited to target EF in treatment studies. Outcome measurement of EF in FXS clinical trials remains in its infancy. Though some traditional neuropsychological measures have demonstrated high test-retest or reproducibility in this patient population (for complete list of measures see [141]), individuals with FXS showed improvement only one measure (i.e., RBANS List Learning) following open label treatment trial with lithium [151]. Still authors even reported that the cognitive battery was too difficult for most FXS participants to complete, suggesting the measures chosen were not appropriate for FXS participants, especially in the context of clinical trials. A recent review of outcome measures suggested potential outcomes measures in FXS treatment studies [152], including specific KiTAP and Woodcock–Johnson subtests. However, these measures only have been used in FXS during initial feasibility studies [56], suggesting studies are needed to confirm their appropriateness in this patient population, especially during treatment trials. Thus, as a field it is important to critically examine the state of literature, and focus future work on identifying (or developing) measures of EF with high test-retest reliability and construct validity that also link to hypothesized neurobiological mechanisms as this would allow for greater potential of success in future clinical trials.

Author Contributions: L.M.S. conceived the idea for the paper, found and interpreted all articles, and wrote the paper. R.C.S., D.R.H., and C.E. assisted with interpretation and paper editing.

Funding: This review was supported by the NIMH/NICHD U54 Fragile X Center.

Acknowledgments: I would like to thank J.N.B. for recommendations on the initial draft.

Conflicts of Interest: R.C.S. receives funding from Fulcrum Therapeutics. D.H receives compensation for consulting to Zynerba, Autifony, and Ovid pharmaceutical companies regarding Fragile X clinical trials. C.E. has received current or past funding from Confluence Pharmaceuticals, Novartis, F. Hoffmann-La Roche Ltd., Seaside Therapeutics, Riovant Sciences, Inc., Fulcrum Therapeutics, Neuren Pharmaceuticals Ltd., Alcobra Pharmaceuticals, Neurotrope, Zynerba Pharmaceuticals, Inc., and Ovid Therapeutics Inc. to consult on trial design or development strategies and/or conduct clinical trials in FXS or other neurodevelopmental disorders. C.E. is additionally the inventor or co-inventor on several patents held by Cincinnati Children's Hospital Medical Center or Indiana University School of Medicine describing methods of treatment in FXS or other neurodevelopmental disorders. Funding source had no role in the conceptualization of the review, the writing of the manuscript, or in the decision to publish the results.

References

1. Kao, D.I.; Aldridge, G.M.; Weiler, I.J.; Greenough, W.T. Altered mRNA transport, docking, and protein translation in neurons lacking fragile X mental retardation protein. *Proc. Natl. Acad. Sci. USA* **2010**, *107*, 15601–15606. [CrossRef] [PubMed]
2. Fu, Y.H.; Kuhl, D.P.; Pizzuti, A.; Pieretti, M.; Sutcliffe, J.S.; Richards, S.; Verkerk, A.J.; Holden, J.J.; Fenwick, R.G.; Warren, S.T. Variation of the CGG repeat at the fragile X site results in genetic instability: Resolution of the Sherman paradox. *Cell* **1991**, *67*, 1047–1058. [CrossRef]
3. Pieretti, M.; Zhang, F.P.; Fu, Y.H.; Warren, S.T.; Oostra, B.A.; Caskey, C.T.; Nelson, D.L. Absence of expression of the FMR-1 gene in fragile X syndrome. *Cell* **1991**, *66*, 817–822. [CrossRef]
4. Park, S.; Park, J.M.; Kim, S.; Kim, J.A.; Shepherd, J.D.; Smith-Hicks, C.L.; Chowdhury, S.; Kaufmann, W.; Kuhl, D.; Ryazanov, A.G.; et al. Elongation factor 2 and fragile X mental retardation protein control the dynamic translation of Arc/Arg3.1 essential for mGluR-LTD. *Neuron* **2008**, *59*, 70–83. [CrossRef] [PubMed]
5. McNaughton, C.H.; Moon, J.; Strawderman, M.S.; Maclean, K.N.; Evans, J.; Strupp, B.J. Evidence for social anxiety and impaired social cognition in a mouse model of fragile X syndrome. *Behav. Neurosci.* **2008**, *122*, 293–300. [CrossRef] [PubMed]
6. Antar, L.N.; Bassell, G.J. Sunrise at the synapse: The FMRP mRNP shaping the synaptic interface. *Neuron* **2003**, *37*, 555–558. [CrossRef]
7. Bassell, G.J.; Warren, S.T. Fragile X syndrome: Loss of local mRNA regulation alters synaptic development and function. *Neuron* **2008**, *60*, 201–214. [CrossRef]
8. Heine-Suñer, D.; Torres-Juan, L.; Morlà, M.; Busquets, X.; Barceló, F.; Picó, G.; Bonilla, L.; Govea, N.; Bernués, M.; Rosell, J. Fragile-X syndrome and skewed X-chromosome inactivation within a family: A female member with complete inactivation of the functional X chromosome. *Am. J. Med. Genet. A* **2003**, *122*, 108–114. [CrossRef]
9. Backes, M.; Genç, B.; Schreck, J.; Doerfler, W.; Lehmkuhl, G.; von Gontard, A. Cognitive and behavioral profile of fragile X boys: Correlations to molecular data. *Am. J. Med. Genet.* **2000**, *95*, 150–156. [CrossRef]
10. Kaufmann, W.E.; Abrams, M.T.; Chen, W.; Reiss, A.L. Genotype, molecular phenotype, and cognitive phenotype: Correlations in fragile X syndrome. *Am. J. Med. Genet.* **1999**, *83*, 286–295. [CrossRef]
11. Tassone, F.; Hagerman, R.J.; Iklé, D.N.; Dyer, P.N.; Lampe, M.; Willemsen, R.; Oostra, B.A.; Taylor, A.K. FMRP expression as a potential prognostic indicator in fragile X syndrome. *Am. J. Med. Genet.* **1999**, *84*, 250–261. [CrossRef]
12. Dykens, E.M. Psychopathology in children with intellectual disability. *J. Child Psychol. Psychiatry* **2000**, *41*, 407–417. [CrossRef]
13. Friedman, N.P.; Miyake, A. Unity and diversity of executive functions: Individual differences as a window on cognitive structure. *Cortex* **2017**, *86*, 186–204. [CrossRef]
14. Tonnsen, B.L.; Grefer, M.L.; Hatton, D.D.; Roberts, J.E. Developmental trajectories of attentional control in preschool males with fragile X syndrome. *Res. Dev. Disabil.* **2015**, *36C*, 62–71. [CrossRef] [PubMed]
15. Cornish, K.; Munir, F.; Wilding, J. A neuropsychological and behavioural profile of attention deficits in fragile X syndrome. *Rev. Neurol.* **2001**, *33*, S24–S29. [PubMed]
16. Munir, F.; Cornish, K.M.; Wilding, J. A neuropsychological profile of attention deficits in young males with fragile X syndrome. *Neuropsychologia* **2000**, *38*, 1261–1270. [CrossRef]
17. Munir, F.; Cornish, K.M.; Wilding, J. Nature of the working memory deficit in fragile-X syndrome. *Brain Cogn.* **2000**, *44*, 387–401. [CrossRef] [PubMed]

18. Wilding, J.; Cornish, K.; Munir, F. Further delineation of the executive deficit in males with fragile-X syndrome. *Neuropsychologia* **2002**, *40*, 1343–1349. [CrossRef]

19. Scerif, G.; Cornish, K.; Wilding, J.; Driver, J.; Karmiloff-Smith, A. Visual search in typically developing toddlers and toddlers with Fragile X or Williams syndrome. *Dev. Sci.* **2004**, *7*, 116–130. [CrossRef]

20. Baker, P.M.; Thompson, J.L.; Sweeney, J.A.; Ragozzino, M.E. Differential effects of 5-HT(2A) and 5-HT(2C) receptor blockade on strategy-switching. *Behav. Brain Res.* **2011**, *219*, 123–131. [CrossRef]

21. Lanfranchi, S.; Cornoldi, C.; Drigo, S.; Vianello, R. Working memory in individuals with fragile X syndrome. *Child Neuropsychol.* **2009**, *15*, 105–119. [CrossRef]

22. Hooper, S.R.; Hatton, D.; Sideris, J.; Sullivan, K.; Hammer, J.; Schaaf, J.; Mirrett, P.; Ornstein, P.A.; Bailey, D.P. Executive functions in young males with fragile X syndrome in comparison to mental age-matched controls: Baseline findings from a longitudinal study. *Neuropsychology* **2008**, *22*, 36–47. [CrossRef] [PubMed]

23. Loesch, D.Z.; Bui, Q.M.; Grigsby, J.; Butler, E.; Epstein, J.; Huggins, R.M.; Taylor, A.K.; Hagerman, R.J. Effect of the fragile X status categories and the fragile X mental retardation protein levels on executive functioning in males and females with fragile X. *Neuropsychology* **2003**, *17*, 646–657. [CrossRef] [PubMed]

24. Lachiewicz, A.M. Abnormal behaviors of young girls with fragile X syndrome. *Am. J. Med. Genet.* **1992**, *43*, 72–77. [CrossRef] [PubMed]

25. Sullivan, K.; Hatton, D.; Hammer, J.; Sideris, J.; Hooper, S.; Ornstein, P.; Bailey, D. ADHD symptoms in children with FXS. *Am. J. Med. Genet. A* **2006**, *140*, 2275–2288. [CrossRef] [PubMed]

26. Menon, V.; Leroux, J.; White, C.D.; Reiss, A.L. Frontostriatal deficits in fragile X syndrome: Relation to FMR1 gene expression. *Proc. Natl. Acad. Sci. USA* **2004**, *101*, 3615–3620. [CrossRef] [PubMed]

27. Lewis, P.; Abbeduto, L.; Murphy, M.; Richmond, E.; Giles, N.; Bruno, L.; Schroeder, S.; Anderson, J.; Orsmond, G. Psychological well-being of mothers of youth with fragile X syndrome: Syndrome specificity and within-syndrome variability. *J. Intellect. Disabil. Res.* **2006**, *50*, 894–904. [CrossRef]

28. Bishop, S.L.; Richler, J.; Cain, A.C.; Lord, C. Predictors of perceived negative impact in mothers of children with autism spectrum disorder. *Am. J. Ment. Retard.* **2007**, *112*, 450–461. [CrossRef]

29. Swanson, H.; Alloway, T. *Working Memory, Learning, and Academic Achievement*, 1st ed.; American Psychological Association: Washington, DC, USA, 2012; pp. 327–366.

30. Diamond, A. Executive functions. *Annu. Rev. Psychol.* **2013**, *64*, 135–168. [CrossRef]

31. Hooper, S.R.; Hatton, D.; Sideris, J.; Sullivan, K.; Ornstein, P.A.; Bailey, D.B. Developmental trajectories of executive functions in young males with fragile X syndrome. *Res. Dev. Disabil.* **2018**, *81*, 73–88. [CrossRef]

32. Van der Molen, M.J.; Huizinga, M.; Huizenga, H.M.; Ridderinkhof, K.R.; Van der Molen, M.W.; Hamel, B.J.; Curfs, L.M.; Ramakers, G.J. Profiling Fragile X Syndrome in males: Strengths and weaknesses in cognitive abilities. *Res. Dev. Disabil.* **2010**, *31*, 426–439. [CrossRef]

33. Quintin, E.M.; Jo, B.; Hall, S.S.; Bruno, J.L.; Chromik, L.C.; Raman, M.M.; Lightbody, A.A.; Martin, A.; Reiss, A.L. The cognitive developmental profile associated with fragile X syndrome: A longitudinal investigation of cognitive strengths and weaknesses through childhood and adolescence. *Dev. Psychopathol.* **2016**, *28*, 1457–1469. [CrossRef]

34. Bennetto, L.; Pennington, B.F.; Porter, D.; Taylor, A.K.; Hagerman, R.J. Profile of cognitive functioning in women with the fragile X mutation. *Neuropsychology* **2001**, *15*, 290–299. [CrossRef] [PubMed]

35. Hessl, D.; Sansone, S.M.; Berry-Kravis, E.; Riley, K.; Widaman, K.F.; Abbeduto, L.; Schneider, A.; Coleman, J.; Oaklander, D.; Rhodes, K.C.; et al. The NIH Toolbox Cognitive Battery for intellectual disabilities: Three preliminary studies and future directions. *J. Neurodev. Disord.* **2016**, *8*, 35. [CrossRef] [PubMed]

36. Scherr, J.F.; Hahn, L.J.; Hooper, S.R.; Hatton, D.; Roberts, J.E. HPA axis function predicts development of working memory in boys with FXS. *Brain Cogn.* **2016**, *102*, 80–90. [CrossRef] [PubMed]

37. Johnson-Glenberg, M.C. Fragile X syndrome: Neural network models of sequencing and memory. *Cogn. Syst. Res.* **2008**, *9*, 274–292. [CrossRef] [PubMed]

38. Miezejeski, C.M.; Jenkins, E.C.; Hill, A.L.; Wisniewski, K.; French, J.H.; Brown, W.T. A profile of cognitive deficit in females from fragile X families. *Neuropsychologia* **1986**, *24*, 405–409. [CrossRef]

39. Loesch, D.Z.; Huggins, R.M.; Bui, Q.M.; Epstein, J.L.; Taylor, A.K.; Hagerman, R.J. Effect of the deficits of fragile X mental retardation protein on cognitive status of fragile X males and females assessed by robust pedigree analysis. *J. Dev. Behav. Pediatr.* **2002**, *23*, 416–423. [CrossRef]

40. Loesch, D.Z.; Huggins, R.M.; Hagerman, R.J. Phenotypic variation and FMRP levels in fragile X. *Ment. Retard. Dev. Disabil. Res. Rev.* **2004**, *10*, 31–41. [CrossRef]

41. Schapiro, M.B.; Murphy, D.G.; Hagerman, R.J.; Azari, N.P.; Alexander, G.E.; Miezejeski, C.M.; Hinton, V.J.; Horwitz, B.; Haxby, J.V.; Kumar, A. Adult fragile X syndrome: Neuropsychology, brain anatomy, and metabolism. *Am. J. Med. Genet.* **1995**, *60*, 480–493. [CrossRef]

42. Baker, S.; Hooper, S.; Skinner, M.; Hatton, D.; Schaaf, J.; Ornstein, P.; Bailey, D. Working memory subsystems and task complexity in young boys with Fragile X syndrome. *J. Intellect. Disabil. Res.* **2011**, *55*, 19–29. [CrossRef] [PubMed]

43. Mazzocco, M.M. Math learning disability and math LD subtypes: Evidence from studies of Turner syndrome, fragile X syndrome, and neurofibromatosis type 1. *J. Learn. Disabil.* **2001**, *34*, 520–533. [CrossRef] [PubMed]

44. Mazzocco, M.M.; Hagerman, R.J.; Cronister-Silverman, A.; Pennington, B.F. Specific frontal lobe deficits among women with the fragile X gene. *J. Am. Acad. Child Adolesc. Psychiatry* **1992**, *31*, 1141–1148. [CrossRef] [PubMed]

45. Pierpont, E.I.; Richmond, E.K.; Abbeduto, L.; Kover, S.T.; Brown, W.T. Contributions of phonological and verbal working memory to language development in adolescents with fragile X syndrome. *J. Neurodev. Disord.* **2011**, *3*, 335–347. [CrossRef] [PubMed]

46. Van der Molen, M.J.; Van der Molen, M.W.; Ridderinkhof, K.R.; Hamel, B.C.; Curfs, L.M.; Ramakers, G.J. Attentional set-shifting in fragile X syndrome. *Brain Cogn.* **2012**, *78*, 206–217. [CrossRef] [PubMed]

47. Cornish, K.M.; Munir, F.; Cross, G. Spatial cognition in males with Fragile-X syndrome: Evidence for a neuropsychological phenotype. *Cortex* **1999**, *35*, 263–271. [CrossRef]

48. Kogan, C.S.; Boutet, I.; Cornish, K.; Graham, G.E.; Berry-Kravis, E.; Drouin, A.; Milgram, N.W. A comparative neuropsychological test battery differentiates cognitive signatures of Fragile X and Down syndrome. *J. Intellect. Disabil. Res.* **2009**, *53*, 125–142. [CrossRef]

49. Kwon, H.; Menon, V.; Eliez, S.; Warsofsky, I.S.; White, C.D.; Dyer-Friedman, J.; Taylor, A.K.; Glover, G.H.; Reiss, A.L. Functional neuroanatomy of visuospatial working memory in fragile X syndrome: Relation to behavioral and molecular measures. *Am. J. Psychiatry* **2001**, *158*, 1040–1051. [CrossRef]

50. Cornish, K.; Cole, V.; Longhi, E.; Karmiloff-Smith, A.; Scerif, G. Mapping developmental trajectories of attention and working memory in fragile X syndrome: Developmental freeze or developmental change? *Dev. Psychopathol.* **2013**, *25*, 365–376. [CrossRef]

51. Freund, L.S.; Reiss, A.L. Cognitive profiles associated with the fra (X) syndrome in males and females. *Am. J. Med. Genet.* **1991**, *38*, 542–547. [CrossRef]

52. Cornish, K.; Scerif, G.; Karmiloff-Smith, A. Tracing syndrome-specific trajectories of attention across the lifespan. *Cortex* **2007**, *43*, 672–685. [CrossRef]

53. Scerif, G.; Karmiloff-Smith, A.; Campos, R.; Elsabbagh, M.; Driver, J.; Cornish, K. To look or not to look? Typical and atypical development of oculomotor control. *J. Cogn. Neurosci.* **2005**, *17*, 591–604. [CrossRef]

54. Hoeft, F.; Hernandez, A.; Parthasarathy, S.; Watson, C.L.; Hall, S.S.; Reiss, A.L. Fronto-striatal dysfunction and potential compensatory mechanisms in male adolescents with fragile X syndrome. *Hum. Brain Mapp.* **2007**, *28*, 543–554. [CrossRef]

55. Woodcock, K.A.; Oliver, C.; Humphreys, G.W. Task-switching deficits and repetitive behaviour in genetic neurodevelopmental disorders: Data from children with Prader-Willi syndrome chromosome 15 q11–q13 deletion and boys with Fragile X syndrome. *Cogn. Neuropsychol.* **2009**, *26*, 172–194. [CrossRef] [PubMed]

56. Knox, A.; Schneider, A.; Abucayan, F.; Hervey, C.; Tran, C.; Hessl, D.; Berry-Kravis, E. Feasibility, reliability, and clinical validity of the Test of Attentional Performance for Children (KiTAP) in Fragile X syndrome (FXS). *J. Neurodev. Disord.* **2012**, *4*, 2. [CrossRef] [PubMed]

57. Kirk, J.W.; Mazzocco, M.M.; Kover, S.T. Assessing executive dysfunction in girls with fragile X or Turner syndrome using the Contingency Naming Test (CNT). *Dev. Neuropsychol.* **2005**, *28*, 755–777. [CrossRef] [PubMed]

58. Lightbody, A.A.; Hall, S.S.; Reiss, A.L. Chronological age, but not FMRP levels, predicts neuropsychological performance in girls with fragile X syndrome. *Am. J. Med. Genet. B Neuropsychiatr. Genet.* **2006**, *141*, 468–472. [CrossRef]

59. Martin, A.; Quintin, E.M.; Hall, S.S.; Reiss, A.L. The Role of Executive Function in Independent Living Skills in Female Adolescents and Young Adults With Fragile X Syndrome. *Am. J. Intellect. Dev. Disabil.* **2016**, *121*, 448–460. [CrossRef] [PubMed]

60. Tamm, L.; Menon, V.; Johnston, C.K.; Hessl, D.R.; Reiss, A.L. fMRI study of cognitive interference processing in females with fragile X syndrome. *J. Cogn. Neurosci.* **2002**, *14*, 160–171. [CrossRef] [PubMed]

61. Garner, C.; Callias, M.; Turk, J. Executive function and theory of mind performance of boys with fragile-X syndrome. *J. Intellect. Disabil. Res.* **1999**, *43*, 466–474. [CrossRef]
62. Scerif, G.; Longhi, E.; Cole, V.; Karmiloff-Smith, A.; Cornish, K. Attention across modalities as a longitudinal predictor of early outcomes: The case of fragile X syndrome. *J. Child Psychol. Psychiatry* **2012**, *53*, 641–650. [CrossRef]
63. Sullivan, K.; Hatton, D.D.; Hammer, J.; Sideris, J.; Hooper, S.; Ornstein, P.A.; Bailey, D.B. Sustained attention and response inhibition in boys with fragile X syndrome: Measures of continuous performance. *Am. J. Med. Genet. B Neuropsychiatr. Genet.* **2007**, *144*, 517–532. [CrossRef]
64. Cornish, K.; Cole, V.; Longhi, E.; Karmiloff-Smith, A.; Scerif, G. Does attention constrain developmental trajectories in fragile x syndrome? A 3-year prospective longitudinal study. *Am. J. Intellect. Dev. Disabil.* **2012**, *117*, 103–120. [CrossRef] [PubMed]
65. Scerif, G.; Cornish, K.; Wilding, J.; Driver, J.; Karmiloff-Smith, A. Delineation of early attentional control difficulties in fragile X syndrome: Focus on neurocomputational changes. *Neuropsychologia* **2007**, *45*, 1889–1898. [CrossRef] [PubMed]
66. Cognition, C. *CANTAB Test Administration Guide*; Cambridge Cognition: Cambridge, UK, 2002.
67. Anderson, P.; Anderson, V.; Northam, E.; Taylor, H. Standardization of the contingency naming test (CNT) for school-ages children: A measure of reactive flexibility. *Clin. Neuropsychol. Rehabil.* **2000**, *8*, 247–273.
68. Keith, R. *The Auditory Continuous Performance Test*; Psychological Corporation: San Antonio, TX, USA, 1994.
69. Kaufman, A.; Kaufman, N.L. *Kaufman Assessment Battery for Children*, 2nd ed.; American Guidance Service: Circle Pines, MN, USA, 1983.
70. Goldsmith, H.; Rothbart, M. *The Laboratory Temperament Assessment Battery*; University of Wisconsin: Madison, WI, USA, 1996.
71. Wilson, B.; Cockburn, J.; Baddeley, A. *The Rivermead Behavioral Memory Test Manual*, Thames Valley: Suffolk, UK, 1985.
72. Wilson, B.; Ivani-Chalin, R.; Aldrich, F. *The Rivermead Behavioural Memory Test for Children Aged 5 to 10 Years*; Thames Valley Test Co.: Bury St. Edmunds, UK, 1991.
73. Thorndike, R.; Hagen, E.; Sattler, J. *Stanford-Binet Intelligence Scale*, 4th ed.; Riverside: Chicago, IL, USA, 1986.
74. Manly, T.; Roberston, I.; Anderson, V.; Nimmo-Smith, I. *The Test of Everyday Attention for Children: TEA-Ch*; Thames Valley Test Company: Bury St. Edmunds, UK, 1999.
75. Wilding, J.; Munir, F.; Cornish, K. The nature of attention differences between group of children differentiated by teacher ratings of attention and hyperactivity. *Br. J. Psychol.* **2001**, *92*, 357–371. [CrossRef] [PubMed]
76. Wechsler, D. *Wechsler Intelligence Scale for Children-III Revised*, 3rd ed.; The Psychological Corporation, Harcourt Brace: New York, NY, USA, 1992.
77. Woodcock, R.; Johnson, M.E.B. *Woodcock–Johnson Tests of Cognitive Ability-III*; DLM: Allen, TX, USA, 2001.
78. Wechsler, D. *The Wechsler Memory Scale–Revised*; Psychological Corporation: San Diego, CA, USA, 1987.
79. Pickering, S.; Gathercole, S. *The Working Memory Test Battery for Children*; The Psychological Corporation: London, UK, 2001.
80. Baddeley, A. Working memory: Looking back and looking forward. *Nat. Rev. Neurosci.* **2003**, *4*, 829–839. [CrossRef] [PubMed]
81. Finestack, L.H.; Abbeduto, L. Expressive language profiles of verbally expressive adolescents and young adults with Down syndrome or fragile X syndrome. *J. Speech Lang. Hear. Res.* **2010**, *53*, 1334–1348. [CrossRef]
82. Friedman, N.P.; Miyake, A. The relations among inhibition and interference control functions: A latent-variable analysis. *J. Exp. Psychol. Gen.* **2004**, *133*, 101–135. [CrossRef] [PubMed]
83. Cornish, K.M.; Munir, F.; Cross, G. Differential impact of the FMR-1 full mutation on memory and attention functioning: A neuropsychological perspective. *J. Cogn. Neurosci.* **2001**, *13*, 144–150. [CrossRef]
84. Huddleston, L.B.; Visootsak, J.; Sherman, S.L. Cognitive aspects of Fragile X syndrome. *Wiley Interdiscip. Rev. Cogn. Sci.* **2014**, *5*, 501–508. [CrossRef]
85. Cepeda, N.J.; Blackwell, K.A.; Munakata, Y. Speed isn't everything: Complex processing speed measures mask individual differences and developmental changes in executive control. *Dev. Sci.* **2013**, *16*, 269–286. [CrossRef] [PubMed]
86. Gathercole, S.E.; Baddeley, A.D. Evaluation of the role of phonological STM in the development of vocabulary in children: A longitudinal study. *J. Mem. Lang.* **1989**, *28*, 200–213. [CrossRef]

87. Conners, F.A.; Moore, M.S.; Loveall, S.J.; Merrill, E.C. Memory profiles of Down, Williams, and fragile X syndromes: Implications for reading development. *J. Dev. Behav. Pediatr.* **2011**, *32*, 405–417. [CrossRef] [PubMed]

88. South, M.; Rodgers, J. Sensory, Emotional and Cognitive Contributions to Anxiety in Autism Spectrum Disorders. *Front. Hum. Neurosci.* **2017**, *11*, 20. [CrossRef] [PubMed]

89. Peng, D.X.; Kelley, R.G.; Quintin, E.M.; Raman, M.; Thompson, P.M.; Reiss, A.L. Cognitive and behavioral correlates of caudate subregion shape variation in fragile X syndrome. *Hum. Brain Mapp.* **2014**, *35*, 2861–2868. [CrossRef] [PubMed]

90. Bray, S.; Hirt, M.; Jo, B.; Hall, S.S.; Lightbody, A.A.; Walter, E.; Chen, K.; Patnaik, S.; Reiss, A.L. Aberrant frontal lobe maturation in adolescents with fragile X syndrome is related to delayed cognitive maturation. *Biol. Psychiatry* **2011**, *70*, 852–858. [CrossRef] [PubMed]

91. Eliez, S.; Blasey, C.M.; Freund, L.S.; Hastie, T.; Reiss, A.L. Brain anatomy, gender and IQ in children and adolescents with fragile X syndrome. *Brain* **2001**, *124*, 1610–1618. [CrossRef]

92. Gothelf, D.; Furfaro, J.A.; Hoeft, F.; Eckert, M.A.; Hall, S.S.; O'Hara, R.; Erba, H.W.; Ringel, J.; Hayashi, K.M.; Patnaik, S.; et al. Neuroanatomy of fragile X syndrome is associated with aberrant behavior and the fragile X mental retardation protein (FMRP). *Ann. Neurol.* **2008**, *63*, 40–51. [CrossRef]

93. Hazlett, H.C.; Poe, M.D.; Lightbody, A.A.; Gerig, G.; Macfall, J.R.; Ross, A.K.; Provenzale, J.; Martin, A.; Reiss, A.L.; Piven, J. Teasing apart the heterogeneity of autism: Same behavior, different brains in toddlers with fragile X syndrome and autism. *J. Neurodev. Disord.* **2009**, *1*, 81–90. [CrossRef]

94. Hessl, D.; Rivera, S.M.; Reiss, A.L. The neuroanatomy and neuroendocrinology of fragile X syndrome. *Ment. Retard. Dev. Disabil. Res. Rev.* **2004**, *10*, 17–24. [CrossRef] [PubMed]

95. Hoeft, F.; Lightbody, A.A.; Hazlett, H.C.; Patnaik, S.; Piven, J.; Reiss, A.L. Morphometric spatial patterns differentiating boys with fragile X syndrome, typically developing boys, and developmentally delayed boys aged 1 to 3 years. *Arch. Gen. Psychiatry* **2008**, *65*, 1087–1097. [CrossRef] [PubMed]

96. Hoeft, F.; Carter, J.C.; Lightbody, A.A.; Cody Hazlett, H.; Piven, J.; Reiss, A.L. Region-specific alterations in brain development in one- to three-year-old boys with fragile X syndrome. *Proc. Natl. Acad. Sci. USA* **2010**, *107*, 9335–9339. [CrossRef]

97. Reiss, A.L.; Abrams, M.T.; Greenlaw, R.; Freund, L.; Denckla, M.B. Neurodevelopmental effects of the FMR-1 full mutation in humans. *Nat. Med.* **1995**, *1*, 159–167. [CrossRef] [PubMed]

98. Hallahan, B.P.; Craig, M.C.; Toal, F.; Daly, E.M.; Moore, C.J.; Ambikapathy, A.; Robertson, D.; Murphy, K.C.; Murphy, D.G. In vivo brain anatomy of adult males with Fragile X syndrome: An MRI study. *Neuroimage* **2011**, *54*, 16–24. [CrossRef] [PubMed]

99. Kates, W.R.; Folley, B.S.; Lanham, D.C.; Capone, G.T.; Kaufmann, W.E. Cerebral growth in Fragile X syndrome: Review and comparison with Down syndrome. *Microsc. Res. Tech.* **2002**, *57*, 159–167. [CrossRef]

100. Grahn, J.A.; Parkinson, J.A.; Owen, A.M. The cognitive functions of the caudate nucleus. *Prog. Neurobiol.* **2008**, *86*, 141–155. [CrossRef]

101. Friedlander, L.; Desrocher, M. Neuroimaging studies of obsessive-compulsive disorder in adults and children. *Clin. Psychol. Rev.* **2006**, *26*, 32–49. [CrossRef]

102. Rotge, J.Y.; Guehl, D.; Dilharreguy, B.; Tignol, J.; Bioulac, B.; Allard, M.; Burbaud, P.; Aouizerate, B. Meta-analysis of brain volume changes in obsessive-compulsive disorder. *Biol. Psychiatry* **2009**, *65*, 75–83. [CrossRef]

103. Levitt, J.J.; McCarley, R.W.; Dickey, C.C.; Voglmaier, M.M.; Niznikiewicz, M.A.; Seidman, L.J.; Hirayasu, Y.; Ciszewski, A.A.; Kikinis, R.; Jolesz, F.A.; et al. MRI study of caudate nucleus volume and its cognitive correlates in neuroleptic-naive patients with schizotypal personality disorder. *Am. J. Psychiatry* **2002**, *159*, 1190–1197. [CrossRef]

104. Mostofsky, S.H.; Mazzocco, M.M.; Aakalu, G.; Warsofsky, I.S.; Denckla, M.B.; Reiss, A.L. Decreased cerebellar posterior vermis size in fragile X syndrome: Correlation with neurocognitive performance. *Neurology* **1998**, *50*, 121–130. [CrossRef] [PubMed]

105. Reiss, A.L.; Aylward, E.; Freund, L.S.; Joshi, P.K.; Bryan, R.N. Neuroanatomy of fragile X syndrome: The posterior fossa. *Ann. Neurol.* **1991**, *29*, 26–32. [CrossRef] [PubMed]

106. Schmahmann, J.D. Disorders of the cerebellum: Ataxia, dysmetria of thought, and the cerebellar cognitive affective syndrome. *J. Neuropsychiatry Clin. Neurosci.* **2004**, *16*, 367–378. [CrossRef] [PubMed]

107. Alvarez, J.A.; Emory, E. Executive function and the frontal lobes: A meta-analytic review. *Neuropsychol. Rev.* **2006**, *16*, 17–42. [CrossRef] [PubMed]

108. D'Cruz, A.M.; Mosconi, M.W.; Ragozzino, M.E.; Cook, E.H.; Sweeney, J.A. Alterations in the functional neural circuitry supporting flexible choice behavior in autism spectrum disorders. *Transl. Psychiatry* **2016**, *6*, e916. [CrossRef]

109. Yerys, B.E.; Antezana, L.; Weinblatt, R.; Jankowski, K.F.; Strang, J.; Vaidya, C.J.; Schultz, R.T.; Gaillard, W.D.; Kenworthy, L. Neural Correlates of Set-Shifting in Children with Autism. *Autism Res.* **2015**, *8*, 386–397. [CrossRef]

110. Barnea-Goraly, N.; Eliez, S.; Hedeus, M.; Menon, V.; White, C.D.; Moseley, M.; Reiss, A.L. White matter tract alterations in fragile X syndrome: Preliminary evidence from diffusion tensor imaging. *Am. J. Med. Genet. B Neuropsychiatr. Genet.* **2003**, *118*, 81–88. [CrossRef]

111. Yuan, P.; Raz, N. Prefrontal cortex and executive functions in healthy adults: A meta-analysis of structural neuroimaging studies. *Neurosci. Biobehav. Rev.* **2014**, *42*, 180–192. [CrossRef]

112. Haas, B.W.; Barnea-Goraly, N.; Lightbody, A.A.; Patnaik, S.S.; Hoeft, F.; Hazlett, H.; Piven, J.; Reiss, A.L. Early white-matter abnormalities of the ventral frontostriatal pathway in fragile X syndrome. *Dev. Med. Child Neurol.* **2009**, *51*, 593–599. [CrossRef] [PubMed]

113. Kemper, M.B.; Hagerman, R.J.; Altshul-Stark, D. Cognitive profiles of boys with the fragile X syndrome. *Am. J. Med. Genet.* **1988**, *30*, 191–200. [CrossRef]

114. Rolls, E.T. The orbitofrontal cortex and reward. *Cereb. Cortex* **2000**, *10*, 284–294. [CrossRef] [PubMed]

115. Huber, K.M.; Gallagher, S.M.; Warren, S.T.; Bear, M.F. Altered synaptic plasticity in a mouse model of fragile X mental retardation. *Proc. Natl. Acad. Sci. USA* **2002**, *99*, 7746–7750. [CrossRef] [PubMed]

116. Comery, T.A.; Harris, J.B.; Willems, P.J.; Oostra, B.A.; Irwin, S.A.; Weiler, I.J.; Greenough, W.T. Abnormal dendritic spines in fragile X knockout mice: Maturation and pruning deficits. *Proc. Natl. Acad. Sci. USA* **1997**, *94*, 5401–5404. [CrossRef]

117. Wang, H.; Ku, L.; Osterhout, D.J.; Li, W.; Ahmadian, A.; Liang, Z.; Feng, Y. Developmentally-programmed FMRP expression in oligodendrocytes: A potential role of FMRP in regulating translation in oligodendroglia progenitors. *Hum. Mol. Genet.* **2004**, *13*, 79–89. [CrossRef] [PubMed]

118. Hinds, H.L.; Ashley, C.T.; Sutcliffe, J.S.; Nelson, D.L.; Warren, S.T.; Housman, D.E.; Schalling, M. Tissue specific expression of FMR-1 provides evidence for a functional role in fragile X syndrome. *Nat. Genet.* **1993**, *3*, 36–43. [CrossRef] [PubMed]

119. Abitbol, M.; Menini, C.; Delezoide, A.L.; Rhyner, T.; Vekemans, M.; Mallet, J. Nucleus basalis magnocellularis and hippocampus are the major sites of FMR-1 expression in the human fetal brain. *Nat. Genet.* **1993**, *4*, 147–153. [CrossRef]

120. Gibson, J.R.; Bartley, A.F.; Hays, S.A.; Huber, K.M. Imbalance of neocortical excitation and inhibition and altered UP states reflect network hyperexcitability in the mouse model of fragile X syndrome. *J. Neurophysiol.* **2008**, *100*, 2615–2626. [CrossRef]

121. Ronesi, J.A.; Collins, K.A.; Hays, S.A.; Tsai, N.P.; Guo, W.; Birnbaum, S.G.; Hu, J.H.; Worley, P.F.; Gibson, J.R.; Huber, K.M. Disrupted Homer scaffolds mediate abnormal mGluR5 function in a mouse model of fragile X syndrome. *Nat. Neurosci.* **2012**, *15*, 431–440. [CrossRef]

122. Westmark, C.J.; Chuang, S.C.; Hays, S.A.; Filon, M.J.; Ray, B.C.; Westmark, P.R.; Gibson, J.R.; Huber, K.M.; Wong, R.K. APP Causes Hyperexcitability in Fragile X Mice. *Front. Mol. Neurosci.* **2016**, *9*, 147. [CrossRef]

123. Lovelace, J.W.; Ethell, I.M.; Binder, D.K.; Razak, K.A. Translation-relevant EEG phenotypes in a mouse model of Fragile X Syndrome. *Neurobiol. Dis.* **2018**, *115*, 39–48. [CrossRef]

124. Wang, J.; Ethridge, L.E.; Mosconi, M.W.; White, S.P.; Binder, D.K.; Pedapati, E.V.; Erickson, C.A.; Byerly, M.J.; Sweeney, J.A. A resting EEG study of neocortical hyperexcitability and altered functional connectivity in fragile X syndrome. *J. Neurodev. Disord.* **2017**, *9*, 11. [CrossRef] [PubMed]

125. Sohal, V.S.; Zhang, F.; Yizhar, O.; Deisseroth, K. Parvalbumin neurons and gamma rhythms enhance cortical circuit performance. *Nature* **2009**, *459*, 698–702. [CrossRef] [PubMed]

126. Mathalon, D.H.; Sohal, V.S. Neural Oscillations and Synchrony in Brain Dysfunction and Neuropsychiatric Disorders: It's About Time. *JAMA Psychiatry* **2015**, *72*, 840–844. [CrossRef] [PubMed]

127. Bosman, C.A.; Lansink, C.S.; Pennartz, C.M. Functions of gamma-band synchronization in cognition: From single circuits to functional diversity across cortical and subcortical systems. *Eur. J. Neurosci.* **2014**, *39*, 1982–1999. [CrossRef] [PubMed]

128. Cho, K.K.; Hoch, R.; Lee, A.T.; Patel, T.; Rubenstein, J.L.; Sohal, V.S. Gamma rhythms link prefrontal interneuron dysfunction with cognitive inflexibility in Dlx5/6(+/−) mice. *Neuron* **2015**, *85*, 1332–1343. [CrossRef]

129. Lewis, D.A.; Curley, A.A.; Glausier, J.R.; Volk, D.W. Cortical parvalbumin interneurons and cognitive dysfunction in schizophrenia. *Trends Neurosci.* **2012**, *35*, 57–67. [CrossRef]

130. Lewis, D.A. Inhibitory neurons in human cortical circuits: Substrate for cognitive dysfunction in schizophrenia. *Curr. Opin. Neurobiol.* **2014**, *26*, 22–26. [CrossRef]

131. Moran, L.V.; Hong, L.E. High vs. low frequency neural oscillations in schizophrenia. *Schizophr. Bull.* **2011**, *37*, 659–663. [CrossRef]

132. Reilly, J.L.; Lencer, R.; Bishop, J.R.; Keedy, S.; Sweeney, J.A. Pharmacological treatment effects on eye movement control. *Brain Cogn.* **2008**, *68*, 415–435. [CrossRef]

133. Hill, E.L. Executive dysfunction in autism. *Trends Cogn. Sci.* **2004**, *8*, 26–32. [CrossRef]

134. Craig, F.; Margari, F.; Legrottaglie, A.R.; Palumbi, R.; de Giambattista, C.; Margari, L. A review of executive function deficits in autism spectrum disorder and attention-deficit/hyperactivity disorder. *Neuropsychiatr. Dis. Treat.* **2016**, *12*, 1191–1202.

135. Constantino, J.N. Deconstructing autism: From unitary syndrome to contributory developmental endophenotypes. *Int. Rev. Psychiatry* **2018**, *30*, 18–24. [CrossRef]

136. Moreno-De-Luca, A.; Evans, D.W.; Boomer, K.B.; Hanson, E.; Bernier, R.; Goin-Kochel, R.P.; Myers, S.M.; Challman, T.D.; Moreno-De-Luca, D.; Slane, M.M.; et al. The role of parental cognitive, behavioral, and motor profiles in clinical variability in individuals with chromosome 16p11.2 deletions. *JAMA Psychiatry* **2015**, *72*, 119–126. [CrossRef]

137. Mervis, C.B.; Klein-Tasman, B.P. Methodological issues in group-matching designs: Alpha levels for control variable comparisons and measurement characteristics of control and target variables. *J. Autism Dev. Disord.* **2004**, *34*, 7–17. [CrossRef] [PubMed]

138. Sansone, S.M.; Schneider, A.; Bickel, E.; Berry-Kravis, E.; Prescott, C.; Hessl, D. Improving IQ measurement in intellectual disabilities using true deviation from population norms. *J. Neurodev. Disord.* **2014**, *6*, 16. [CrossRef] [PubMed]

139. Hessl, D.; Nguyen, D.V.; Green, C.; Chavez, A.; Tassone, F.; Hagerman, R.J.; Senturk, D.; Schneider, A.; Lightbody, A.; Reiss, A.L.; et al. A solution to limitations of cognitive testing in children with intellectual disabilities: The case of fragile X syndrome. *J. Neurodev. Disord.* **2009**, *1*, 33–45. [CrossRef] [PubMed]

140. Au, J.; Berkowitz-Sutherland, L.; Schneider, A.; Schweitzer, J.B.; Hessl, D.; Hagerman, R. A feasibility trial of Cogmed working memory training in fragile X syndrome. *J. Pediatr. Genet.* **2014**, *3*, 147–156.

141. Berry-Kravis, E.; Sumis, A.; Kim, O.K.; Lara, R.; Wuu, J. Characterization of potential outcome measures for future clinical trials in fragile X syndrome. *J. Autism Dev. Disord.* **2008**, *38*, 1751–1757. [CrossRef] [PubMed]

142. Yan, Q.J.; Asafo-Adjei, P.K.; Arnold, H.M.; Brown, R.E.; Bauchwitz, R.P. A phenotypic and molecular characterization of the fmr1-tm1Cgr fragile X mouse. *Genes Brain Behav.* **2004**, *3*, 337–359. [CrossRef] [PubMed]

143. Durtch-Belgium Fragile X Consortium. Fmr1 knockout mice: A model to study fragile X mental retardation. The Dutch-Belgian Fragile X Consortium. *Cell* **1994**, *78*, 23–33.

144. Peier, A.M.; McIlwain, K.L.; Kenneson, A.; Warren, S.T.; Paylor, R.; Nelson, D.L. (Over) correction of FMR1 deficiency with YAC transgenics: Behavioral and physical features. *Hum. Mol. Genet.* **2000**, *9*, 1145–1159. [CrossRef] [PubMed]

145. D'Hooge, R.; Nagels, G.; Franck, F.; Bakker, C.E.; Reyniers, E.; Storm, K.; Kooy, R.F.; Oostra, B.A.; Willems, P.J.; De Deyn, P.P. Mildly impaired water maze performance in male Fmr1 knockout mice. *Neuroscience* **1997**, *76*, 367–376. [CrossRef]

146. Robbins, T.W. The 5-choice serial reaction time task: Behavioural pharmacology and functional neurochemistry. *Psychopharmacology* **2002**, *163*, 362–380. [CrossRef] [PubMed]

147. Kramvis, I.; Mansvelder, H.D.; Loos, M.; Meredith, R. Hyperactivity, perseveration and increased responding during attentional rule acquisition in the Fragile X mouse model. *Front. Behav. Neurosci.* **2013**, *7*, 172. [CrossRef] [PubMed]

148. Moon, J.; Beaudin, A.E.; Verosky, S.; Driscoll, L.L.; Weiskopf, M.; Levitsky, D.A.; Crnic, L.S.; Strupp, B.J. Attentional dysfunction, impulsivity, and resistance to change in a mouse model of fragile X syndrome. *Behav. Neurosci.* **2006**, *120*, 1367–1379. [CrossRef]

149. Krueger, D.D.; Osterweil, E.K.; Chen, S.P.; Tye, L.D.; Bear, M.F. Cognitive dysfunction and prefrontal synaptic abnormalities in a mouse model of fragile X syndrome. *Proc. Natl. Acad. Sci. USA* **2011**, *108*, 2587–2592. [CrossRef] [PubMed]

150. D'Cruz, A.M.; Ragozzino, M.E.; Mosconi, M.W.; Shrestha, S.; Cook, E.H.; Sweeney, J.A. Reduced behavioral flexibility in autism spectrum disorders. *Neuropsychology* **2013**, *27*, 152–160. [CrossRef] [PubMed]

151. Berry-Kravis, E.; Sumis, A.; Hervey, C.; Nelson, M.; Porges, S.W.; Weng, N.; Weiler, I.J.; Greenough, W.T. Open-label treatment trial of lithium to target the underlying defect in fragile X syndrome. *J. Dev. Behav. Pediatr.* **2008**, *29*, 293–302. [CrossRef]

152. Budimirovic, D.B.; Berry-Kravis, E.; Erickson, C.A.; Hall, S.S.; Hessl, D.; Reiss, A.L.; King, M.K.; Abbeduto, L.; Kaufmann, W.E. Updated report on tools to measure outcomes of clinical trials in fragile X syndrome. *J. Neurodev. Disord.* **2017**, *9*, 14. [CrossRef] [PubMed]

Review

Early Identification of Fragile X Syndrome through Expanded Newborn Screening

Katherine C. Okoniewski *, Anne C. Wheeler, Stacey Lee, Beth Boyea, Melissa Raspa, Jennifer L. Taylor and Donald B. Bailey, Jr.

RTI International, Research Triangle Park, NC 27709-2194, USA; acwheeler@rti.org (A.C.W.); snlee@rti.org (S.L.); mlincolnboyea@rti.org (B.B.); mraspa@rti.org (M.R.); jltaylor@rti.org (J.L.T.); dbailey@rti.org (D.B.B., Jr.)
* Correspondence: kokoniewski@rti.org; Tel.: +1-919-541-7461

Received: 14 November 2018; Accepted: 24 December 2018; Published: 3 January 2019

Abstract: Over the past 20 years, research on fragile X syndrome (FXS) has provided foundational understanding of the complex experiences of affected individuals and their families. Despite this intensive focus, there has been little progress on earlier identification, with the average age of diagnosis being 3 years. For intervention and treatment approaches to have the greatest impact, they need to begin shortly after birth. To access this critical timespan, differential methods of earlier identification need to be considered, with an emerging focus on newborn screening practices. Currently, barriers exist that prevent the inclusion of FXS on standard newborn screening panels. To address these barriers, an innovative program is being implemented in North Carolina to offer voluntary screening for FXS under a research protocol, called Early Check. This program addresses the difficulties observed in prior pilot studies, such as recruitment, enrollment, lab testing, and follow-up. Early Check provides an opportunity for stakeholders and the research community to continue to gain valuable information about the feasibility and greater impact of newborn screening on the FXS population.

Keywords: fragile X syndrome; newborn screening; early identification

1. Introduction

Although parents, pediatricians, and early educators frequently identify early developmental differences in infants and toddlers with fragile X syndrome (FXS) [1], it often takes up to 2 years between first concern and diagnosis in males. As a result, the average age of diagnosis for a child with FXS is around 36 months [2]. This timeline is even longer for females, who tend to be less severely affected as a result of their second X chromosome and X-inactivation. In addition to causing delays in access to targeted interventions, there are important implications for the family because of this delay in diagnosis. These include increased family emotional and financial stress related to the diagnostic odyssey, as well as implications for reproductive decision making in immediate and extended family, with many families having more than one child with FXS before a diagnosis is made.

There is now accumulating evidence that symptoms in FXS are detectable within the first year of life [3–5]. Both animal and neuroimaging studies suggest that the consequences of FXS begin in the prenatal period with diminished production of fragile X mental retardation protein (FMRP) believed to play a key role in early brain development [6]. Recent findings, suggesting white matter development differences in the brains of infants with FXS as young as 6 months of age [5], confirm that neurological differences are evident before observable symptoms appear.

Therapeutic development has been on a rapid course since the early 2000's, when a theory was proposed suggesting that excessive mGluR5 function was associated with reduced FMRP [7]. This theory led to several studies demonstrating "rescue" of the FXS phenotype in *fmr1* knockout mice [8–10], which subsequently led to an increase in clinical trials for mGluR5 inhibitors [11]. Although a cure has yet to be fully realized, the pace of discovery in the basic science realm has generated excitement within the FX (fragile X) community and has spurred increased discussion of how to maximize the potential benefit of emerging therapeutics. Symptom-based behavioral interventions are most commonly used in this population, incorporating multiple disciplines and techniques to address needs in individuals and their families; however, there is limited knowledge of the effects of these pre-symptomatically or early on in development. Overall, most researchers and clinicians agree that for a treatment to be most effective for improving long-term outcomes for individuals with FXS, it would need to be implemented very early, likely within the first year of life.

As a result of this gap between the potential benefits of earlier diagnosis and the reality of age of diagnosis, there has been increasing interest in earlier identification of FXS. Several solutions have been proposed to facilitate earlier identification [12], including preconception carrier testing, newborn screening, and systematic universal developmental screening of infants and toddlers. Of these, the solution that has received the most attention is newborn screening (NBS). In this paper we discuss current practice for early identification of individuals with FXS, describe possible screening approaches, and outline a new project that offers voluntary newborn screening to all birthing parents in one state, and use lessons learned from prior pilot NBS studies to guide our work.

2. Diagnosis of FXS

FXS is caused by an expansion of over 200 CGG repeats in the *FMR1* gene, resulting in significantly reduced FMRP, which is necessary for healthy brain development. Although FXS is relatively rare (1:4000–6000 male births, 1:6000–8000 female births), it is considered the most common form of inherited intellectual disability and one of the most well-studied genetic causes of autism spectrum disorders. Although there are clear guidelines by groups such as the American Academy of Pediatrics and the National Society of Genetic Counselors regarding focused screening recommendations, FXS still remains under recognized [12]. Current practices for receiving a diagnosis for FXS almost always involves a significant "diagnostic odyssey" on the part of the family [13]. This odyssey may start when parents recognize there are delays in their child's development, usually noticeable by 9 months of age [4], leading them to report their concerns to their child's pediatrician. The pediatrician may respond with a referral for a developmental evaluation and/or early intervention services, or they may suggest taking a "wait and see" approach, further delaying access to treatment. Even when a child receives timely access to early intervention, they will likely first receive a diagnosis of developmental delay or autism, and even this diagnosis can take up to 12 months to obtain. It may take several more years before genetic testing for FXS is recommended. During this time, many families, not knowing their reproductive risk, will go on to have additional children with FXS.

Several solutions to reduce the diagnostic odyssey and allow for earlier identification of FXS have been proposed. Maternal testing for preconception carrier status would allow for more informed reproductive decision making and planning and is reported as the preferred timing by parents who are already caregivers to a child with FXS [14]. However, current practices for preconception genetic testing generally require a family history or other risk factors to trigger testing. Further, universal preconception testing would require that each pregnancy is planned and that potential parents have the resources to seek and receive this testing prior to conception. Another option is pairing genetic testing with systematic universal developmental screening procedures for observed delays. This would refine the current problem-based evaluation of children but would still delay the diagnosis until after the child was symptomatic. The earliest, most universal approach would be to focus on fetal testing or newborn screening. Prenatal testing for conditions like FXS is controversial, especially given the lack of refined prognosis prediction due to the spectrum of phenotypic outcomes, most markedly in females. The use of prenatal testing is becoming more common and allows for reproductive choice, early identification, and access to intervention; however, universal access to prenatal testing is varied throughout the population, posing a barrier to many. NBS therefore has emerged as the solution with the greatest potential to reach the most individuals and with the least potential of bias towards income and access to healthcare [15].

3. Fragile X and Newborn Screening

For a disorder to be included on the Recommended Uniform Screening Panel (RUSP) for NBS, it must meet a specific set of requirements [16]. These factors broadly include overall benefit of screening (e.g., health status and importance of early identification) as well as feasibility and current readiness for state-level implementation (e.g., validated screening assays, state health laboratory capacities). While a disorder may meet some or all of the criteria, it is still within the authority of the Advisory Committee on Heritable Disorders in Newborns and Children and the Secretary of Health and Human Services to make a final recommendation for conditions considered for the RUSP. Implementation of NBS for disorders on the RUSP is overseen by state public health departments that determine which disorders to screen for, streams of financing for screening procedures, and ways in which follow-up and support can be provided to identified infants and their families [13].

Proponents of the addition of FXS cite a high enough prevalence rate and level of impairment along with an array of behavioral and developmental interventions consistent with RUSP guidelines, making it a viable candidate for NBS. However, the lack of an inexpensive and valid screening measure for FXS, no proven medical treatment, and no feasibility studies, have stood as significant barriers [13]. To address these concerns, several studies exploring the feasibility, buy-in, and acceptability of NBS for FXS have been conducted.

In 2008, a multi-site study aiming to identify the extent of acceptance, any adverse experiences that may occur because of early identification to the infant, and a feasible consent process for FXS NBS was executed [17]. Encouraging findings emerged, particularly around parent buy-in and uptake. At the end of the study, screening opportunities had been offered to over 28,000 families and accepted by 62%. Although initial perceptions were positive, difficulties with NBS screening in FXS were also identified, particularly related to recruitment and consent. Since FXS was not yet on the RUSP, a direct consent model needed to be implemented to allow screening for this disorder to occur. In this study in-hospital direct recruitment was implemented. While findings showed feasibility in this approach, accompanying challenges included difficulty with recruiting mothers soon after birth and training of hospital staff to effectively and independently recruit families. Consent was required from both parents, which was an additional challenge for maximizing opportunities for all families [17].

Recently, a comprehensive review of the literature and expert report identified ongoing barriers to implementation of universal NBS for FXS [18]. These barriers include issues related to identification of carriers, varied access to early intervention, no effective medical treatment for FXS, issues related to uncertainty and anxiety for caregivers, and implications for family planning. Until recently, a feasible and affordable screening test was not available. Finally, the capacity for follow-up across states is unknown. An expansion of public and professional education is needed to adequately support identified infants and families and to overcome these barriers.

Fragile X Premutation as a Complicating Factor for Newborn Screening for FXS

One of the more controversial concerns regarding NBS for FXS is the detection and reporting of infants with an *FMR1* premutation (PM). *FMR1* premutations occur when the number of expanded CGG repeats is between 55 and 200, and is much more common (1:200 females, 1:430 males) [19–21] than FXS. Given the high number of infants with the PM that would be identified through NBS, the main challenge would be the large burden on providing genetic counseling to so many families.

In addition, it may be challenging to convey the uncertainty that comes from a PM result. Decades of targeted research have shown that the PM conveys its own set of health risks and phenotypic traits [22–24], although these are often seen in adults, not infants. These include two well-documented conditions; FX-associated primary ovarian insufficiency (FXPOI) [25] and FX-associated tremor ataxia syndrome (FXTAS) [26], as well as a host of other cognitive, emotional, and medical problems. Similarly, individuals with the PM are frequently referred to as "carriers" because of the increased risk for having offspring with FXS in women with a PM. Recent emerging evidence suggests there may be increased risk for developmental delays or differences in a subset of young children with the PM indicating a potential need for early identification of the PM. Furthermore, individuals with a high CGG repeat number in the PM range (e.g. >150) may be at risk for a more similar phenotype to those with a diagnosis of FXS due to repeats above 200 [27–30].

It is important to highlight however that most individuals with a PM will have few to no developmental challenges or health risks. The majority of PM alleles are in the 55–70 range, a range that confers a much lower risk of expansion in the next generation and is believed to have fewer associated health risks than alleles with >70, although there is evidence to suggest this may not always be the case. For example, there are several reports of FXTAS occurring in individuals with CGG repeats in the low PM range [31] and multiple studies suggest a curvilinear pattern of risk with those with mid-range CGG repeats having greater risk for poor outcomes than those with low or high range repeats [32–36].

Without additional biomarkers to help predict risk, conveying information about the PM is complex and challenging. Our limited information about genotype–phenotype associations in the PM is a problem for NBS for FXS; however, it also increases the need for prospective studies examining the natural history of these conditions. Ultimately, the full range of the PM is unlikely to meet the criteria of proven benefit for NBS. However, inclusion of the PM in pilot studies of NBS for FXS allows for the opportunity to identify which infants may be at greatest risk for the spectrum of developmental concerns associated with FXS and can help with the identification of potential biomarkers that can help guide prognosis and treatment.

4. Early Check: Expanded Screening in Newborns

With a better understanding of the barriers and promising evidence for NBS in FXS, a diverse team of researchers, clinicians, public health professionals, advocacy groups, universities, and state institutions have come together to create an innovative program called Early Check (www.EarlyCheck. org). Early Check offers voluntary screening for a second panel of conditions that are not part of standard NBS. One of the goals of Early Check is to facilitate earlier identification of conditions not currently eligible for the RUSP to promote greater understanding of the natural history of the condition and allow for pre-symptomatic treatment studies. More specifically, Early Check aims to address condition-specific questions regarding (1) prevalence rates and medical implications of the disorder on the public health system, (2) practicality and feasibility of affordable screening assays that can be performed using dried blood spot (DBS) specimens, and (3) efficient follow-up practices, connecting identified individuals and families to interventions and treatments, as well as clinical trials.

To address the challenges of providing voluntary newborn screening, Early Check has established collaborations with a state public health laboratory, three local university medical centers, and biotechnological corporations to expand capabilities of the use of DBS specimens. Early Check has a diverse lab team of experts in newborn screening, informatics, neuroscience, chemistry, and molecular biology developing and executing innovative methodologies in the world of newborn screening. Furthermore, execution of various pilot studies for recommended conditions such as X-linked adrenoleukodystrophy (X-ALD), has expanded the expertise of Early Check team members in navigating and better understanding state-level requirements for the implementation of new conditions.

The first step of the Early Check process is to disseminate information through strategic outreach campaigns that are easily accessible to birthing mothers. Early Check developers spent considerable time establishing targeted communication and recruitment procedures as well as an easily accessible online consent module to access, inform, and enroll as many birthing mothers as possible. Formative evaluations utilizing extensive literature reviews, focus groups, and pilots of materials supported and facilitated the creation of a comprehensive campaign to spread the news of Early

Check throughout the state of North Carolina. These efforts are paired with a user-friendly web-based consent portal to achieve a representative comprehensive sample. Through a phased model of roll out, Early Check will access a variety of communication networks to inform and provide birthing mothers the opportunity to access expanded screening for rare disorders. Participants can consent through an electric permissions portal in the pre- and post-natal periods. Once an enrolled infant's newborn screening test is complete, Early Check laboratory staff use the residual DBS specimen to run targeted assays on a panel of conditions.

Following screening, families identified with screen-positive results are provided confirmatory testing and short- and long-term follow-up opportunities. For conditions with limited understanding of the effectiveness of early behavioral intervention or medical treatments, Early Check has developed a systematic follow-up program. Short- and long-term follow-up protocols include standard confirmatory testing, genetic counseling, assessment of developmental functioning and growth, and evaluation of parent well-being. Parents are provided the opportunity to engage in the Early Check registry, allowing for continued dissemination of information on interventions, treatments, and future studies.

5. Early Check and FXS

FXS is one of the introductory conditions included on the Early Check panel. This provides a unique and invaluable opportunity to address the specific challenges faced by the FXS community in attaining earlier identification. Below we outline specific components of Early Check designed to capture the unique issues related to NBS for FXS.

5.1. Consenting

One of the first barriers faced in the pilot study of NBS for FXS were challenges with in-hospital recruitment and achieving optimal uptake rates. To address this, we developed an electronic portal consent aimed to remedy these challenges, such that (1) a person is not responsible for face-to-face contact to recruit a family to enroll, (2) recruitment methodologies and access have been specifically developed to provide a range of information to enhance informed decision making in parents during the prenatal period and for approximately 4 weeks after the infant's birth. These methodologies allow for a differentiated and cost-effective option for wide recruitment and enrollment as compared to previous practices that proved to be a challenge. While a universal system of recruitment would be ideal, since FXS is not yet on the RUSP a voluntary consent-based model must be utilized.

Inclusion of the PM is also a challenge with regard to NBS. Although parents may appreciate and desire knowing early about a condition like FXS, they may feel less sure about wanting to know the PM status of their newborn. With Early Check, parents are offered a second tiered consent for the PM such that they must first consent to screening for FXS. Once they have completed the consent procedures, they are immediately offered the opportunity to also consent to receive information regarding PM findings. This allows us to better understand the desires of parents to know PM status, while also allowing the opportunity to identify and follow a subset of infants with the PM to develop a better understanding of the natural history of the full spectrum of *FMR1* mutations.

5.2. Laboratory Test

The lack of a validated, efficient, affordable screening assay for FXS has been a consistent challenge for NBS. The Early Check laboratory team was able to successfully utilize a custom PCR (Polymerase chain reaction)-based assay and analysis software program developed by Asuragen for a high-throughput sample workflow to provide robust detection of *FMR1* repeat expansions from DBS specimens. This method was characterized by analyzing cell lines and quality control reference material with CGG repeat sizes spanning a range of genotypes, including normal, premutation, and full mutation alleles. The performance of this method was characterized by evaluating assay precision, accuracy, sensitivity, and specificity, with the assay consistently performing within 5% of the expected

CGG repeat requirements, with proficiency testing results in 100% concordance with the results from reference laboratories. In addition, a straightforward sample preparation workflow was utilized for the analysis of 963 de-identified newborn DBS samples from the North Carolina Laboratory of Public Health NBS program to determine preliminary population distributions and to develop a screening algorithm, with results finding 957 normal, 6 premutation, and 0 full mutation specimens.

5.3. Confirmatory Testing and Genetic Counseling

The great majority of mothers who participate in Early Check sign in to the secure portal and read a reassuring, lay-language document indicating that their infant tested normal. They are also provided a downloadable clinical screening report to provide to their child's pediatrician. However, approximately 460 of the 120,000 babies born in North Carolina each year are expected to have either the fragile X full mutation or premutation. Given the sensitivity of the Asuragen assay, virtually all of these cases are potentially identifiable through Early Check [17,37]. Despite the very low false positive rate for the Asuragen assay, confirmatory testing is strongly recommended and is provided free of charge up until the baby's first birthday. Soon after the positive screening result is relayed, parents are sent a cheek swab kit with instructions for collecting a buccal swab, which they send to a local molecular laboratory in a pre-paid mailer for confirmatory testing. Carrier testing is also offered to mothers of babies identified with the full or premutation and to fathers of girls with the premutation whose mothers are found to have normal *FMR1* alleles. Confirmatory test results include CGG repeat number, possible AGG interruptions, and reflexive methylation studies on samples with CGG repeats $\geq$100. Once screening results are confirmed, parents of babies with FXS will be offered in-person genetic counseling at a centralized, easily accessible partner clinic. Parents of babies with the premutation will be offered telegenetic counseling using a HIPAA (Health Insurance Portability and Accountability Act)-compliant, app-based video-conferencing platform accessible through a smartphone or computer.

Genetic counseling for FXS or the PM identified by Early Check is notably complicated by the convergence of an unusual combination of factors. First, the concepts and implications of variable expressivity, reduced penetrance, and an X-linked inheritance pattern nuanced by genetic anticipation and mediated by the number of CGG repeats and AGG interruptions in a female fragile X premutation carrier are far more challenging to explain than the simple genetic mechanism implicated in most single gene conditions. In addition, communication about the uncertain potential effects of the premutation in babies and in parents unexpectedly identified as carriers presents substantial challenges. Further, unlike the typical FXS diagnostic scenario in which parents have often been searching for an explanation for their child's developmental delays, parents whose babies are identified pre-symptomatically through Early Check will typically have had no forewarning about the diagnosis and may well doubt its validity. Finally, North Carolina's population varies widely regarding income, ethnicity, education, and health literacy [38], presenting another challenge to the meaningful communication of these results to parents. Given these confounding factors, it comes as no surprise that parents in the fragile X newborn screening pilot sometimes required multiple conversations with a genetic counselor or medical geneticist [17].

Early Check employs a multidimensional, parent-centered approach to returning results and providing education and support, using an innovative suite of communication technologies including automated emails and texts, carefully crafted educational websites, print materials, and visual aids that reiterate and augment the genetic counseling content. Several metrics will be used to evaluate the use and effectiveness of the genetic counseling intervention, including timing, number, duration, and mode of contact with a genetic counselor, web content, and clicks on links to external resources. Follow-up surveys will assess parent understanding, retention, well-being, and decision regret.

5.4. Treatment and Intervention

The Early Check system provides parents with the opportunity for short- and long-term follow-up options, including surveillance, connections to current clinical trials and treatments available, as well as developmental monitoring and early intervention. All families will be offered the opportunity to have their infant's development assessed at two time points around 3 and 6 months of age. Families can also choose ongoing surveillance via our longitudinal research registry.

All infants with FXS will be referred for community-based early intervention services, a federally funded, state-based program that infants with FXS will be eligible for due to having an established condition. Infants with the PM who agree to participate in our long-term follow-up research registry will receive ongoing developmental monitoring via regular parent surveys and will be referred for early intervention if they show signs of early delay. We have also developed a specialized clinic where children identified with any condition through Early Check can receive ongoing medical and developmental monitoring outside of a research protocol.

Because FXS has traditionally been diagnosed later in early childhood, there is very little known about effective treatments for infants and toddlers with FXS. One case study followed the identification of a child with FXS through a pilot NBS study and found significant positive effects of early intervention practices on cognitive and behavioral functioning [39]. However, this has yet to be replicated. As part of traditional early intervention, a multidisciplinary team approach, including pediatricians, neurologists, and speech, developmental and occupational therapists, is commonly used to address areas that need improvement [40]. Speech/language therapy focuses on expressive, receptive, and pragmatic skills, whereas physical and occupational therapies address motor delays and sensory sensitivities.

Current treatments for older children with FXS include symptom-based behavioral and pharmacological interventions. Pharmacological interventions are available to address comorbid behavioral difficulties such as attention-deficit/hyperactivity, anxiety, sleep, and aggression. Traditional long-acting stimulants in children over age 5 have been effective in addressing attentional difficulties, and selective serotonin reuptake inhibitors show promising results when used to address anxiety. Antipsychotics such as risperidone have also been utilized to address more significant levels of hyperactive and aggressive behaviors [41–43]. For individuals with FXS and comorbid autism spectrum disorder (ASD), evidence-based intervention strategies designed for young children at risk for ASD also prove effective [44]. As of this current review, there are no targeted behaviorally based interventions developed and validated specifically for young individuals with FXS.

While these intervention and treatment practices have been implemented and evaluated in samples of older individuals with FXS, little is known about the earliest developmental trajectories of these children and the lasting implications of intervention implementation pre-symptomatically. A goal of Early Check is to address this gap.

5.5. Long-Term Follow-up and Research

Key to the efficacy of Early Check is the ability to document improved outcomes as a result of earlier identification. To monitor these outcomes, an infrastructure for long-term follow-up and collaborative research is critical. We have developed a long-term research registry, open only to families who participate in the Early Check screening program. By enrolling in this registry, families will have the opportunity to provide ongoing information about their child's development, their family's adaptation to the diagnosis, their access to and satisfaction with treatment options, and an evaluation of the Early Check program. In addition, as new clinical trials or research studies are funded for their child's condition, the registry will provide a portal for notifying families and providing information about how to participate.

Although there are ongoing clinical trials for older children and adults with FXS, at the time of writing there are no clinical trials targeting infants with FXS or the PM. However, targeted treatments for FXS have been supported by studies in animal models of FXS such as the *FMR1* knockout mouse, with particular focus on mGluR [45]. As these trials continue into human application, findings have been varied. For example, initial phases of human trials of implications of the mGluR5 negative modulator AFQ056 showed promising results in decreasing stereotypic and hyperactive behaviors; however, expansion into Phase II did not result in significant findings [46]. Efforts to explore these therapeutic possibilities are ongoing, with a general consensus that earlier implementation is likely to be more effective. Although it is likely to be years before an approved clinical trial for infants with FXS is available, documenting the natural history and establishing procedures for pre-symptomatic identification will set the stage for rapid implementation of those trials when available.

In the meantime, our initial focus on longitudinal outcomes will focus on the potential benefits of early intervention for infants who are pre-symptomatic. This will include monitoring uptake, types, and intensity of community-based early interventions for families who agree to join our research registry. In addition, we will offer families of infants with FXS the opportunity to participate in a trial of a targeted enhanced early intervention program. This program will be overseen by early intervention specialists at a local university and will capitalize on empirically based parent-mediated early intervention programming for infants at risk of developmental differences or ASDs. We will compare the outcomes of these infants to a cohort of young children who have received a diagnosis and early intervention in North Carolina but who were not diagnosed through Early Check and did not receive the enhanced intervention program. Evidence of improved outcomes for those in the Early Check program would provide critical support for universal newborn screening for FXS.

For infants with the PM, lack of knowledge about the relative risk for developmental differences and the biological or environmental predictors of worse outcomes make a natural history study a priority. Because we anticipate identifying more infants with a PM than with FXS, and because we do not expect many to demonstrate overt developmental differences in infancy, we will conduct initial developmental surveillance with identified infants with a PM primarily through parent report. We will also invite parents of infants with a PM to enroll in a pilot study of remote developmental assessment techniques and will pursue additional funding for a more robust natural history study of the PM as more families enroll.

6. Discussion

FXS is one of the most well-characterized neurogenetic conditions. However, diagnosis typically occurs well after the onset of observable delays. To capitalize on critical periods of early development, newborn screening for FXS has been proposed as a method for providing earlier identification. FXS does not currently meet the necessary criteria for consideration for standard NBS. At the time of writing, there are no medical treatments to prevent or reduce the impact of FXS on the developing child, requiring psychoeducational and behavioral interventions as primary treatment. No condition has ever been approved for the RUSP based on psychoeducational or behavioral treatment benefits; therefore, it is unlikely that FXS will meet criteria for standard NBS in the near future. As such, FXS may be better suited for a second-tier voluntary screening panel, which would allow parents to choose expanded screening for their newborn while increasing the number of infants identified with FXS prior to symptom onset.

Early Check is an innovative program designed to provide this expanded screening panel for conditions thought to benefit from earlier identification, but which do not meet current criteria for the RUSP. Access to pre-symptomatic infants and their families will allow for important long-term follow-up and natural history data to be collected that can inform future treatment approaches. With improved knowledge of the early natural history of infants with an *FMR1* expansion will come greater knowledge of the interactions between genetic risk factors and environmental influences in outcomes for affected children and their families. It also provides potential access to pre-symptomatic infants, a population likely to benefit most from emerging therapeutics. Studies that focus on the timing of early intervention for infants and toddlers with FXS, the content, intensity, and approach to behavioral treatments, and the impact of treatments will guide the development of recommended practice for infants identified during the newborn period.

In addition to providing earlier identification and access to a cohort of newborns with FXS and the PM, Early Check will test important concepts and procedures needed to implement voluntary expanded newborn screening for conditions like FXS on a state-wide level. For example, we know that asking eligible parents to provide electronic consent through an opt-in model will not result in universal uptake. The reasons for this are manifold, with the most critical being that many hard-to-reach populations will not have access or the ability to participate. Thus, there is a possibility of missing affected individuals. However, this model allows us to test various outreach efforts to determine which strategies have maximum reach and for whom.

Early Check screening procedures will also provide important information about the feasibility of using a high-throughput FXS assay in an NBS context. Traditional methods for FXS testing have been laborious and unsuitable for high-throughput, rapid screening. The Asuragen assay and accompanying analytical software provide a streamlined screening process that provides results in less than two days.

The development of standard operating procedures for screening and follow-up will provide critical information for states to use in a future appraisal of their readiness to implement NBS for FXS, whether through traditional or expanded protocols. State evaluations of readiness rely on answering important questions, such as what diagnostic confirmation methods are available and whether there exists standard treatment and follow-up protocols to manage the disorder. Having implemented screening on a large scale, we will have detailed information on quality control measures (e.g., timeliness) and other indicators that states can use to assess their ability to add FXS to traditional or expanded newborn screening procedures.

FXS is one of many rare neurogenetic conditions that are believed to benefit greatly from earlier identification and treatment. Inclusion of FXS on the Early Check panel will not only provide a mechanism to identify and test theories about outcomes for individuals with *FMR1* expansions, but will serve as a prototype for expanded NBS for many conditions resulting in intellectual or developmental disabilities. As breakthroughs in understanding of molecular pathways for these rare conditions continue to occur, increasing focus on therapeutic development and a recognition that earlier onset of treatment is critical are likely to come more sharply into focus. Simultaneously, concentrated efforts to demonstrate efficacy of psychoeducational and behavioral early intervention techniques may provide significant benefits for the child and family. These efforts have the potential to result in a paradigm shift in how the benefits of NBS are defined, and could impact NBS policy for future generations.

Author Contributions: K.C.O. led the writing of the paper; A.C.W., M.R., and D.B.B., Jr. conceived of the paper and contributed to specific sections; B.B., S.L., and J.L.T. contributed to specific sections. All authors reviewed and approved the final version.

Funding: This research was funded by the National Center for Advancing Translational Sciences of the National Institutes of Health (Award # U01TR001792); the Eunice Kennedy Shriver National Institute for Child Health and Human Development (Award # HHSN27500003); The John Merck Fund, and Cure SMA. The APC was funded by The John Merk Fund.

Acknowledgments: Research reported in this publication was supported by the National Center for Advancing Translational Sciences of the National Institutes of Health (Award # U01TR001792); the Eunice Kennedy Shriver National Institute for Child Health and Human Development (Award # HHSN27500003); The John Merck Fund, Asuragen, and Cure SMA. The content is solely the responsibility of the authors and does not necessarily represent the official views of the National Institutes of Health or any of the other project funders.

Conflicts of Interest: The authors declare no conflict of interest.

References

1. Bailey, D.B., Jr.; Hatton, D.D.; Skinner, M. Early developmental trajectories of males with fragile X syndrome. *Am. J. Ment. Retard.* **1998**, *103*, 29–39. [CrossRef]
2. Bailey, D.B., Jr.; Raspa, M.; Bishop, E.; Holiday, D. No change in the age of diagnosis for fragile X syndrome: Findings from a national parent survey. *Pediatrics* **2009**, *124*, 527–533. [CrossRef] [PubMed]

3. Burris, J.L.; Barry-Anwar, R.A.; Sims, R.N.; Hagerman, R.J.; Tassone, F.; Rivera, S.M. Children with fragile X syndrome display threat-specific biases toward emotion. *Biol. Psychiatry Cogn. Neurosci. Neuroimaging* **2017**, *2*, 487–492. [CrossRef] [PubMed]

4. Roberts, J.E.; Mankowski, J.B.; Sideris, J.; Goldman, B.D.; Hatton, D.D.; Mirrett, P.L.; Baranek, G.T.; Reznick, J.S.; Long, A.C.; Bailey, D.B., Jr. Trajectories and predictors of the development of very young boys with fragile X syndrome. *J. Pediatr. Psychol.* **2009**, *34*, 827–836. [CrossRef] [PubMed]

5. Swanson, M.R.; Wolff, J.J.; Shen, M.D.; Styner, M.; Estes, A.; Gerig, G.; McKinstry, R.C.; Botteron, K.N.; Piven, J.; Hazlett, H.C.; et al. Development of white matter circuitry in infants with fragile X syndrome. *JAMA Psychiatry* **2018**, *75*, 505–513. [CrossRef] [PubMed]

6. Antar, L.N.; Dictenberg, J.B.; Plociniak, M.; Afroz, R.; Bassell, G.J. Localization of FMRP-associated mRNA granules and requirement of microtubules for activity-dependent trafficking in hippocampal neurons. *Genes Brain Behav.* **2005**, *4*, 350–359. [CrossRef]

7. Bear, M.F.; Huber, K.M.; Warren, S.T. The mGluR theory of fragile X mental retardation. *Trends Neurosci.* **2004**, *27*, 370–377. [CrossRef]

8. Bassell, G.J.; Gross, C. Reducing glutamate signaling pays off in fragile X. *Nat. Med.* **2008**, *14*, 249–250. [CrossRef]

9. Dolen, G.; Osterweil, E.; Rao, B.S.S.; Smith, G.B.; Auerbach, B.D.; Chattarji, S.; Bear, M.F. Correction of fragile X syndrome in mice. *Neuron* **2007**, *56*, 955–962. [CrossRef]

10. Nakamoto, M.; Nalavadi, V.; Epstein, M.P.; Narayanan, U.; Bassell, G.J.; Warren, S.T. Fragile X mental retardation protein deficiency leads to excessive mGluR5-dependent internalization of AMPA receptors. *Proc. Natl. Acad. Sci. USA* **2007**, *104*, 15537–15542. [CrossRef]

11. Braat, S.; Kooy, R.F. Insights into GABAAergic system deficits in fragile X syndrome lead to clinical trials. *Neuropharmacology* **2015**, *88*, 48–54. [CrossRef] [PubMed]

12. Hill, M.K.; Archibald, A.D.; Cohen, J.; Metcalfe, S.A. A systematic review of population screening for fragile X syndrome. *Genet. Med.* **2010**, *12*, 396–410. [CrossRef] [PubMed]

13. Bailey, D.B., Jr. Newborn screening for fragile X syndrome. *Ment. Retard. Dev. Disabil. Res. Rev.* **2004**, *10*, 3–10. [CrossRef] [PubMed]

14. Bailey, D.B., Jr.; Bishop, E.; Raspa, M.; Skinner, D. Caregiver opinions about fragile X population screening. *Genet. Med.* **2012**, *14*, 115–121. [CrossRef] [PubMed]

15. Brosco, J.P.; Grosse, S.D.; Ross, L.F. Universal state newborn screening programs can reduce health disparities. *JAMA Pediatr.* **2015**, *169*, 7–8. [CrossRef] [PubMed]

16. Kemper, A.R.; Green, N.S.; Calonge, N.; Lam, W.K.; Comeau, A.M.; Goldenberg, A.J.; Ojodu, J.; Prosser, L.A.; Tanksley, S.; Bocchini, J.A., Jr. Decision-making process for conditions nominated to the recommended uniform screening panel: Statement of the US Department of Health and Human Services Secretary's Advisory Committee on Heritable Disorders in Newborns and Children. *Genet. Med.* **2014**, *16*, 183–187. [CrossRef]

17. Bailey, D.B., Jr.; Berry-Kravis, E.; Gane, L.W.; Guarda, S.; Hagerman, R.; Powell, C.M.; Tassone, F.; Wheeler, A. Fragile X newborn screening: Lessons learned from a multisite screening study. *Pediatrics* **2017**, *139*, S216–S225. [CrossRef]

18. Riley, C.; Wheeler, A. Assessing the fragile X syndrome newborn screening landscape. *Pediatrics* **2017**, *139*, S207–S215. [CrossRef]

19. Maenner, M.J.; Baker, M.W.; Broman, K.W.; Tian, J.; Barnes, J.K.; Atkins, A.; McPherson, E.; Hong, J.; Brilliant, M.H.; Mailick, M.R. FMR1 CGG expansions: Prevalence and sex ratios. *Am. J. Med. Genet. B* **2013**, *162*, 466–473. [CrossRef]

20. Seltzer, M.M.; Baker, M.W.; Hong, J.; Maenner, M.; Greenberg, J.; Mandel, D. Prevalence of CGG expansions of the FMR1 gene in a US population-based sample. *Am. J. Med. Genet. B* **2012**, *159*, 589–597. [CrossRef]

21. Tassone, F.; Iong, K.P.; Tong, T.H.; Lo, J.; Gane, L.W.; Berry-Kravis, E.; Nguyen, D.; Mu, L.Y.; Laffin, J.; Bailey, D.B.; et al. FMR1 CGG allele size and prevalence ascertained through newborn screening in the United States. *Genome Med.* **2012**, *4*, 100. [CrossRef] [PubMed]

22. Grigsby, J.; Cornish, K.; Hocking, D.; Kraan, C.; Olichney, J.M.; Rivera, S.M.; Schneider, A.; Sherman, S.; Wang, J.Y.; Yang, J.C. The cognitive neuropsychological phenotype of carriers of the FMR1 premutation. *J. Neurodev. Disord.* **2014**, *6*, 28. [CrossRef] [PubMed]

23. Wheeler, A.; Raspa, M.; Hagerman, R.; Mailick, M.; Riley, C. Implications of the FMR1 Premutation for Children, Adolescents, Adults, and Their Families. *Pediatrics* **2017**, *139*, S172–S182. [CrossRef] [PubMed]

24. Wheeler, A.C.; Bailey, D.B., Jr.; Berry-Kravis, E.; Greenberg, J.; Losh, M.; Mailick, M.; Mila, M.; Olichney, J.M.; Rodriguez-Revenga, L.; Sherman, S.; et al. Associated features in females with an FMR1 premutation. *J. Neurodev. Disord.* **2014**, *6*, 30. [CrossRef] [PubMed]

25. Spath, M.A.; Feuth, T.B.; Allen, E.G.; Smits, A.P.; Yntema, H.G.; van Kessel, A.G.; Braat, D.D.; Sherman, S.L.; Thomas, C.M. Intra-individual stability over time of standardized anti-Mullerian hormone in FMR1 premutation carriers. *Hum. Reprod.* **2011**, *26*, 2185–2191. [CrossRef] [PubMed]

26. Leehey, M.A.; Berry-Kravis, E.; Goetz, C.G.; Zhang, L.; Hall, D.A.; Li, L.; Rice, C.D.; Lara, R.; Cogswell, J.; Reynolds, A.; et al. FMR1 CGG repeat length predicts motor dysfunction in premutation carriers. *Neurology* **2008**, *70*, 1397–1402. [CrossRef] [PubMed]

27. Bailey, D.B., Jr.; Raspa, M.; Olmsted, M.; Holiday, D.B. Co-occurring conditions associated with FMR1 gene variations: Findings from a national parent survey. *Am. J Med. Genet. A* **2008**, *146*, 2060–2069. [CrossRef]

28. Gallego, P.K.; Burris, J.L.; Rivera, S.M. Visual motion processing deficits in infants with the fragile X premutation. *J. Neurodev. Disord.* **2014**, *6*, 29. [CrossRef]

29. Raspa, M.; Wylie, A.; Wheeler, A.C.; Kolacz, J.; Edwards, A.; Heilman, K.; Porges, S.W. Sensory difficulties in children with an FMR1 premutation. *Front. Genet.* **2018**, *9*, 351. [CrossRef]

30. Wheeler, A.C.; Sideris, J.; Hagerman, R.; Berry-Kravis, E.; Tassone, F.; Bailey, D.B., Jr. Developmental profiles of infants with an FMR1 premutation. *J. Neurodev. Disord.* **2016**, *8*, 40. [CrossRef]

31. Hagerman, R.; Hagerman, P. Advances in clinical and molecular understanding of the FMR1 premutation and fragile X-associated tremor/ataxia syndrome. *Lancet Neurol.* **2013**, *12*, 786–798. [CrossRef]

32. Ennis, S.; Ward, D.; Murray, A. Nonlinear association between CGG repeat number and age of menopause in FMR1 premutation carriers. *Eur. J. Hum. Genet.* **2006**, *14*, 253–255. [CrossRef] [PubMed]

33. Mailick, M.R.; Hong, J.; Greenberg, J.; Smith, L.; Sherman, S. Curvilinear association of CGG repeats and age at menopause in women with FMR1 premutation expansions. *Am. J. Med. Genet. B* **2014**, *165*, 705–711. [CrossRef] [PubMed]

34. Roberts, J.E.; Tonnsen, B.L.; McCary, L.M.; Ford, A.L.; Golden, R.N.; Bailey, D.B., Jr. Trajectory and predictors of depression and anxiety disorders in mothers with the FMR1 premutation. *Biol. Psychiatry* **2016**, *79*, 850–857. [CrossRef] [PubMed]

35. Seltzer, M.M.; Barker, E.T.; Greenberg, J.S.; Hong, J.; Coe, C.; Almeida, D. Differential sensitivity to life stress in FMR1 premutation carrier mothers of children with fragile X syndrome. *Health Psychol.* **2012**, *31*, 612–622. [CrossRef] [PubMed]

36. Sullivan, A.K.; Marcus, M.; Epstein, M.P.; Allen, E.G.; Anido, A.E.; Paquin, J.J.; Yadav-Shah, M.; Sherman, S.L. Association of FMR1 repeat size with ovarian dysfunction. *Hum. Reprod.* **2005**, *20*, 402–412. [CrossRef]

37. March of Dimes PeriStats. Available online: www.marchofdimes.org/peristats (accessed on 30 October 2018).

38. Health Literacy Data Map. Available online: http://healthliteracymap.unc.edu/ (accessed on 26 October 2018).

39. Winarni, T.I.; Schneider, A.; Borodyanskara, M.; Hagerman, R.J. Early intervention combined with targeted treatment promotes cognitive and behavioral improvements in young children with fragile X syndrome. *Case Rep. Genet.* **2012**, *2012*, 280813. [CrossRef]

40. Oostra, B.A.; Hoogeveen, A. FMR1 Protein Studies and Animal Model for Fragile X Syndrome. In *Fragile X Syndrome: Diagnosis, Treatment and Research*; Hagerman, R.J., Hagerman, P.J., Eds.; Johns Hopkins Univ. Press: Baltimore, MD, USA, 2002; pp. 169–190.

41. Erickson, C.A.; Stigler, K.A.; Posey, D.J.; McDougle, C.J. Aripiprazole in autism spectrum disorders and fragile X syndrome. *Neurotherapeutics* **2010**, *7*, 258–263. [CrossRef]

42. Greiss Hess, L.; Fitzpatrick, S.E.; Nguyen, D.V.; Chen, Y.; Gaul, K.N.; Schneider, A.; Lemons Chitwood, K.; Eldeeb, M.A.; Polussa, J.; Hessl, D.; et al. A randomized, double-blind, placebo-controlled trial of low-dose sertraline in young children with fragile X syndrome. *J. Dev. Behav. Pediatr.* **2016**, *37*, 619–628. [CrossRef]

43. Hagerman, R.J.; Berry-Kravis, E.; Kaufmann, W.E.; Ono, M.Y.; Tartaglia, N.; Lachiewicz, A.; Kronk, R.; Delahunty, C.; Hessl, D.; Visootsak, J.; et al. Advances in the treatment of fragile X syndrome. *Pediatrics* **2009**, *123*, 378–390. [CrossRef]

44. Rogers, S.J.; Hayden, D.; Hepburn, S.; Charlifue-Smith, R.; Hall, T.; Hayes, A. Teaching young nonverbal children with autism useful speech: A pilot study of the Denver Model and PROMPT interventions. *J. Autism Dev. Disord.* **2006**, *36*, 1007–1024. [CrossRef] [PubMed]

45. Hagerman, R.J.; Berry-Kravis, E.; Hazlett, H.C.; Bailey, D.B., Jr.; Moine, H.; Kooy, R.F.; Tassone, F.; Gantois, I.; Sonenberg, N.; Mandel, J.L.; et al. Fragile X syndrome. *Nat. Rev. Dis. Primers* **2017**, *3*, 17065. [CrossRef] [PubMed]
46. Berry-Kravis, E.; Des Portes, V.; Hagerman, R.; Jacquemont, S.; Charles, P.; Visootsak, J.; Brinkman, M.; Rerat, K.; Koumaras, B.; Zhu, L.; et al. Mavoglurant in fragile X syndrome: Results of two randomized, double-blind, placebo-controlled trials. *Sci. Transl. Med.* **2016**, *8*, 321. [CrossRef] [PubMed]

Review

Clinical Development of Targeted Fragile X Syndrome Treatments: An Industry Perspective

Anna W. Lee [1],*, Pamela Ventola [2], Dejan Budimirovic [3], Elizabeth Berry-Kravis [4] and Jeannie Visootsak [1]

[1] Ovid Therapeutics Inc., New York, NY 10036, USA; jvisootsak@ovidrx.com
[2] Child Study Center, Yale University, New Haven, CT 06520, USA; pamela.ventola@yale.edu
[3] Departments of Psychiatry and Behavioral Sciences, Kennedy Krieger Institute and Child Psychiatry, Johns Hopkins University, Baltimore, MD 21205, USA; budimirovic@kennedykrieger.org
[4] Departments of Pediatrics, Neurological Sciences, Biochemistry, Rush University Medical Center, Chicago, IL 60612, USA; Elizabeth_Berry-Kravis@rush.edu
* Correspondence: ALee@ovidrx.com; Tel.: +1-646-350-0369

Received: 14 November 2018; Accepted: 30 November 2018; Published: 5 December 2018

Abstract: Fragile X syndrome (FXS) is the leading known cause of inherited intellectual disability and autism spectrum disorder. It is caused by a mutation of the fragile X mental retardation 1 (*FMR1*) gene, resulting in a deficit of fragile X mental retardation protein (FMRP). The clinical presentation of FXS is variable, and is typically associated with developmental delays, intellectual disability, a wide range of behavioral issues, and certain identifying physical features. Over the past 25 years, researchers have worked to understand the complex relationship between FMRP deficiency and the symptoms of FXS and, in the process, have identified several potential targeted therapeutics, some of which have been tested in clinical trials. Whereas most of the basic research to date has been led by experts at academic institutions, the pharmaceutical industry is becoming increasingly involved with not only the scientific community, but also with patient advocacy organizations, as more promising pharmacological agents are moving into the clinical stages of development. The objective of this review is to provide an industry perspective on the ongoing development of mechanism-based treatments for FXS, including identification of challenges and recommendations for future clinical trials.

Keywords: fragile X syndrome; clinical trials; targeted treatments; drug development

---

1. Introduction

Fragile X syndrome (FXS) is the most common inherited cause of intellectual disability and the second most common cause of intellectual disability after Down syndrome [1,2]. It is also the most common known monogenetic cause of autism spectrum disorder (ASD), with approximately 30% to 54% of males and 16% to 20% of females with FXS meeting the diagnostic criteria for ASD by direct assessment [3]. FXS is a genetic disorder caused by the expansion of over 200 cytosine–guanine–guanine (CGG) triplet repeats in the fragile X mental retardation 1 (*FMR1*) gene on the X chromosome [4]. The normal range of CGG repeats varies from 5 to 44 [4]. The full mutation, defined as over 200 CGG repeats, results in the hypermethylation and silencing of the promoter region of the *FMR1* gene, and the absence or reduction of fragile X mental retardation protein (FMRP). FMRP is an RNA-binding protein, which regulates the synthesis of many synaptic proteins. FMRP is required for normal neural development and its absence leads to abnormalities in brain development and function [5].

Individuals with 55–200 CGG repeats in the *FMR1* gene are said to be premutation carriers [4]. In contrast to full mutation, premutation is associated with a functional gene that transcribes the mutation and is associated with certain clinical features that do not occur in those with the full mutation. Males with premutation may develop fragile X-associated tremor/ataxia syndrome (FXTAS), an adult onset neurodegenerative disorder. This phenotype usually manifests late in life (early 60s) and is characterized by action tremor, gait ataxia, and impaired cognition. Parkinsonism-like symptoms, lower extremity neuropathy, dysautonomia, anxiety and depression, and behavioral problems may also be present. Females also may develop FXTAS, in about 10% of cases [6], and symptoms tend to be milder [4]. Females with the premutation can also be affected by fragile X-associated primary ovarian insufficiency (FXPOI) [6].

FMR1 genes with CGG repeat lengths of 45–54 are classified as intermediate or grey zone alleles. It is not yet clear whether these small expansions are related to an increased risk of disease [4]. Recent reports suggest that intermediate alleles may be associated with the development of mild FXTAS, FXPOI, Parkinson disease, ataxia, or multiple system atrophy; however, more research is needed for a clarification of definite risks. Many intermediate alleles are stable and do not change over generations. Individuals with intermediate alleles are not at risk for FXS or to have children with FXS. However, in some families, intermediate alleles with zero or one AGG interspersions can show instability and can expand to a premutation in future generations.

The clinical phenotype of FXS is more varied than basic descriptions suggest. Some individuals express the full mutation in some cells and premutation in others [6]. This "size mosaicism" can result in a milder and even mixed clinical picture, with the presence of both FXS and FXTAS having been described in a very small number of cases of older individuals with FXS. A complicated clinical picture may also be present in individuals with "methylation mosaicism", which occurs when only some of the full mutation alleles are methylated. Furthermore, some individuals may have deletions or point mutations [7] in *FMR1* rather than—or in addition to—repeat expansion [6]. Clinical presentation in such individuals may or may not be typical.

There have been a number of studies aimed at determining the prevalence of FXS in males and females. FXS affects all ethnic and socioeconomic backgrounds. Current estimates suggest that it affects between 1 in 4000 to 5000 males and 1 in 6000 to 8000 females worldwide [1]. The prevalence of premutation carriers is more common, affecting an estimated 1 in 300 to 450 males and 1 in 150 to 200 females. These estimates translate to approximately 430,000 male and close to 1 million female premutation carriers in the US.

2. FXS Phenotypes

Individuals with FXS present with a broad range of physical features, symptoms, and limitations (Table 1) [4,8–10]. The subtle physical features and variability in the clinical presentation of FXS make diagnosis dependent on molecular confirmation. Individuals with FXS can have characteristic physical features that become more apparent with increasing age [8]. These features include a long, narrow face, tall forehead, large prominent ears, high arched palate, unusually hyperflexible fingers, and flat feet. Males also have macro-orchidism after puberty. However, in some individuals with FXS, there may be minimal or no obvious physical features.

Table 1. Common features of fragile X syndrome (FXS).

Neurocognitive [8]	Developmental Delays (Motor and/or Language) Cognitive Deficits/Intellectual Disabilities
Behavioral [8]	Hand flapping and/or biting Gaze avoidance Tactile defensiveness Hyperarousal to sensory stimuli Impaired social skills Social anxiety and mood disorders Hyperactivity Impulsivity Aggression Perseverative behavior
Physical Features [8,9]	Large ears Long, narrow face Prominent forehead or chin Large testicles in teen/adults High palate Flat feet Hyperflexible joints
Other [9]	Recurrent otitis media Strabismus Sleep disorders Gastroesophageal reflux Seizures Weight gain

Physical features of FXS become more apparent with increasing age [10]. Physical features are often not apparent in younger individuals [8,10]. The newborn may appear entirely normal and the only indicator of FXS in infancy may be some degree of reduced muscle tone. At this age, feeding problems with gastroesophageal refluxes (often resulting in emesis) and poor latch or suck with breastfeeding, as well as chronic otitis media are relatively common [10]. Delays in developmental and speech milestones usually become evident before 2 years of age and are the most common symptom leading to diagnosis in patients with FXS. Motor delays and variable hypotonia are present in some infants with FXS, with delays in crawling and walking. Hypotonia and motor coordination deficits become less evident as children become older. Behavioral symptoms become evident in early childhood. In older patients, the facial phenotype may become more accentuated, facilitating clinical diagnosis, but this is not uniform. Joint laxity becomes less prominent after childhood, but for many patients persists throughout life.

There is a wide range of behavioral symptoms in FXS (Table 1), although these tend to be more severe in males than in females [10]. Individuals with FXS can experience problems in one or more behavioral domains. Many individuals with FXS have features of ASD; 30% to 54% of males and 16% to 20% of females meet diagnostic criteria for ASD [3]. Early language and motor milestones [11], and language–communication and social interaction (reviewed in Budimirovic and Kaufmann, 2011) are more affected in patients with ASD and FXS than in those with FXS only. Anxiety is one of the most common reported behavioral symptoms [12,13], with specific symptoms of shyness and social anxiety especially common as patients go into adolescence. Hyperarousal is common at younger ages in particular, and stereotypic behavior such as hand flapping and hand biting are often seen. Many of these behaviors are thought to result from a high underlying level of social anxiety, poor flexibility in response to unexpected situations, and general over-responsiveness to sensory stimuli. Social anxiety in individuals with FXS can be a significant issue for families, interfering with daily activities, education, and socialization, and it may be associated to some extent with high parenting stress [14,15]. Sleep issues are commonly reported by parents of children with FXS as well.

Children with FXS may experience more problems falling asleep, frequent nighttime awakenings, early awakenings, restless sleep, and daytime sleepiness than typically developing children [16].

Females with FXS usually present with less severe symptoms compared to males [6]. In some cases, females with FXS do not appear to display any obvious symptoms. This likely reflects the presence of the normal *FMR1* allele on the active X chromosome in a proportion of cells in females. Most males with FXS have moderate to severe intellectual disabilities, while only about a quarter of females with FXS have intellectual disability (an IQ < 70) [17]. Adult males have an average IQ of about 40 and a mental age of about 5–6 years [18], whereas most females typically have cognitive abilities in the borderline to low–normal range (25% IQ < 70; 28% IQ 70–84) [19]. However, there is large variability in cognitive ability, in both males and females, due to a number of factors yet to be fully understood. Additionally, IQ scores decline with age, with adolescents and adults consistently scoring lower than young children [20]. Floor effects further complicate the interpretation of test results [21]. However, individuals with FXS do not regress in their cognitive functioning; rather, they do not progress at the same rate as their peers, resulting in the lower cognitive scores over time. It is estimated that approximately 60% to 75% of females and nearly 100% of males with FXS have significant language deficits [6]. Females also are prone to the development of emotional problems, specifically internalizing disorders, such as social anxiety and depression [8].

3. Current State of FXS Treatment

Currently, there is no FDA-approved treatment for FXS. The associated (as opposed to the core) symptoms of FXS are typically managed using pharmacologic interventions, such as stimulants for attention deficit and hyperactivity, selective serotonin reuptake inhibitors (SSRIs) for anxiety, antipsychotic drugs for aggression and mood instability, and melatonin for sleep [22,23]. These pharmacologic treatments target only the behavioral symptoms and not the cognitive/language impairments or the underlying brain deficits. Interventional services such as speech–language therapy, occupational therapy, physical therapy, special education services, and behavior management are commonly utilized to address specific behaviors and developmental issues, and comorbid conditions such as ASD, anxiety, attention deficit/hyperactivity disorder (ADHD), seizures, frequent otitis media, strabismus, sleep difficulties, and gastrointestinal problems, which may require specialized medical care [9,22,24]. Even with substantial support and therapies, individuals with FXS continue to present with significant impairments in their functioning throughout their life.

In recent years, an effort to improve the lives of individuals with FXS has driven increased research into the pathophysiologic causes of FXS manifestations and new therapeutic approaches to manage them. The results of these studies have not only demonstrated the complexity of the relationship between FMRP deficiency, downstream neurobiological abnormalities, and FXS symptoms, but also have identified a vast array of potential interventional targets.

4. Development of Targeted Therapies

4.1. Preclinical Rationale

The relationship between FMRP deficiency and aberrant synaptic function is thought to underlie symptoms of FXS and has been the focus of intense research, using animal models of FXS. FMRP is widely expressed in mammals, particularly in the brain and gonads [25–27]. A primary function of FMRP appears to be the regulation of mRNA translation and the synthesis of proteins essential for normal dendritic spine morphology and neuronal signaling [5], as demonstrated by widespread structural and functional abnormalities in its absence [28–30]. In the brain, altered signaling when FMRP is absent is believed to be a result of excessive ribosomal activity and protein synthesis, causing a shift in the excitatory/inhibitory balance towards excessive excitation [31]. Considerable research has focused on the effects of excess group I metabotropic glutamate receptor (mGluR) signaling (excitatory), in particular mGluR5 [32] and deficient γ-aminobutyric acid (GABA) signaling (inhibitory) [33]. Research into these abnormalities and related/downstream effects has resulted in the identification of a large number of potential therapeutic targets, with the most data having been collected on effects of mGluR5 antagonists (2-methyl-6-(phenylethynyl)pyridine (MPEP), fenobam, 2-chloro-4-((2,5-dimethyl-1-(4-(trifluoromethoxy)phenyl)-1H-imidazol-4-yl)ethynyl)pyridine (CTEP), AFQ056, STX107, RO4917526), and GABA receptor activators (baclofen, arbaclofen, acamprosate, ganaxolone, gaboxadol, metadoxine) [10,18]. Other mechanisms targeted for FXS include the inhibition of pathways downstream of group I mGluR signaling (lithium, serine/threonine-protein kinase (PAK) inhibitors, lovastatin, glycogen synthase kinase-3β (GSK3β) inhibitors, PI3K enhancer (PIKE) inhibitors, P13K inhibitors, ERK/Akt inhibitors), blocking excess activity of overproduced proteins (minocycline, striatum-enriched protein-tyrosine phosphatase (STEP) inhibitors, rolipram, phosphodiesterase (PDE) inhibitors); increasing deficient α-amino-3-hydroxy-5-methyl-4-isoxazolepropionic acid (AMPA) receptor activity (CX516); blocking excess synthesis of specific proteins with micro RNAs (miR125A); regulating abnormal channel activity (BK channel blockers, KCNQ1 (Slack) channel blockers); regulating abnormal insulin signaling (metformin); inhibition of endocannabinoid signaling (cannabidiol, endocannabinoid blockers); and inhibition of acetyl cholinesterase (donepezil).

4.2. Clinical Development

FXS has been at the forefront of efforts to test preclinical evidence for targeted interventions in clinical studies. However, to date, the clinical translation of new FXS-specific agents has failed to meet primary endpoints in trials. For example, recent notable compounds demonstrating great preclinical promise, but failing to move forward after early phase 2 trials in the clinic include the selective GABA$_B$ agonist arbaclofen (STX209 (Seaside)), the selective GABA$_A$ modulator ganaxolone, the monoamine-independent GABA transmission modulator metadoxine, and the noncompetitive mGluR5 antagonists mavoglurant (AFQ056 (Novartis)) and basimglurant (RO4917523 (Roche)) [34]. It is unclear whether the failure of these compounds to show benefit in clinical trials reflects the inadequacy of the compounds themselves or the design of the studies devised to evaluate them. Some of these studies were phase 2 studies (ganaxolone, metadoxine) or closed without full enrollment due to financial issues (arbaclofen) and, thus, were statistically underpowered to detect between-group treatment differences at expected effect sizes [18], but most also relied on outcome measures that were not well matched to the expected effects of the drug and/or inadequate to detect meaningful improvements in FXS populations tested. Some studies which showed potential beneficial effects of the drug have simply not moved forward due to the financial position of the sponsor (arbaclofen, metadoxine). With respect to the latter, the Outcome Measures Working Groups convened by the National Institutes of Health to examine endpoints used in FXS clinical trials identified a number of shortcomings, including the dearth of measures validated for FXS, the failure of investigators to use a common set of measures (no consensus), the inability of available measures to assess a broad range of function, limited standardization and floor effects, and the absence of direct-observation measures or validated biological markers (biomarkers) [34].

Despite the history of several "negative" FXS trials, the search for effective FXS-targeted therapies continues. As of October 2018, ClinicalTrials.gov listed 71 FXS studies in various stages of activity, including those completed [35]. While the majority of currently active studies (planned, recruiting/enrolling, or active; Table 2) are investigator-initiated trials, the pharmaceutical industry is visibly involved in the development of at least four promising targeted agents (OV101/gaboxadol, ZYN002/cannabidiol, BPN14770, and Bryostatin-1).

Table 2. Ongoing* clinical trials of pharmacologic interventions for the treatment of FXS [35].

Phase	Compound	Identifier	Industry-Sponsored? (Y/N)
1	AZD7325	NCT03140813	N
2	Acamprosate, lovastatin, minocycline	NCT02998151	N
2	AFQ056	NCT02920892	N
2	BPN14770	NCT03569631	Y (Tetra Discovery Partners/FRAXA/RUMC)
2	Metformin	NCT03722290	N
2	OV101/gaboxadol	NCT03697161/ROCKET	Y (Ovid Therapeutics Inc.)
2/3	Acamprosate	NCT01911455	N
2/3	Metformin	NCT03479476	N
2/3	ZYN002	NCT03614663/CONNECT-FX	Y (Zynerba Pharmaceuticals Inc.)
4	Lovastatin	NCT02642653	N
4	Methylphenidate, fluoxetine, risperidone	NCT00768820	N

* Planned, recruiting/enrolling, or active; Y: Yes; N: No

4.2.1. BPN14770

BPN14770 (Tetra Discovery Partners Inc., Grand Rapids, MI, USA) is a phosphodiesterase-4D negative allosteric modulator (PDE4D-NAM) [36]. It has been proposed that inhibition of PDE4 can prevent the degradation of cAMP, which is reduced in the presence of FMRP deficiency [37–39]. Inhibition of PDE4 may attenuate associated effects, as demonstrated by the rescue of behavioral and structural deficits in the *Drosophila* FXS model [40,41]. In a preclinical study conducted in adult male *Fmr1* KO mice, BPN14770 significantly improved behavioral phenotypes (reduced hyperarousal in the open field, increased frequency of social interaction, and improved natural behaviors (nestin and marble burying)) vs. vehicle administration, with effects persisting after a 2-week drug washout period [36]. In humanized PDE4D mice, single doses of BPN14770 increased brain cAMP, augmented the extended phase of long-term potentiation (LTP), reversed scopolamine-induced short-term memory impairment, and improved long-term memory [42].

The safety and tolerability of BPN14770 were established in two phase 1 studies conducted in healthy young and elderly subjects [43]. A phase 2 two-period crossover study (NCT03569631; target N, 30) is underway to assess the safety and tolerability of BPN14770 in individuals with FXS and explore the effects of BPN14770 on cognitive function and behavior in adult males with FXS [35]. Planned outcome assessment measures include the NIH-Toolbox cognitive battery (NIH-TCB) modified for intellectual disabilities; the KiTAP test of attentional performance; the clinical global impression–severity (CGI–S); the CGI–I; a visual analog rating scale assessment of behavior problems, language abilities, and eating behavior; the aberrant behavior checklist community edition (ABC–C); the anxiety, depression, and mood scale (ADAMS); the Vineland-3 adaptive behavior scale; electroencephalogram assessment of event-related potentials; and eye tracking.

4.2.2. OV101/Gaboxadol

OV101 (gaboxadol or 4,5,6,7-tetrahydroisoxazolo-[5,4-c]pyridine-3-ol (THIP); Ovid Therapeutics Inc.) is a highly selective GABA receptor agonist. [44]. In vitro binding studies have shown that OV101/gaboxadol binds selectively to the extrasynaptic $\alpha 4\delta$-containing $GABA_A$ receptor subpopulation [44–46]. In contrast to OV101/gaboxadol, benzodiazepines or other agents acting at the benzodiazepine binding sites do not act on these same receptors [47]. OV101/gaboxadol, through its actions on extrasynaptic $\alpha 4\delta$-containing $GABA_A$ receptors, is believed to restore tonic inhibition and contribute to sleep induction and maintenance.

The therapeutic potential of OV101 and other GABA agonists has been tested in genetically engineered animal models of FXS that recapitulate the human phenotype of FXS. In the *Fmr1* knockout (KO) mouse model [10], FMRP deficiency is thought to lead to reductions in $GABA_A$ receptors and enzymes necessary for GABA production [48] as well as defects in phasic (synaptic) and tonic (extrasynaptic) inhibitory signaling [10]. Disruptions to GABA signaling are associated with excessive neural excitation [33], a common feature of FXS. In animal models, compounds targeting the GABA$_A$ receptor (agonists) were shown to improve behavioral characteristics, such as hyperactivity, auditory/audiogenic seizures, and repetitive and/or perseverative behaviors [10]. A transient induction of the excitatory–inhibitory switch was also shown to improve hyperactivity and autistic behaviors in offspring in the *Fmr1* KO mouse model. In the *Fmr1* KO, OV101 reduced sensory hypersensitivity and motor hyperactivity improved prepulse inhibition (signal to noise ratio) and enhanced tonic inhibition in the amygdala, a region of the brain thought to be associated with behavioral abnormalities in individuals with FXS [49,50]. In a separate study conducted in *Fmr1* KO mice, OV101 normalized behavioral abnormalities relevant to hyperactivity, anxiety, irritability/aggression, and repetitive behavior [51].

In a phase 1 PK clinical trial, OV101 was well tolerated in adolescents age 13–17 years with Angelman syndrome or FXS (*n* = 12) (NCT03109756) [35]. A phase 2 three-arm, double-blind clinical study (ROCKET; NCT03697161) is ongoing [35]. ROCKET is enrolling adolescent and young adult males with FXS aged 13 to 22 years. It was designed primarily to evaluate the safety of OV101 in FXS, and also includes secondary measures to assess changes in behavior and functioning using the aberrant behavior checklist community edition (ABC–C) and the clinical global impressions–improvement (CGI–I) scale.

4.2.3. ZYN002

ZYN002 (Zynerba Pharmaceuticals, Inc.) is a transdermal cannabidiol gel formulation that targets the dysregulation of the endocannabinoid system [52]. Cannabidiol is a non-intoxicating cannabinoid with numerous molecular targets [53,54]. It has been reported to potentially demonstrate a broad range of activity, including analgesic, anti-inflammatory, antioxidant, antiemetic, antianxiety, antipsychotic, anticonvulsant, and selective cytotoxic effects [53,55] and has shown efficacy as an antispasticity and antiepileptic agent. It is proposed that transdermal administration will reduce the risk for psychomimetic adverse effects versus oral administration by circumventing the conversion to psychoactive components (i.e., Δ-9 tetrahydrocannabinol (THC), Δ-8 THC) in the acidic environment of the gastrointestinal tract [56].

Preclinical evidence indicates that FMRP deficiency enhances mGluR1-dependent endocannabinoid mobilization and subsequent synaptic effects [57–59]. This potentiates inhibitory short- and long-term depression and excitatory postsynaptic potential (EPSP)-spike coupling in the hippocampus [57]. The effects of endocannabinoid modulators in the *Fmr1* KO have been complex, with both activators and blockers of endocannabinoid activity showing phenotype reversal in different brain areas and neural cell types. A blockade of presynaptic cannabinoid type 1 receptors (CB1) normalized several FXS phenotypes in the mouse model [58], as did a blockade of the degradation of the predominant endocannabinoid, 2-arachidonoylglycerol (2-AG) [60]. A blockade of the degradation of the endocannabinoid anandamide also improved performance on tests of learning and memory in the mouse model [61].

ZYN002 was demonstrated to be safe and well tolerated in phase 1 studies conducted in healthy subjects and patients with epilepsy [62,63]. It was also well tolerated in a phase 2 open-label study and extension conducted in 20 children and adolescents with FXS, with the most common treatment-emergent adverse events being gastroenteritis and upper respiratory tract infection [52]. Improvements in behavioral symptoms were noted and are being further evaluated in an ongoing phase 3 double-blind placebo-controlled clinical trial (CONNECT-FX; NCT03614663) [35]. CONNECT-FX plans to utilize the ABC–C with the Fragile X Factor Structure and the CGI–I scale to evaluate drug effects [35].

4.2.4. AFQ056

AFQ056 is an mGluR5 negative modulator (NAM). This class of drugs has shown reversal of a comprehensive list of synaptic, molecular, electrophysiological, morphological, behavioral, and cognitive phenotypes in the fragile X mouse, fly, and rat models, with consistent results from four different mGluR5 NAMs (including AFQ056) published in over 50 scientific papers. Six phenotypes in multiple categories in the fragile X mouse are also corrected by genetic reduction of mGluR5 receptors. The preclinical body of evidence surrounding this treatment target is certainly the largest in any genetic neurodevelopmental disorder [18,64]. A phase 1 trial of fenobam, another mGluR5 NAM, showed target engagement with prepulse inhibition in participants with FXS [65], yet subsequent trials failed to show a behavioral benefit across the trial cohort in phase 2a and b trials in adolescents and adults [66,67]. This suggests that the preclinical data from models are not helpful in choosing agents for human trials or that the trials may not have been targeting the correct outcome at the appropriate age for a developmental disorder. Because the mGluR5 NAMs have synaptic and learning effects in the animal models not measured in the human trials, the ability of these drugs to affect learning and cognition in FXS has not yet been truly tested.

In order to determine if this class of drugs can help to accelerate learning in FXS, a novel trial is currently enrolling 3–6-year-old children with FXS in a double-blind placebo-controlled trial, in which subjects are randomized to AFQ056 or placebo and then receive a 6-month intensive language intervention to assess whether language learning can be enhanced by the drug. This study is also using a novel objective videotaped measure of communication during play tasks to assess improvements in language and communication, as well as standard objective measures of language and cognition, and eye tracking and auditory event related potential biomarkers, tests which should not be prone to placebo effects. This trial is being conducted through a Novartis independent investigator program project in partnership with an NIH grant to run the trial through the NIH-funded NeuroNext network as an exploratory new trial design to test targets in neurodevelopmental disorders with strong preclinical effects on synaptic and learning function in model systems.

5. Considerations for Future Preclinical Research

The failure of animal models to predict the efficacy of potential therapies in psychiatric diseases and neurologic disorders, including FXS, is a central problem in drug development. The reproducibility of preclinical results remains a critical issue for translation. The industry has the resources to validate published results rapidly, but the issues with reproducibility of published results from one lab to another often hinder efforts and have a net impact of slowing the speed of drug development. Experience with the *Fmr1* KO model in predicting clinical success has demonstrated their value as experimental systems for proof-of-principle assessments of new interventions. However, this work has also shown that phenotypes that are improved in mice do not necessarily translate directly onto affected individuals. The translation of behaviors is especially difficult, and indeed, the available preclinical data suggest that the behavioral phenotypes in *Fmr1* KO mice do not translate well to behavioral symptoms measured in clinical trials.

In terms of progress, there are now electrophysiological measures that show similar abnormalities in *Fmr1* KO mice and individuals with FXS. Therefore, the field needs to emphasize the development of preclinical animal testing that can be evaluated in a similar manner in humans.

These markers need to be employed to help with target engagement and show direct translation from mouse to man. Furthermore, it would strengthen preclinical work to systematically validate the data across multiple assays from behavioral to electrophysiological to molecular, in at least two labs and in multiple species as per the NIH's published guidelines.

Another barrier to translatability stems from the premature publication of results that haven't been validated, and a publication culture focused on "positive" preclinical data. Understanding which particular treatments may not improve animal phenotypes may be as important as understanding the aspects of success with a specific drug. Thus, the industry would encourage the publication of negative preclinical data. These "negative" results would be critical to inform study design, outcome measure selection, and execution of future such studies in humans.

6. Considerations for Conducting Clinical Trials

As previous experience has shown, clinical trials to evaluate potential FXS pharmacologic treatments are challenging. From an industry perspective, there is much to be learned from previous clinical study failures, but also a wealth of resources upon which to draw in order to improve the study implementation/execution to increase the likelihood of developing safe and effective targeted treatment options for the FXS population.

Patient advocacy groups are uniquely positioned to provide insight into the needs of individuals with FXS and their families/caregivers, which can then be incorporated into trial design. Moving forward, it is important that the pharmaceutical industry collaborate with organizations such as the National Fragile X Foundation (NFXF) (www.fragilex.org) and FRAXA (www.fraxa.org) to broaden our collective understanding of outcomes that are pertinent and important to this community. Specifically, such relationships will help to ensure outcomes are relevant and clinically meaningful. Additionally, partnerships will help with recruitment and engagement in trials; an alliance with patient registries (e.g., the Fragile X Research Registry (www.fragilexregistry.org)) and the Fragile X Online Registry with Accessible Research Database (FORWARD) can also be considered as a mechanism to identify individuals and families affected by FXS that may be interested in study participation.

The industry must also reach out to the scientific community, which collectively harbors a wealth of knowledge from decades of research and clinical experience. It is notable that the US-based FRAXA Research Foundation (www.fraxa.org) not only funds research grants and fellowships at top universities, but also partners with biomedical and pharmaceutical companies to "bridge the gap between research discoveries and actual treatments".

With respect to study design, some key considerations include the heterogeneity of the condition and the lack of sensitive, validated clinical outcome measures. For example, in a 2017 commentary published in *Translational Neuroscience*, Duy and Budimirovic [68] highlighted the need to utilize strategies to detect potential differences in response based on both heterogeneity of phenotype and baseline symptom severity. Additionally, outcome measures need to be able to capture the wide range of symptom presentations in the syndrome, and they need to be sensitive enough to detect potentially small changes, as in FXS, even small gains can be quite meaningful and impactful on individuals' daily lives.

Desirable attributes of clinical outcome assessment measures are summarized in Table 3. Unfortunately, no reliable clinical outcomes' assessment measures that fulfill these criteria have been established. In 2017, an expert working group review of tools used to quantify outcomes in FXS clinical trials did identify several promising measures, including the KiTAP [69,70], http://www.psytest.net/index.php?page=KiTAP, Expressive Language Sampling (ELS) [71], and the NIH-TCB [72,73]; (http://www.healthmeasures.net/explore-measurement-systems/nih-toolbox) for assessing cognition [34]; independent living scales (ILS; https://www.pearsonclinical.com/therapy/products/100000181/independent-living-scales-ils.html) and the Waisman activities of daily living scale (W-ADL; www.waisman.wisc.edu/family/WADL) [74] for evaluating adaptive behavior [34]; and the fragile X syndrome rating scale (FXSRS) for evaluating changes in behavior and emotion [34]. Additional studies, such as the ongoing SKYROCKET trial (http://www.ovidrx.com/our-pipeline/)—a nondrug study of males with FXS aged 5 to 30 years examining the suitability of scales for the measurement of behavior, sleep, and functioning in individuals with FXS (OVID website)—are important to validate the appropriateness of these measures for inclusion in future FXS treatment trials.

Data quality and accuracy, overall, also need to be specifically considered in FXS trials. Given the nature of the condition, caregiver-report measures are critical to capture information about the subjects' emotional and behavioral presentation, particularly. Many mothers of children with FXS tend to be generally anxious [75], in part related to the high likelihood of being a premutation carrier, and potentially caring for more than one child with FXS, and this heightened general anxiety may affect their ability to objectively recall and report on their child's presentation. Therefore, training on how to complete measures accurately is of critical importance. Caregivers need direct instructions, clear examples, and concrete and specific anchors for reporting on behavior change. Similarly, in multisite trials, rater training is of the utmost importance. As with supporting caregivers in accurate reporting, trials need to provide instructions for clinicians on how to ensure reliability across raters and sites. Examples include a strong study-specific training curriculum, ongoing data quality assurances, and the use of objective outcomes when possible. For more subjective outcome assessments that rely on clinical judgment, trials need to employ highly expert raters and utilize training methods to ensure the clinicians are reliably and uniformly coding the subjects' presentations.

In 2006, the NFXF brought together the network of majority of Fragile X clinics in the US into a "Fragile X Clinical and Research Consortium (FXCRC)". The Foundation provides its administrative structure, such as meeting planning, material development, publicity, and internal communication. The Consortium's members led by the Clinical Committee have written expert-level consensus documents (i.e., ASD, pharmacological, toileting, medical issues, etc.) primarily aimed at properly informing and empowering care providers and families of individuals with *FMR1* mutations. The Clinical Trials Committee (CTC) was formed in 2015 to try to help guide the industry with trial design and drug development planning to optimize the process and ensure stakeholder input. The CTC consists of the top clinical investigators in the FXS field who donate time to provide free consultation and input on any interventional trial to be performed on patients with FXS.

To integrate the work of basic and clinical intramural scientists and to facilitate progress in clinical trials, the NIH established a Bench-to-Bedside program in 1999. The program has continued to grow and in 2006, it opened to partnerships between intramural and extramural programs, which further enabled fruitful collaborations; for example, the investigation of signaling pathways of antipsychotics, which are widely prescribed in individuals with FXS [18].

Due to the intellectual disabilities in the FXS population, the clinical trial endpoints are heavily based on caregiver reports. These questionnaires are performed by caregivers based on their perception on the subject's behavior, and these are highly susceptible to large placebo effects. The placebo response in FXS clinical trials is strong, as is the case in other CNS indications, and this may certainly contribute to the failure of many past studies to show a drug effect. Clinical trials in FXS often fail to show a statistically significant difference between the treatment and placebo control groups. In these cases, it is unclear whether the treatment was truly not effective or whether the larger than expected placebo response masked the treatment effect. The presence of a significant placebo effect makes it more difficult to observe a treatment effect because the treatment will need to result in a much larger improvement in order to see a positive response. Thus, it is important to include assessments that are less susceptible to placebo responses, such as direct subject assessments of cognition and language [18], and not rely on caregiver assessments alone.

At present, behavioral and cognitive measures are the most directly clinically meaningful outcome assessments. As reviewed in Budimirovic and colleagues (2017) [34], however, biomarkers are also critical to consider in clinical trials. Examples of biomarkers in FXS include blood and tissue markers (e.g., mitogen-activated protein kinases/extracellular signal regulated kinases (MAPK/ERKs), the brain-derived neurotrophic factor (BDNF), the amyloid precursor protein (APP)), neurophysiological measures (e.g., prepulse inhibition (PPI), eye tracking and pupillometry, event-related potentials (ERPs), electroencephalography (EEG) spectral analyses), and neuroimaging. As the field progresses, biomarkers will likely provide evidence for target engagement and possibly will serve as early efficacy indicators. Changes in neural signals or neurophysiological measures, for example, may be able to serve as indicators of treatment response prior to overt or measurable change in behaviors. Additionally, biomarkers would ideally be able to provide objective measures of treatment response as well as predictors of outcome. Given the heterogeneity in FXS despite the common etiology of the expansion mutation in 99% of cases, biomarkers may even be able to inform novel and personalized therapeutics and approaches. Further work, though, needs to be completed to both identify and validate potential biomarkers. At present, measurements of biomarkers tend to have limited feasibility due to cost, availability of equipment, and subjects' ability to tolerate the procedures. Therefore, overall, the benefits of biomarkers on investigational FXS therapies remain to be established [34], but there are significant potential benefits of the approach, as the field progresses.

Table 3. Desirable clinical outcome measure attributes [76].

Tests a broad range of ability
Overcomes cooperation/variable performance problems
Results can be reproduced
Quantifies core defects
Correlates with quality of life/true functional improvement

7. Conclusions

There remains a great need for safe and effective treatments for FXS, particularly for targeted treatments that surpass symptom management and address the pathophysiologic abnormalities that underlie the most common manifestations. The pharmaceutical industry can potentially aid this effort by taking on a more prominent role in both the preclinical and clinical phases of FXS drug development. It is imperative that the industry collaborate with both the research and advocacy communities to develop well-designed clinical studies that can produce meaningful results. This effort should include a leading role in the identification and validation of practical and reliable clinical outcomes assessments and biomarkers for use in future drug evaluation studies.

Author Contributions: All authors (A.W.L., P.V., D.B., E.B.-K., J.V.) wrote, reviewed, and edited the paper.

Funding: This research received no external funding.

Conflicts of Interest: A.W.L. and J.V. are employed by Ovid Therapeutics Inc. P.V. (Yale University) is also employed by Cogstate and has received funding from Roche Pharmaceuticals and Takeda Pharmaceutical Company to consult on clinical trial design in neurodevelopmental disorders. D.B. has received support for clinical trials in FXS from Seaside Therapeutics, Ovid Therapeutics Inc (and consult on trial design), and also NIH-research funding through Asuragen Inc. Within the last three years, D.B. has done ad hoc consulting work (rarely) for the American Academy of Child and Adolescent Psychiatry, Ironshore, MEDACorp, Guidepoint, Ovid Therapeutics, and Sunovion. E.B.-K. (Rush University) has received funding from Seaside Therapeutics, Novartis, Roche, Alcobra, Neuren, Cydan, Fulcrum, GW, Neurotrope, Marinus, BioMarin, Ovid Therapeutics, Zynerba, Tetra, and Yamo Pharmaceuticals to consult on trial design or development strategies and/or conduct clinical trials in FXS and other neurogenetic and neurodevelopmental disorders, from Vtesse/Sucampo/Mallinkcrodt to consult on trial design and results, and conduct clinical trials in NP-C, and from Asuragen Inc to do test validation studies and develop testing standards for FMR1 testing.

References

1. Riley, C.; Mailick, M.; Berry-Kravis, E.; Bolen, J. The future of fragile X syndrome: CDC Stakeholder Meeting summary. *Pediatrics* **2017**, *139*, S147–S152. [CrossRef] [PubMed]
2. Thurman, A.J.; McDuffie, A.; Hagerman, R.; Abbeduto, L. Psychiatric symptoms in boys with fragile X syndrome: A comparison with nonsyndromic autism spectrum disorder. *Res. Dev. Disabil.* **2014**, *35*, 1072–1086. [CrossRef] [PubMed]
3. Kaufmann, W.E.; Kidd, S.A.; Andrews, H.F.; Budimirovic, D.B.; Esler, A.; Haas-Givler, B.; Stackhouse, T.; Riley, C.; Peacock, G.; Sherman, S.L.; et al. Autism Spectrum Disorder in fragile X syndrome: Cooccurring conditions and current treatment. *Pediatrics* **2017**, *139*, S194–S206. [CrossRef] [PubMed]
4. Grigsby, J. The fragile X mental retardation 1 gene (*FMR1*): Historical perspective, phenotypes, mechanism, pathology, and epidemiology. *Clin. Neuropsychol.* **2016**, *30*, 815–833. [CrossRef] [PubMed]
5. Banerjee, A.; Ifrim, M.F.; Valdez, A.N.; Raj, N.; Bassell, G.J. Aberrant RNA translation in fragile X syndrome: From FMRP mechanisms to emerging therapeutic strategies. *Brain Res.* **2018**, *1693*, 24–36. [CrossRef] [PubMed]
6. Rajaratnam, A.; Shergill, J.; Salcedo-Arellano, M.; Saldarriaga, W.; Duan, X.; Hagerman, R. Fragile X syndrome and fragile X-associated disorders. *F1000Research* **2017**, *6*, 2112. [CrossRef] [PubMed]

7. Coffee, B.; Ikeda, M.; Budimirovic, D.B.; Hjelm, L.N.; Kaufmann, W.E.; Warren, S.T. MosaicFMR1 deletion causes fragile X syndrome and can lead to molecular misdiagnosis: A case report and review of the literature. *Am. J. Med. Genet. A* **2008**, *146A*, 1358–1367. [CrossRef] [PubMed]

8. Garber, K.B.; Visootsak, J.; Warren, S.T. Fragile X syndrome. *Eur. J. Hum. Genet.* **2008**, *16*, 666–672. [CrossRef]

9. Kidd, S.A.; Lachiewicz, A.; Barbouth, D.; Blitz, R.K.; Delahunty, C.; McBrien, D.; Visootsak, J.; Berry-Kravis, E. Fragile X syndrome: A review of associated medical problems. *Pediatrics* **2014**, *134*, 995–1005. [CrossRef]

10. Hagerman, R.J.; Berry-Kravis, E.; Hazlett, H.C.; Bailey, D.B.; Moine, H.; Kooy, R.F.; Tassone, F.; Gantois, I.; Sonenberg, N.; Mandel, J.L.; et al. Fragile X syndrome. *Nat. Rev. Dis. Primers* **2017**, *3*, 17065. [CrossRef]

11. Hinton, R.; Budimirovic, D.B.; Marschik, P.B.; Talisa, V.B.; Einspieler, C.; Gipson, T.; Johnston, M.V. Parental reports on early language and motor milestones in fragile X syndrome with and without autism spectrum disorders. *Dev. Neurorehabil.* **2013**, *16*, 58–66. [CrossRef] [PubMed]

12. Cordeiro, L.; Ballinger, E.; Hagerman, R.; Hessl, D. Clinical assessment of DSM-IV anxiety disorders in fragile X syndrome: Prevalence and characterization. *J. Neurodev. Disord.* **2011**, *3*, 57–67. [CrossRef] [PubMed]

13. Kaufmann, W.E.; Capone, G.; Clarke, M.; Budimirovic, D.B. Autism in genetic intellectual disability: Insights into idiopathic autism. In *Autism: Current Theories and Evidence*; The Humana Press Inc.: Totowa, NJ, USA, 2008; pp. 81–108.

14. Smith, L.E.; Seltzer, M.M.; Greenberg, J.S. Daily health symptoms of mothers of adolescents and adults with fragile X syndrome and mothers of adolescents and adults with autism spectrum disorder. *J. Autism Dev. Disord.* **2012**, *42*, 1836–1846. [CrossRef] [PubMed]

15. Abbeduto, L.; Seltzer, M.M.; Shattuck, P.; Krauss, M.W.; Orsmond, G.; Murphy, M.M. Psychological well-being and coping in mothers of youths with autism, down syndrome, or fragile X syndrome. *Am. J. Ment. Retard.* **2004**, *109*, 237–254. [CrossRef]

16. Kronk, R.; Bishop, E.E.; Raspa, M.; Bickel, J.O.; Mandel, D.A.; Bailey, D.B. Prevalence, nature, and correlates of sleep problems among children with fragile X syndrome based on a large scale parent survey. *Sleep* **2010**, *33*, 679–687. [CrossRef] [PubMed]

17. Gallagher, A.; Hallahan, B. Fragile X-associated disorders: A clinical overview. *J. Neurol.* **2012**, *259*, 401–413. [CrossRef]

18. Berry-Kravis, E.M.; Lindemann, L.; Jønch, A.E.; Apostol, G.; Bear, M.F.; Carpenter, R.L.; Crawley, J.N.; Curie, A.; Des Portes, V.; Hossain, F.; et al. Drug development for neurodevelopmental disorders: Lessons learned from fragile X syndrome. *Nat. Rev. Drug Discov.* **2017**, *17*, 280–299. [CrossRef]

19. Hagerman, R.J.; Jackson, C.; Amiri, K.; Silverman, A.; O-Connor, R.; Sobesky, W. Girls with fragile X syndrome: Physical and neurocognitive status and outcome. *Pediatrics* **1992**, *89*, 395–400.

20. Wright-Talamante, C.; Cheema, A.; Riddle, J.E.; Luckey, D.W.; Taylor, A.K.; Hagerman, R.J. A controlled study of longitudinal IQ changes in females and males with fragile X syndrome. *Am. J. Med. Genet.* **1996**, *64*, 350–355. [CrossRef]

21. Sansone, S.M.; Schneider, A.; Bickel, E.; Berry-Kravis, E.; Prescott, C.; Hessl, D. Improving IQ measurement in intellectual disabilities using true deviation from population norms. *J. Neurodev. Disord.* **2014**, *6*, 16. [CrossRef]

22. Raspa, M.; Wheeler, A.C.; Riley, C. Public health literature review of fragile X syndrome. *Pediatrics* **2017**, *139*, S153–S171. [CrossRef]

23. Hagerman, R.J.; Berry-Kravis, E.; Kaufmann, W.E.; Ono, M.Y.; Tartaglia, N.; Lachiewicz, A.; Kronk, R.; Delahunty, C.; Hessl, D.; Visootsak, J.; et al. Advances in the treatment of fragile X syndrome. *Pediatrics* **2009**, *123*, 378–390. [CrossRef] [PubMed]

24. Lozano, R.; Azarang, A.; Wilaisakditipakorn, T.; Hagerman, R.J. Fragile X syndrome: A review of clinical management. *Intractable Rare Dis. Res.* **2016**, *5*, 145–157. [CrossRef] [PubMed]

25. Devys, D.; Lutz, Y.; Bellocq, J.-P.; Mandel, J.-L. The FMR-1 protein is cytoplasmic, most abundant in neurons and appears normal in carriers of a fragile X premutation. *Nat. Genet.* **1993**, *4*, 335–340. [CrossRef] [PubMed]

26. Tamanini, F. Differential expression of FMR1, FXR1 and FXR2 proteins in human brain and testis. *Hum. Mol. Genet.* **1997**, *6*, 1315–1322. [CrossRef] [PubMed]

27. Bakker, C.E.; de Diego Otero, Y.; Bontekoe, C.; Raghoe, P.; Luteijn, T.; Hoogeveen, A.T.; Oostra, B.A.; Willemsen, R. Immunocytochemical and biochemical characterization of FMRP, FXR1P, and FXR2P in the mouse. *Exp. Cell Res.* **2000**, *258*, 162–170. [CrossRef] [PubMed]

28. Erickson, C.A.; Davenport, M.H.; Schaefer, T.L.; Wink, L.K.; Pedapati, E.V.; Sweeney, J.A.; Fitzpatrick, S.E.; Brown, W.T.; Budimirovic, D.; Hagerman, R.J.; et al. Fragile X targeted pharmacotherapy: Lessons learned and future directions. *J. Neurodev. Disord.* **2017**, *9*, 7. [CrossRef]

29. Busquets-Garcia, A.; Maldonado, R.; Ozaita, A. New insights into the molecular pathophysiology of fragile X syndrome and therapeutic perspectives from the animal model. *Int. J. Biochem. Cell Biol.* **2014**, *53*, 121–126. [CrossRef]

30. He, C.X.; Portera-Cailliau, C. The trouble with spines in fragile X syndrome: Density, maturity and plasticity. *Neuroscience* **2013**, *251*, 120–128. [CrossRef]

31. Budimirovic, D.; Duy, P. Neurobehavioral features and targeted treatments in fragile X syndrome: Current insights and future directions. *Engrami* **2015**, *37*, 5–26. [CrossRef]

32. Bear, M.F.; Huber, K.M.; Warren, S.T. The mGluR theory of fragile X mental retardation. *Trends Neurosci.* **2004**, *27*, 370–377. [CrossRef] [PubMed]

33. Paluszkiewicz, S.M.; Martin, B.S.; Huntsman, M.M. Fragile X syndrome: The GABAergic system and circuit dysfunction. *Dev. Neurosci.* **2011**, *33*, 349–364. [CrossRef] [PubMed]

34. Budimirovic, D.B.; Berry-Kravis, E.; Erickson, C.A.; Hall, S.S.; Hessl, D.; Reiss, A.L.; King, M.K.; Abbeduto, L.; Kaufmann, W.E. Updated report on tools to measure outcomes of clinical trials in fragile X syndrome. *J. Neruodev. Disord.* **2017**, *9*, 14. [CrossRef] [PubMed]

35. US National Library of Medicine Fragile X Syndrome. Available online: https://clinicaltrials.gov (accessed on 11 October 2018).

36. Gurney, M.E.; Cogram, P.; Deacon, R.M.; Rex, C.; Tranfaglia, M. Multiple behavior phenotypes of the fragile-X syndrome mouse model respond to chronic inhibition of phosphodiesterase-4D (PDE4D). *Sci. Rep.* **2017**, *7*, 14653. [CrossRef] [PubMed]

37. Kelley, D.J.; Davidson, R.J.; Elliott, J.L.; Lahvis, G.P.; Yin, J.C.P.; Bhattacharyya, A. The cyclic AMP cascade is altered in the fragile X nervous system. *PLoS ONE* **2007**, *2*, e931. [CrossRef] [PubMed]

38. Berry-Kravis, E.; Huttenlocher, P.R. Cyclic AMP metabolism in fragile X syndrome. *Ann. Neurol.* **1992**, *31*, 22–26. [CrossRef]

39. Berry-Kravis, E.; Sklena, P. Demonstration of abnormal cyclic AMP production in platelets from patients with fragile X syndrome. *Am. J. Med. Genet.* **1993**, *45*, 81–87. [CrossRef]

40. Choi, C.H.; Schoenfeld, B.P.; Weisz, E.D.; Bell, A.J.; Chambers, D.B.; Hinchey, J.; Choi, R.J.; Hinchey, P.; Kollaros, M.; Gertner, M.J.; et al. PDE-4 Inhibition rescues aberrant synaptic plasticity in drosophila and mouse models of fragile X syndrome. *J. Neurosci.* **2015**, *35*, 396–408. [CrossRef]

41. Kanellopoulos, A.K.; Semelidou, O.; Kotini, A.G.; Anezaki, M.; Skoulakis, E.M.C. Learning and memory deficits consequent to reduction of the fragile X mental retardation protein result from metabotropic glutamate receptor-mediated inhibition of cAMP signaling in drosophila. *J. Neurosci.* **2012**, *32*, 13111–13124. [CrossRef]

42. Zhang, C.; Xu, Y.; Chowdhary, A.; Fox, D.; Gurney, M.E.; Zhang, H.-T.; Auerbach, B.D.; Salvi, R.J.; Yang, M.; Li, G.; et al. Memory enhancing effects of BPN14770, an allosteric inhibitor of phosphodiesterase-4D, in wild-type and humanized mice. *Neuropsychopharmacology* **2018**, *43*, 2299–2309. [CrossRef]

43. Prickaerts, J.; Heckman, P.R.A.; Blokland, A. Investigational phosphodiesterase inhibitors in phase I and phase II clinical trials for Alzheimer's disease. *Expert Opin. Investig. Drugs* **2017**, *26*, 1033–1048. [CrossRef] [PubMed]

44. Meera, P.; Wallner, M.; Otis, T.S. Molecular basis for the high THIP/gaboxadol sensitivity of extrasynaptic GABA $_A$ receptors. *J. Neurophysiol.* **2011**, *106*, 2057–2064. [CrossRef] [PubMed]

45. Belelli, D. Extrasynaptic GABAA Receptors of Thalamocortical Neurons: A Molecular Target for Hypnotics. *J. Neurosci.* **2005**, *25*, 11513–11520. [CrossRef] [PubMed]

46. Brown, N.; Kerby, J.; Bonnert, T.P.; Whiting, P.J.; Wa, K.A. Pharmacological characterization of a novel cell line expressing human a4b3d GABAA receptors. *Br. J. Pharmacol.* **2002**, *136*, 965–974. [CrossRef] [PubMed]

47. Wafford, K.A.; Macaulay, A.J.; Fradley, R.; O'Meara, G.F.; Reynolds, D.S.; Rosahl, T.W. Differentiating the role of γ-aminobutyric acid type A (GABAA) receptor subtypes. *Biochem. Soc. Trans.* **2004**, *32*, 553–556. [CrossRef] [PubMed]

48. D'Hulst, C.; Heulens, I.; Brouwer, J.R.; Willemsen, R.; De Geest, N.; Reeve, S.P.; De Deyn, P.P.; Hassan, B.A.; Kooy, R.F. Expression of the GABAergic system in animal models for fragile X syndrome and fragile X associated tremor/ataxia syndrome (FXTAS). *Brain Res.* **2009**, *1253*, 176–183. [CrossRef] [PubMed]

49. Olmos-Serrano, J.L.; Paluszkiewicz, S.M.; Martin, B.S.; Kaufmann, W.E.; Corbin, J.G.; Huntsman, M.M. Defective GABAergic neurotransmission and pharmacological rescue of neuronal hyperexcitability in the amygdala in a mouse model of fragile X syndrome. *J. Neurosci.* **2010**, *30*, 9929–9938. [CrossRef]

50. Olmos-Serrano, J.L.; Corbin, J.G. Amygdala regulation of fear and emotionality in fragile X syndrome. *Dev. Neurosci.* **2011**, *33*, 365–378. [CrossRef]

51. Cogram, P.; Deacon, R.J.; von Schimmelmann, M.J.; During, M.J.; Abrahams, B.S. Gaboxadol normalizes behavioral abnormalities in a mouse model of fragile X syndrome (P1.323). *Neurology* **2018**, *19*, 323.

52. Heussler, H.S.; Cohen, J.; Silove, N.; Tich, N.; O'Neill, C.; Bonn-Miller, M.O. *Transdermal Cannabidiol (CBD) Gel for the Treatment of Fragile X Syndrome (FXS)*; Zynerba Pharmaceuticals: Devon, PA, USA, 2018.

53. Russo, E.B. Cannabidiol claims and misconceptions. *Trends Pharmacol. Sci.* **2017**, *38*, 198–201. [CrossRef]

54. McPartland, J.M.; Duncan, M.; Di Marzo, V.; Pertwee, R.G. Are cannabidiol and Δ^9-tetrahydrocannabivarin negative modulators of the endocannabinoid system? A systematic review: Mechanistic studies (in vivo and ex vivo) of CBD and THCV. *Br. J. Pharmacol.* **2015**, *172*, 737–753. [CrossRef] [PubMed]

55. Russo, E.B. Taming THC: Potential cannabis synergy and phytocannabinoid-terpenoid entourage effects: Phytocannabinoid-terpenoid entourage effects. *Br. J. Pharmacol.* **2011**, *163*, 1344–1364. [CrossRef] [PubMed]

56. Merrick, J.; Lane, B.; Sebree, T.; Yaksh, T.; O'Neill, C.; Banks, S.L. Identification of psychoactive degradants of cannabidiol in simulated gastric and physiological fluid. *Cannabis Cannabiniod Res.* **2016**, *1*, 102–112. [CrossRef] [PubMed]

57. Zhang, L.; Alger, B.E. Enhanced endocannabinoid signaling elevates neuronal excitability in fragile X syndrome. *J. Neurosci.* **2010**, *30*, 5724–5729. [CrossRef] [PubMed]

58. Busquets-Garcia, A.; Gomis-González, M.; Guegan, T.; Agustín-Pavón, C.; Pastor, A.; Mato, S.; Pérez-Samartín, A.; Matute, C.; de la Torre, R.; Dierssen, M.; et al. Targeting the endocannabinoid system in the treatment of fragile X syndrome. *Nat. Med.* **2013**, *19*, 603–607. [CrossRef] [PubMed]

59. Maccarrone, M.; Rossi, S.; Bari, M.; De Chiara, V.; Rapino, C.; Musella, A.; Bernardi, G.; Bagni, C.; Centonze, D. Abnormal mGlu 5 receptor/endocannabinoid coupling in mice lacking FMRP and BC1 RNA. *Neuropsychopharmacology* **2010**, *35*, 1500–1509. [CrossRef] [PubMed]

60. Jung, K.-M.; Sepers, M.; Henstridge, C.M.; Lassalle, O.; Neuhofer, D.; Martin, H.; Ginger, M.; Frick, A.; DiPatrizio, N.V.; Mackie, K.; et al. Uncoupling of the endocannabinoid signalling complex in a mouse model of fragile X syndrome. *Nat. Commun.* **2012**, *3*, 1080. [CrossRef] [PubMed]

61. Qin, M.; Zeidler, Z.; Moulton, K.; Krych, L.; Xia, Z.; Smith, C.B. Endocannabinoid-mediated improvement on a test of aversive memory in a mouse model of fragile X syndrome. *Behav. Brain Res.* **2015**, *291*, 164–171. [CrossRef] [PubMed]

62. Bonn-Miller, M.; Sebree, T.; O'Neill, C.; Messenheimer, J. *Neuropsychological Effects of ZYN002 (Synthetic Cannabidiol) Transdermal Gel in Healthy Subjects and Patients with Epilepsy: Phase 1, Randomized, Double-Blind, Placebo-Controlled Studies*; Zynerba Pharmaceuticals: Devon, PA, USA, 2016.

63. Sebree, T.; O'Neill, C.; Messenheimer, J.; Gutterman, D. *Safety and Tolerability of ZYN002 (Synthetic Cannabidiol) Transdermal Permeation-Enhanced Gel in Healthy Subjects and Patients with Epilepsy: Three Phase 1, Randomized, Double-Blind, Placebo-Controlled Studies*; Zynerba Pharmaceuticals: Devon, PA, USA, 2016.

64. Gross, C.; Hoffmann, A.; Bassell, G.J.; Berry-Kravis, E.M. Therapeutic strategies in fragile X syndrome: From bench to bedside and back. *Neurotherapeutics* **2015**, *12*, 584–608. [CrossRef] [PubMed]

65. Berry-Kravis, E.; Hessl, D.; Coffey, S.; Hervey, C.; Schneider, A.; Yuhas, J.; Hutchison, J.; Snape, M.; Tranfaglia, M.; Nguyen, D.V.; et al. A pilot open label, single dose trial of fenobam in adults with fragile X syndrome. *J. Med. Genet.* **2009**, *46*, 266–271. [CrossRef] [PubMed]

66. Jacquemont, S.; Curie, A.; des Portes, V.; Torrioli, M.G.; Berry-Kravis, E.; Hagerman, R.J.; Ramos, F.J.; Cornish, K.; He, Y.; Paulding, C.; et al. Epigenetic modification of the FMR1 gene in fragile X syndrome is associated with differential response to the mGluR5 antagonist AFQ056. *Sci. Transl. Med.* **2011**, *3*, 64ra1. [CrossRef]

67. Berry-Kravis, E.; des Portes, V.; Hagerman, R.; Jacquemont, S.; Charles, P.; Visootsak, J.; Brinkman, M.; Rerat, K.; Koumaras, B.; Zhu, L.; et al. Mavoglurant in fragile X syndrome: Results of two randomized, double-blind, placebo-controlled trials. *Sci. Transl. Med.* **2016**, *8*, 321ra5. [CrossRef] [PubMed]

68. Duy, P.Q.; Budimirovic, D.B. Fragile X syndrome: Lessons learned from the most translated neurodevelopmental disorder in clinical trials. *Transl. Neurosci.* **2017**, *8*, 7–8. [CrossRef] [PubMed]

69. Zimmermann, P. A test battery for attentional performance. In *Applied Neuropsychology of Attention Theory, Diagnosis and Rehabilatation*; Leclercq, M., Zimmermann, P., Eds.; Psychology Press: London, UK, 2004.

70. Knox, A.; Schneider, A.; Abucayan, F.; Hervey, C.; Tran, C.; Hessl, D.; Berry-Kravis, E. Feasibility, reliability, and clinical validity of the Test of Attentional Performance for Children (KiTAP) in Fragile X syndrome (FXS). *J. Neurodev. Disord.* **2012**, *4*, 2. [CrossRef] [PubMed]

71. Berry-Kravis, E.; Doll, E.; Sterling, A.; Kover, S.T.; Schroeder, S.M.; Mathur, S.; Abbeduto, L. Development of an expressive language sampling procedure in fragile X syndrome: A pilot study. *J. Dev. Behav. Pediatr.* **2013**, *34*, 245–251. [CrossRef] [PubMed]

72. Weintraub, S.; Dikmen, S.S.; Heaton, R.K.; Tulsky, D.S.; Zelazo, P.D.; Bauer, P.J.; Carlozzi, N.E.; Slotkin, J.; Blitz, D.; Wallner-Allen, K.; et al. Cognition assessment using the NIH Toolbox. *Neurology* **2013**, *80*, S54–S64. [CrossRef]

73. Hessl, D.; Sansone, S.M.; Berry-Kravis, E.; Riley, K.; Widaman, K.F.; Abbeduto, L.; Schneider, A.; Coleman, J.; Oaklander, D.; Rhodes, K.C.; et al. The NIH Toolbox Cognitive Battery for intellectual disabilities: Three preliminary studies and future directions. *J. Neurodev. Disord.* **2016**, *8*, 35. [CrossRef]

74. Maenner, M.J.; Smith, L.E.; Hong, J.; Makuch, R.; Greenberg, J.S.; Mailick, M.R. Evaluation of an activities of daily living scale for adolescents and adults with developmental disabilities. *Disabil. Health J.* **2013**, *6*, 8–17. [CrossRef]

75. Hauser, C.T.; Kover, S.T.; Abbeduto, L. Maternal well-being and child behavior in families with fragile X syndrome. *Res. Dev. Disabil.* **2014**, *35*, 2477–2486. [CrossRef]

76. Gross, C.; Berry-Kravis, E.M.; Bassell, G.J. Therapeutic strategies in fragile X syndrome: Dysregulated mGluR signaling and beyond. *Neuropsychpharmacology* **2012**, *37*, 178–195. [CrossRef]

MDPI

St. Alban-Anlage 66

4052 Basel

Switzerland

Tel. +41 61 683 77 34

Fax +41 61 302 89 18

www.mdpi.com

Brain Sciences Editorial Office

E-mail: brainsci@mdpi.com

www.mdpi.com/journal/brainsci